SACRAMENTO PUBLIC LIBRARY
3 3029 05449 3234

Global communications

THE NEW

HOW THINGS WORK

EVERYDAY
TECHNOLOGY
EXPLAINED

BY JOHN LANGONE

ART BY PETE SAMEK,
ANDY CHRISTIE,
AND BRYAN CHRISTIE

NATIONAL GEOGRAPHIC

WASHINGTON, D.C.

CONTENTS

Preceding pages: Laser light from the Starfire Telescope in New Mexico pierces the night sky.

INTRODUCTION

It has been said that whenever Thomas Edison showed visitors the numerous inventions and gadgets that filled his home, someone would invariably ask why they still had to enter through an old-fashioned turnstile. The master inventor had an easy and altogether practical answer: "Because every single soul who forces his way through that old stile pumps three gallons of water up from my well and into my water tank."

Versatility, as well as the ability to perform useful work, is what machines and mechanisms are often all about, and while a device may have a specific use, chances are that its operating principles and many of its parts, if not the device itself, can be put to work in a variety of ways. The compact disc, for example, fills our homes with music, but it also can provide the images, data, and sound that emerge from a personal computer. Propellers drive planes through the air, but they also help generate electricity in a hydroelectric power plant. Water irrigates soil-grown crops, and it can itself become the medium for growth on a soilless hydroponics farm. Fabrics let air in but also keep it out; lenses allow us to look deep inside our bodies and far into the universe. A home's furnace can also run a central air-conditioning system; an electron microscope works more like a television set than like a conventional microscope; the telephone expands into the Internet; the steam engine lends a hand to drive a nuclear-powered submarine; the levers that move a piano's keys are akin to those that run a typewriter.

The machines and other technologies that have extended the range of human capability are all products of a human ingenuity that has successfully utilized

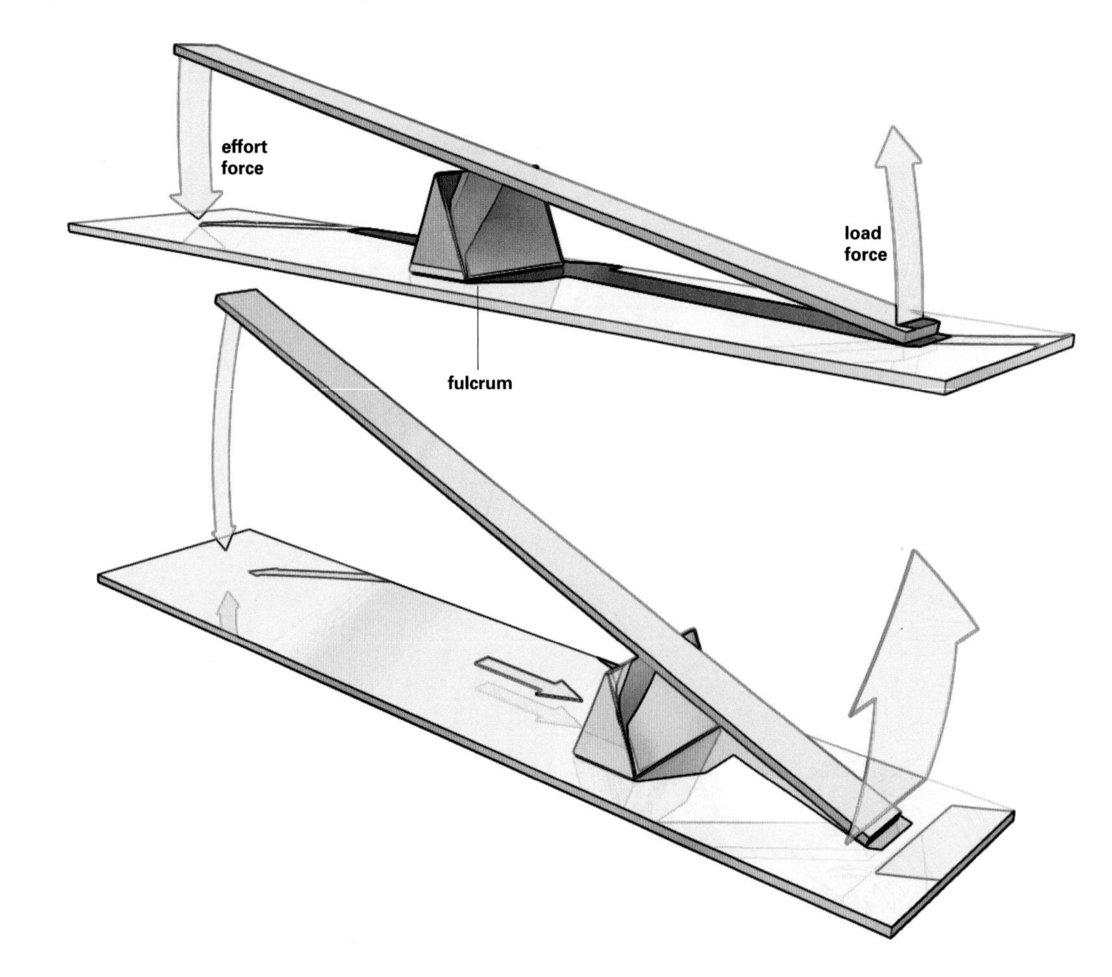

LEVERS

These simplest of mechanical contrivances require effort at one end to raise a load on the other, and the effort required to lift the load depends on the position of the fulcrum. With equal amounts of effort and load at its ends, a lever balances when its fulcrum occupies a position at the center. Moving the fulcrum farther from the load (top) makes the load harder to raise. Placing it closer to the load (bottom) reduces the amount of effort needed for lifting.

the science behind motion, forces, and various forms of energy. One can look under the hood of an automobile or unscrew the back of a television set to check out the working parts, but to truly understand how a machine works—and a machine can be something as simple as a lever, an inclined plane, a wheel and axle, a pulley, or a screw—one has to have at least a casual acquaintance with nature's forces and how they are modified, transformed, transmitted, and otherwise adapted to do work.

In general terms, machines work because some kind of force, or energy that affects motion, is applied. Put another way, machines are devices that overcome resistance at one point by the application of a force at some other point. Various movable mechanical parts, such as levers, gears, and springs—or electrical wires, transistors, and electromagnets—manipulate and transmit the force exerted by a "prime mover," sending it where it is needed and getting the most out of it. This is where efficiency comes in. Always less than 100 percent

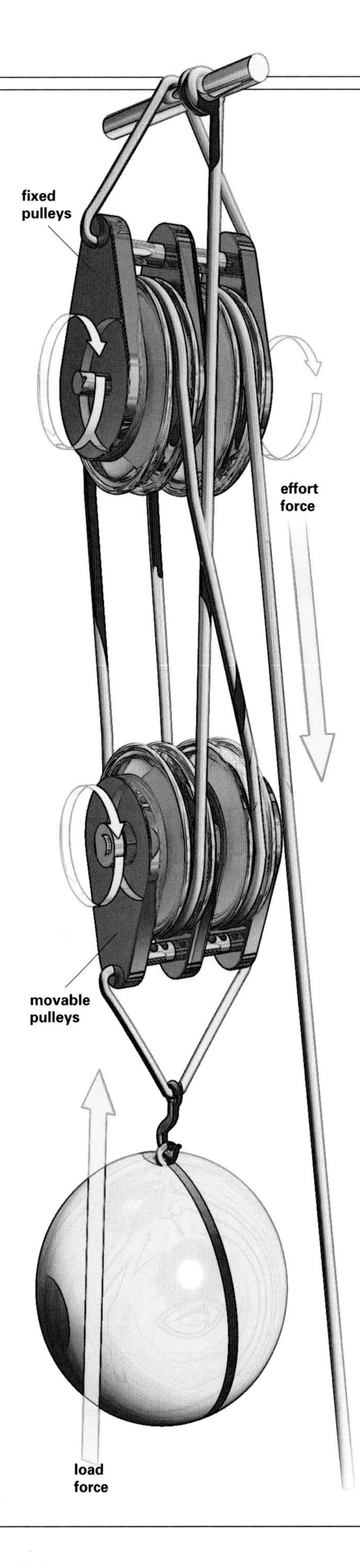

PULLEYS

Another basic machine, a pulley may consist of grooved wheels mounted on blocks, with a rope running in the grooves. A fixed pulley changes the direction of force applied to a rope. If you run the end of that rope around an unfixed pulley and then pass it back to the fixed one, you can raise a load attached to the rope with half the effort. Various combinations of ropes and pulleys help distribute the weight of a load, easing the strain on workers who must lift heavy objects.

because some energy is lost to friction as heat, a machine's efficiency is the ratio of useful work to the energy put into it, and it varies enormously depending on the machine. Only about 6 to 8 percent of the energy from burned coal went into the work the old steam locomotive did to haul the train. On the other hand, a gasoline engine and a steam turbine may have a thermal efficiency of around 25 percent, a diesel oil burner about 35 percent, a hoist some 60 percent, and an electric motor around 95 percent. The process through which a machine works often involves interconnections: As a force moves one component of a machine, that part exerts force at another point, moving some other part, and so on. Thus, energy is handed along by the machine in a sort of relay. In a nuclear power plant, for example, the force of heat from split atoms vaporizes water, creating the steam that turns the turbine that runs

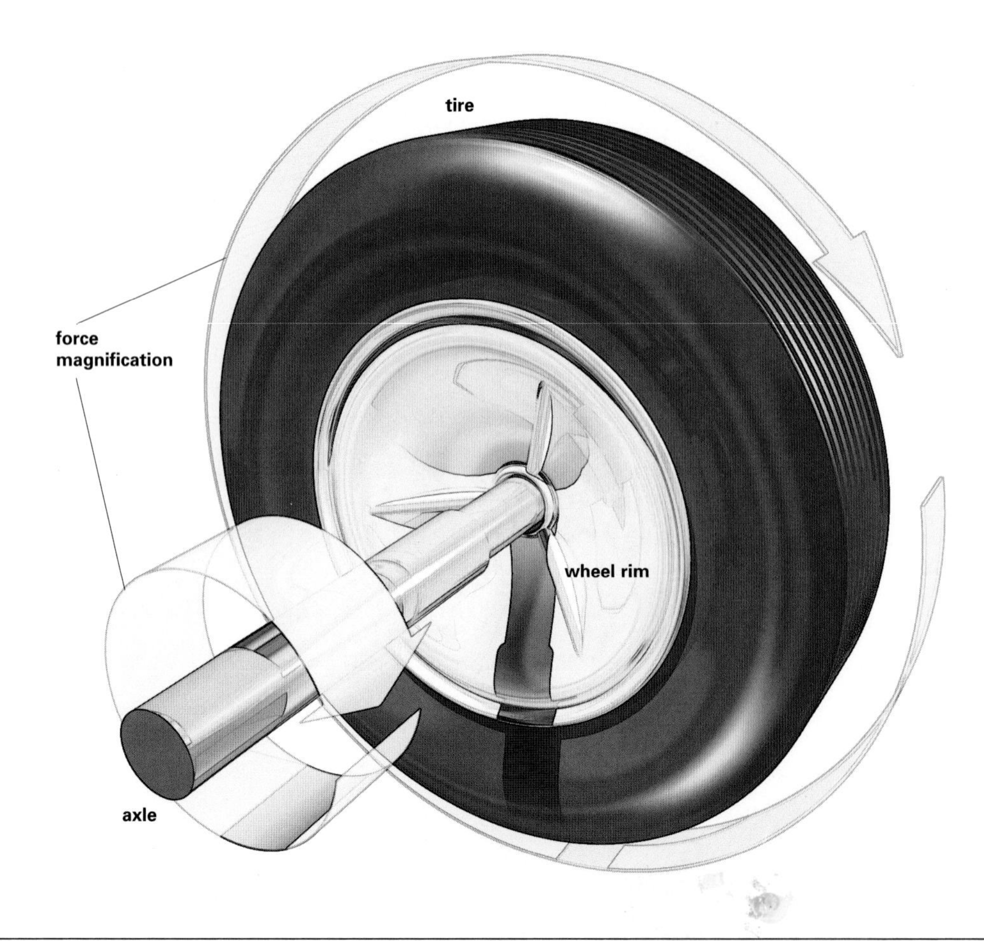

HOW A WHEEL WORKS

With its familiar circular frame, hub, and axle, a wheel (left) effectively transmits power and motion, moving everything from toys and vehicles to houses and the hands of a watch. Whether spoked, smooth, toothed, or flanged—or used in a Ferris wheel (above) or a wheelbarrow—the wheel provides the rotary motion essential to countless machines. Its center acts as a fulcrum, making it, in fact, a rotating lever.

the electrical generator. In an automobile, vaporized gasoline ignites and burns inside the cylinders, transmitting motion to the wheels.

But the laws of force and motion do not apply only to running a steam engine or driving a vehicle. In the human body, chemical energy moves the muscles that do the work when we swing axes or hit golf balls, while the bones act as levers, and the joints are controlled by the body's equivalent of belts and pulleys. Fired bullets and rockets follow the force-motion rules. The water that rushes from a faucet, the furnace that forces heat into a room, and the electrical pulses that actuate a loudspeaker do so as well.

The simplest of our machines is probably the lever, a rigid beam or rod pivoted at a fulcrum point. Used in one basic form for centuries, at first to lift heavy stones or tree trunks, it is now used in crowbars, wheelbarrows, nutcrackers, typewriter bars, pliers, wrenches, nail clippers, kitchen scales, and even the links within a piano that move the hammers that strike the tightened wires to sound the notes. A pulley, which is a wheel—another simple machine—on a shaft, is also a type of lever, as are the gear wheels of a clock, which form a system of levers whose lengths are the radii of the wheels. The inclined plane is a simple machine that enabled the ancient Egyptians to haul up stones to build the pyramids; its spiral cousins include the screw and the automobile jack. The operating principle of an inclined plane (effort to move an object to a higher plane is reduced by extending the distance the object must travel) also governs the action of the wedge, which translates work at one end to force at the sides, as in splitting wood.

Last, but certainly not least, are wheels and axles, which take advantage of the principle that rolling

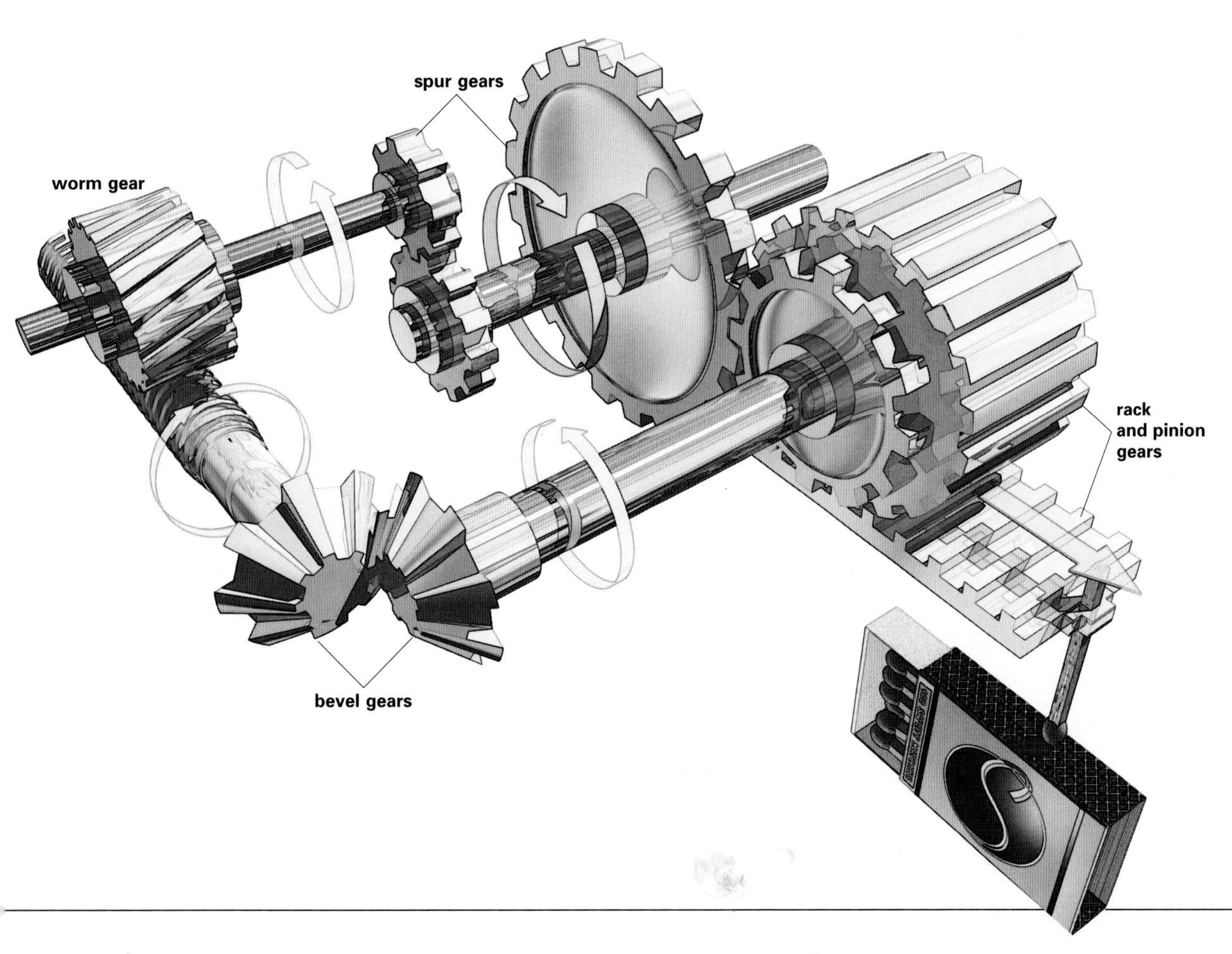

GEARS

The shape and size of a gear's teeth control the number of rotations, the direction of motion, the speed, and the amount of force exerted. Worm gears link shafts that have axes at right angles but that do not intersect; bevel gears connect shafts at an angle; rack and pinion gears have pinion wheels that mesh with sliding, toothed racks.

friction is less of a drag than sliding friction. Shoving a heavy box across a floor, for example, may be impossible without rollers of some kind beneath; with them, however, friction is dissipated and relatively little force is required. A power and motion transmitter par excellence, the wheel mediates between rotary and linear motion, stores energy as in a flywheel, and shows up in gears, winches, capstans, turbines, and pulleys.

The mechanisms and principles behind simple machines, and the complex ones that they create when mixed and matched, is what this book is all about. *The New How Things Work: Everyday Technology Explained* is not, however, a home repair book. It will not tell you how to fix a laser printer or tune a piano, for example. Nor is it a manual that will teach you how to pilot a plane or stitch a hem with a sewing machine. What it will do instead is satisfy your curiosity about how and why these machines work. It will also tell you about a host of other technologies that affect our lives in various ways.

The pages ahead include more than information on conventional machinery. Other technologies—those that appear to be "nonmachines," for want of a better phrase—are equally interesting, and while they are not gadgets in the classic sense, they still have a place here because they, too, rely on scientific principles to function. Thus, the book not only explores what goes on inside a refrigerator, a computer, a jet engine, and most of our other familiar machines, but it also examines the technology and the principles behind such achievements as building arches, genetically altered animals and crops, clarinets, fireworks displays, and how detergents work.

On one page you will see inside a baseball and a golf ball; another page will let you peer inside a gigantic magnetic bottle that contains the awesome

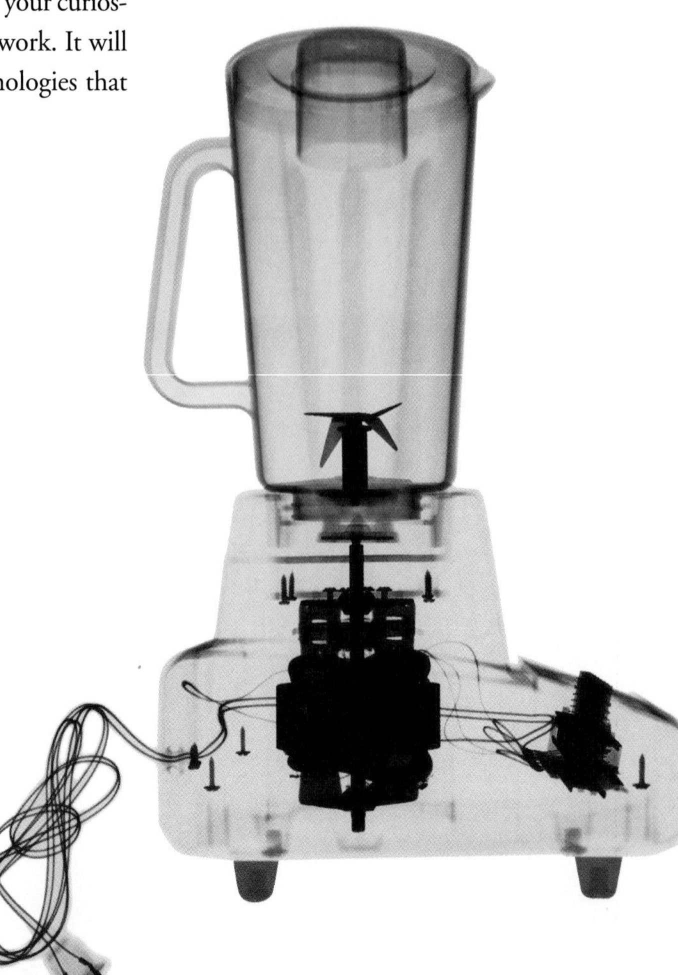

A MOTOR

A motor (right) converts electrical into mechanical energy—just the reverse of what a generator, or dynamo, does—and it works on the principle that an electrical conductor moves when in the presence of a magnetic field at right angles to the current. Current enters the motor via a commutator and its brushes, and it turns a laminated steel rotor that contains conducting coils. The commutator reverses the flow of current every half turn of the coils, and this action reverses the magnetic field, thus ensuring that the coil continues moving. As the rotor turns, a shaft delivers mechanical power sufficient to drive an electric blender (left) and every other household appliance.

spawn of fusion; a computer, logic gates, and the Internet are stripped to their bare essentials; the hooks and loops of a Velcro fastener are unraveled. Household plumbing and electrical systems emerge skeleton-like when walls are removed. The mysterious mechanism beneath a clock's face and the powerhouse inside a nuclear submarine are opened to view. From an air bag to a climbing crane to kimono patterning to a Zamboni, everyday technology is not only explained but also rendered easy on the eyes. Within each chapter, sidebars give you insight into such topics as the pendulum, earthquake protection, the Polaroid, hearing aids, environmental cleanup, the zipper, processing film, satellite radio, and the electronic-music keyboard. They also help to highlight the application of science to specific technologies and will provide a grace note or two in the midst of some of the more technical discussions. You can open *The New How Things Work* at any place and read each single-topic spread as if it were a magazine article.

Moreover, because technology overlaps so many fields, the interdependence of the scientific, industrial, and technical disciplines involved in creating and utilizing our machines, equipment, and systems is always hovering implicitly or stated explicitly in the text. The overlap becomes especially obvious in the section describing the MRI scanner, which is a wedding of medicine, electronics, computer technology, and nuclear physics. You can also see it in the explanation of the kidney dialysis machine, which is a product of physics and bioengineering; and in the discussions of such industrial processes as the manufacture of steel, plastics, and glass, all of which involve the blending of basic science and technology.

Inventions and innovations are, of course, the creations of the human mind. Where possible, *The New How Things Work: Everyday Technology*

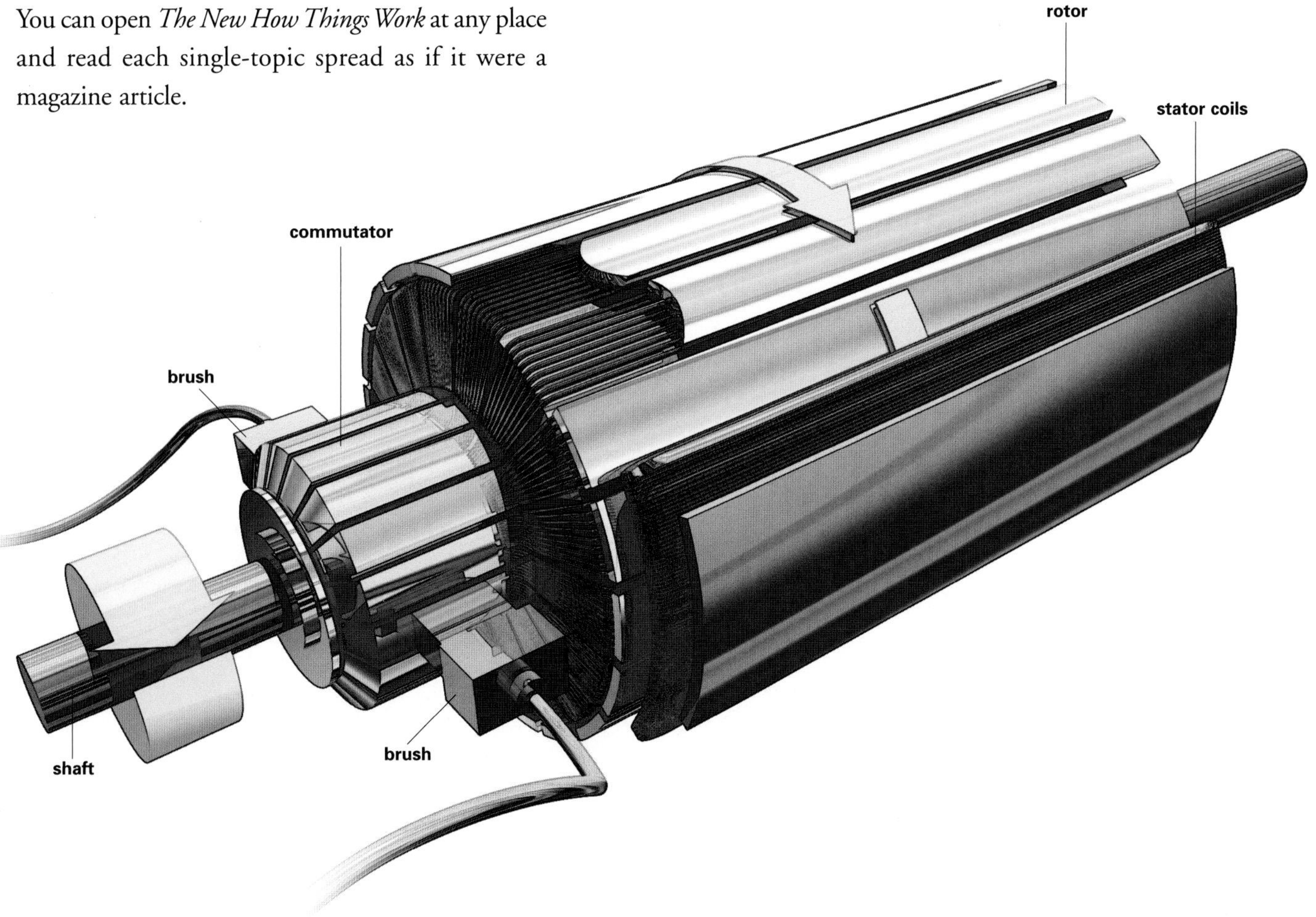

Explained includes historical information about a gadget or a process, relating how, when, and why it came to be and describing its impact on society. We learn, for instance, that clocks may have been in use in 2000 B.C., that Leonardo da Vinci designed a steam-powered vehicle, that Queen Elizabeth I refused to grant a patent to the inventor of a knitting machine, and that miners were once paid by the tons of coal they loaded by hand.

The book's focus on technology should not, however, be construed as our endowing its intricate and awe-inspiring workings with the same divine right as kings or implying that it is omnipotent and infallible. Red flags certainly abound. Computerization, our newest symbol for automation and mechanization, has become, for some people, the cyber-devil that drains away human reasoning and dumbs us down, makes us slaves to speed and materialism, and increasingly isolates us. Sperm and eggs are now routinely manipulated outside their natural hosts, embryos are transplanted, sheep and other animals are cloned, all of which raise moral and ethical questions.

POWER STROKES

In a four-stroke internal combustion engine (below), vaporized gasoline ignites and burns inside the cylinders. The downward induction stroke opens an inlet valve, sucking in fuel and air; on compression, an upward stroke compresses the air and fuel mixture; in the downward power stroke, a spark plug ignites the fuel; on exhaust, a valve opens, expelling gases. Variations of the internal combustion engine can power a wide variety of things, including water craft like the one being enjoyed by two children plying Caribbean waters (far right).

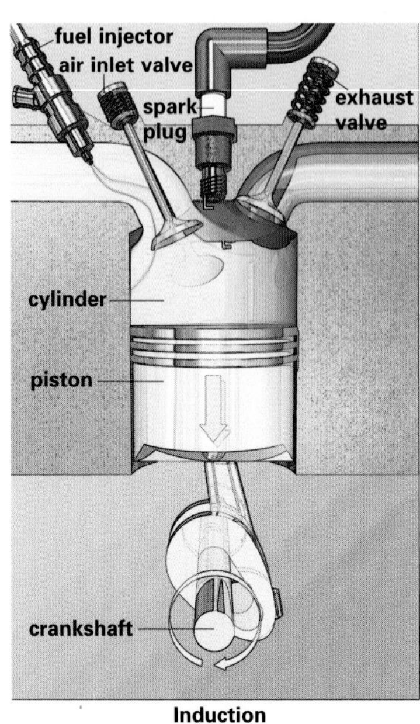

Induction

Compression

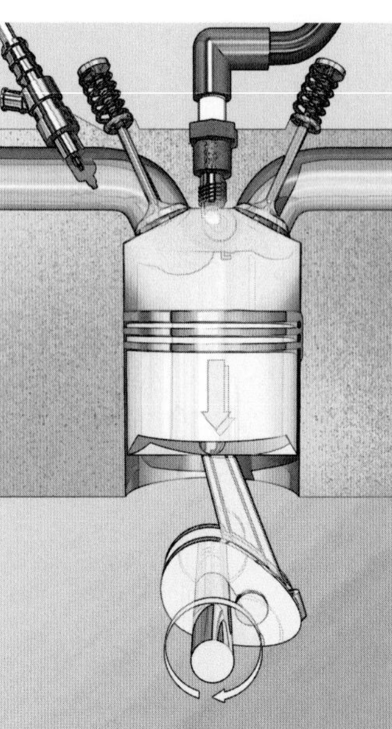

Ignition

Exhaust

The mechanization of medical care may well have distanced doctors and made caregivers seem less compassionate. A generation that knows almost no other way relies on a digital language to perform countless tasks that once took hands-on effort, as well as human innovation, intuition, and imagination. Military technology with its almost unbelievably expert arsenal of "smart" weapons and systems has a purpose that the civilized world would rather not have to boast about.

In light of these caveats, it is no wonder that science and technology have had their detractors. Mark Twain, acknowledging that science was fascinating, mused, "One gets such wholesome returns of conjectures out of such trifling investment of fact." For George Bernard Shaw, science was always wrong. "It never solves a problem without creating ten new ones," he said. For Oliver Wendell Holmes, it was "a first rate piece of furniture for a man's upper chamber, if he has common sense on the ground floor." More to the point may be a remark by Canadian physician Wilder Penfield. "The trouble is not science," he observed, "but in the uses men make of it. Doctor and laymen alike must learn wisdom in their employment of science, whether this applies to atom bombs or blood transfusions."

Perhaps science and technology will never solve all of nature's mysteries, and maybe have given us too many things, some of which are not always desirable. But without science and technology, we would have very few things at all.

CHAPTER 1

AT HOME: RUNNING THE HOUSE

BE IT EVER SO HUMBLE, THERE IS NO PLACE LIKE a modern home. Furnaces, fans, refrigerators, stoves, washing machines—today's comfortable, efficient structures are pierced by a circuitous system of wires, pipes, vents, and ducts. They rely on electricity and the transformer, a 19th-century invention that made it possible to transmit power for domestic and industrial use. Because we have electricity, houses are filled with "home basics" people once considered luxuries. Some homes even have "smart" digitally connected environments that link television sets, personal computers, programmable thermostats, lighting, and security systems. Eventually, most of the pieces of equipment in a home will be "talking" electronically among themselves. The average home contains a hundred pieces of powered equipment—machines that can reduce our physical labors and free up time.

Despite nature's backdrop, each of these homes in St. George, Utah, is undoubtedly ruled by an array of high-tech appliances.

COOKING WITHOUT GUESSWORK

IN 1900, THE SEARS, ROEBUCK AND COMPANY catalog offered a top-of-the-line Acme sterling steel, nickel-plated range that burned coal or wood and sold for a then whopping $26.50 to $31.05. Today, a high-tech home cookstove with computerized precision controls can cost thousands of dollars, but a person doesn't have to pay anywhere near that amount to cook like Fannie Farmer or Julia Child.

Standard electric and gas ranges are efficient heat producers, while microwave ovens, toaster ovens, electric skillets, and slow cookers are good alternatives. Although each of the electrical cooking appliances has its own size, shape, and purpose, all but the microwave do their jobs by converting electricity into heat. As electricity passes through insulated wires inside metal coils or loops, called heating elements, the electrical resistance heats the outer metal. Manual controls regulate the heat by adjusting voltage in the wires, or a thermostat makes adjustments automatically.

In a gas range, natural or bottled gas flows in and is mixed with air in a chamber; then the mixture is ignited by a spark, the flame of a pilot light, or an electric heating coil, firing the burners. Some modern ranges are duel-fuel appliances, utilizing both electric ovens and gas tops. Another variation is the convection oven, and in this appliance a fan circulates heated air uniformly and continuously around food for faster and more even cooking. Still another is the induction range—also called cool electric induction—which uses a magnetic field, generated electronically, under a ceramic range-top. Instant heat is generated through resistance when magnetic pans are set on the ceramic plate.

A microwave oven produces high-frequency electromagnetic waves. Passing through food, the waves reverse polarity billions of times a second. The food's water molecules also have polarity, and they react to each change by rapidly reversing themselves. Friction results, heating the water and cooking the food.

SUSPENDED ANIMATION

(Right) Heat, latches, timers, and springs pop toasted bread out of a toaster.

MICROWAVE OVEN

(Left) This oven uses the same electromagnetic radiation as radar to magnetically agitate water molecules in food (left inset), causing them to heat faster than in a conventional oven. Produced electrically in a magnetron (right inset), the high-frequency microwaves pass through a wave-guide, encounter a stirrer-fan, and reflect into the oven. Invisible and seemingly benign, microwaves can injure human tissue.

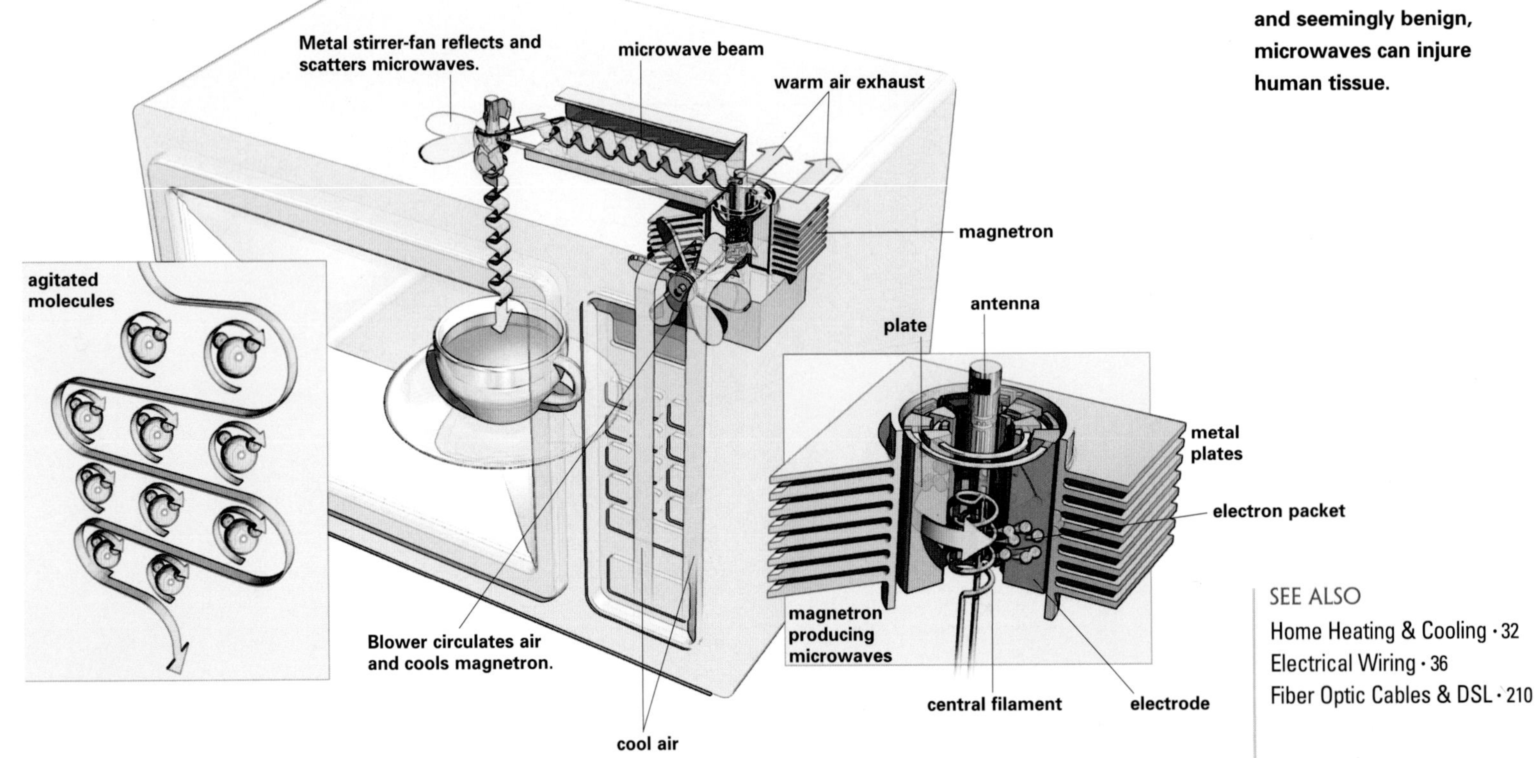

SEE ALSO
Home Heating & Cooling · 32
Electrical Wiring · 36
Fiber Optic Cables & DSL · 210

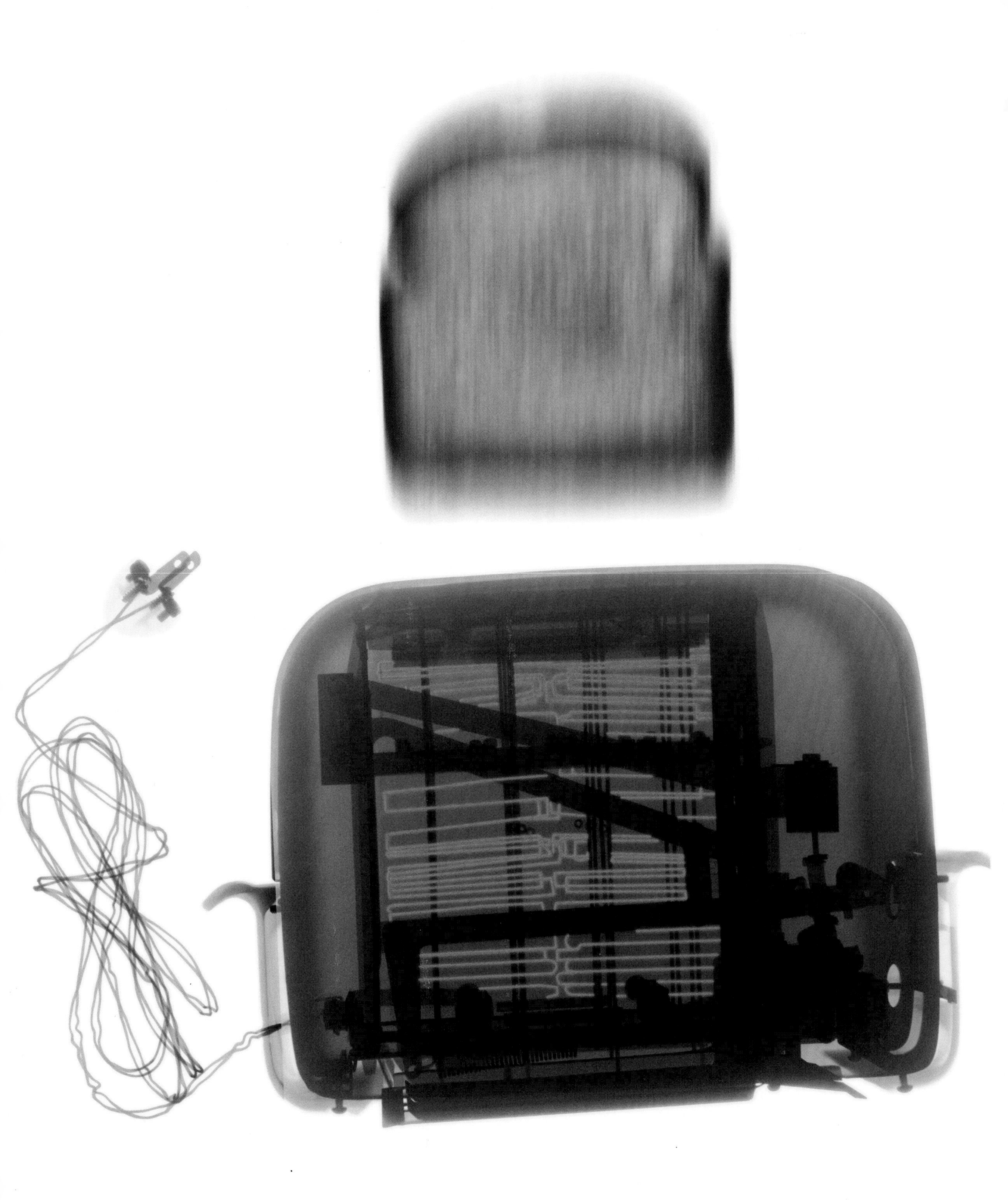

REFRIGERATORS

IN THE DAYS OF THE EARLY ICEBOX, THE SIMPLE process of ice melting inside the enclosure took up a certain amount of heat and kept the icebox cool. If someone covered the block of ice with paper to conserve the coolant, the box would not function properly because the paper kept the ice from melting. The icebox was quickly replaced when the first Frigidaire came off a 1921 assembly line at the Delco Light Plant, a subsidiary of General Motors. Eventually, Frigidaire products found a place in one out of four U.S. homes.

The modern refrigerator also cools by extracting heat, but it does so in a more complicated way. A compressor pumps coolant vapor through sealed tubes; after increasing its pressure and temperature, it routes the vapor outside the box and into a condenser where the coolant releases heat and becomes a liquid. From there, pipes lead the liquid back into the box, through a control valve, and into the evaporator, or freezing unit; it vaporizes in the coils surrounding the unit and absorbs heat, thereby cooling the unit. The warmed vapor returns to the compressor, and the cycle begins again. In a frost-free refrigerator, a fan in the freezer compartment circulates air cooled by the evaporator coils, helping to prevent condensation on the freezer walls. Two or three times a day, a timer activates a defrost heater that melts frost on evaporator coils. Water drains into a pan at the bottom of the refrigerator and slowly evaporates into the room.

Of course, the reason for all of this is to preserve food. Although the cold air does not kill bacteria and molds that are present, it checks their growth and slows the chemical breakdown of food. A freezer compartment is typically 0°F to 10°F, cold enough to preserve food up to a year, and the refrigerator enclosure itself is usually between 32°F and 40°F.

Another addition, the automatic icemaker, replaces the trays in which water is frozen into ice cubes. All that an icemaker generally needs to go through its cycle and make cubes is an electric motor and heating unit, a connection to a water line, a water valve to fill plastic molds, and a thermostat to regulate the water in the molds. When the molds are filled and frozen, a heating coil warms their underside, and ejector blades attached to the motor free the cubes and spew them into a bin.

COLD STORAGE

Refrigerated food is kept from deteriorating by chilled air. It is a principle recognized long ago when people stored foodstuffs in cool caves. Cool air slows the growth of bacteria but, contrary to popular belief, does not kill them.

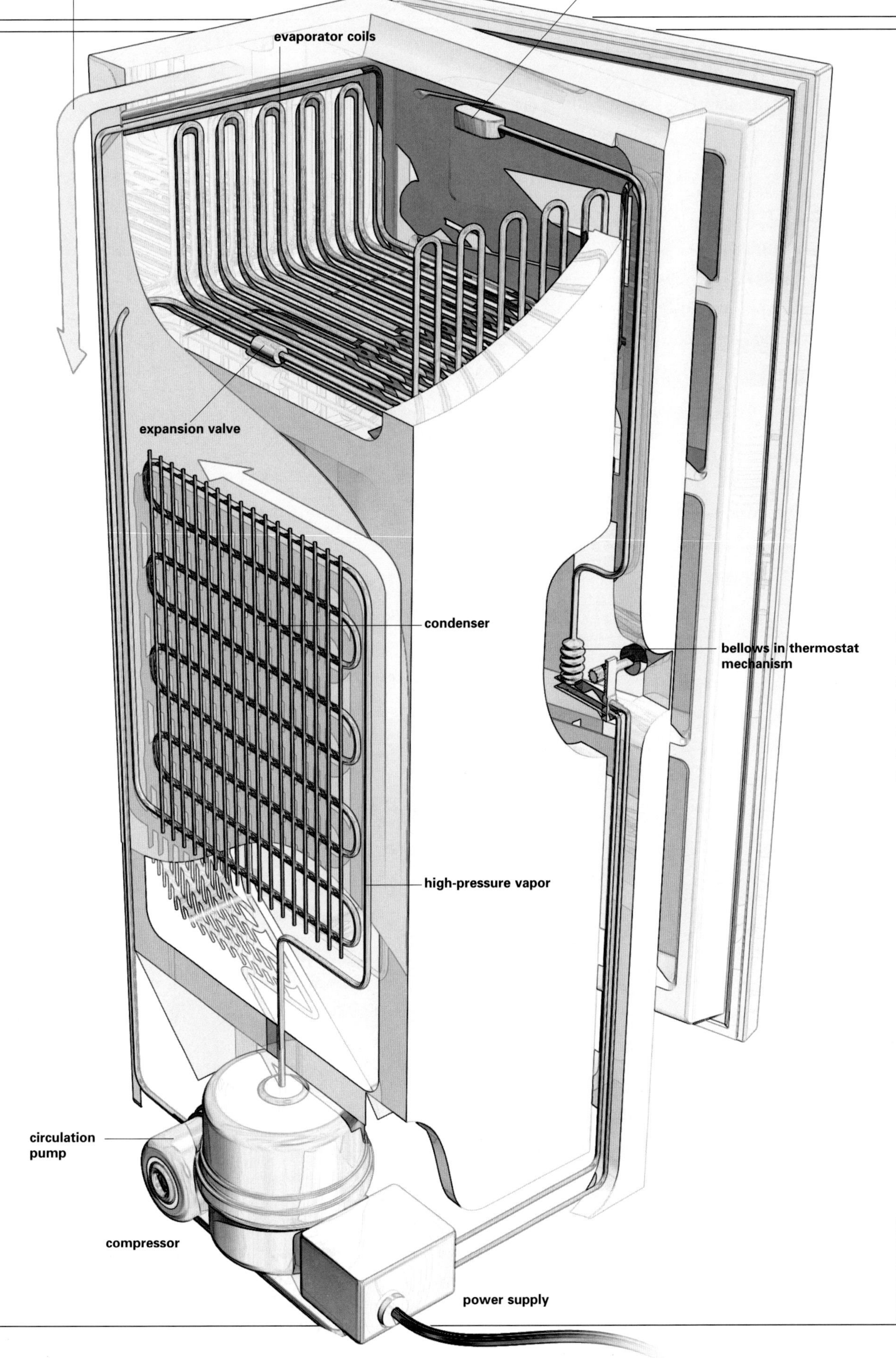

COLD STORAGE

Like an old-fashioned icebox, a refrigerator also keeps food cold. But this common modern appliance does not require a big block of ice melting inside to help it accomplish its task. Instead, a refrigerator relies on coolant circulated through a sealed system of tubes by a compressor. As the coolant evaporates, it absorbs heat from food stored inside the refrigerator compartments. The warmed gas returns to the compressor, which sends it to the condenser, where it cools and becomes a liquid. Then the coolant recirculates through the sealed system.

SEE ALSO
Home Heating & Cooling · 32

VACUUM CLEANERS

THE CARPET SWEEPER USED TO BE (AND STILL IS in some homes) a good way to pick lint and crumbs off floors and rugs. A simple device, it consists of an enclosed pan on wheels pushed about at the end of a long stick; small bits of debris are scooped up into the pan by the roller action of a cylindrical brush mounted on the underside. Although the sweeper is not much better than an ordinary dustpan and brush, it does not require stooping and kneeling.

Electric vacuum cleaners, which made their debut early in the 20th century, use suction, not gathering, to pick up particles sweepers just run over. Three basic systems—upright, canister, and central—all work in essentially the same way: A motor-driven fan creates a partial vacuum that sucks up dirt loosened by a beater brush and deposits it in a bag or another type of collector. (Vacuum cleaner fan blades spin as much as 18,000 times a minute; jet engine blades reach 7,000 to 8,000 a minute.) An upright vacuum is a single unit with a bag usually attached along the backside of the handle. The canister type has a long, detachable cleaning wand attached to a rolling unit that houses a more powerful motor, the fan, and the dust bag.

A central vacuum cleaner has a power unit and a collection canister in the basement, garage, or spare room that is connected to metal or plastic tubes in walls and under floors. To activate the system, a cleaning hose and its attachment-tool are plugged into inlet valves in the walls. Dust and debris are sucked through the tubes and sent to the dirt-collection canister, which has to be emptied only a few times a year. More powerful and easier to maneuver than a portable vacuum cleaner, a central unit is also quieter, and it doesn't spray out any of the fine dust particles that flow through a bag's porous walls.

Yet another variation on vacuum cleaner technology is the battery-operated robot cleaner that scoots over floors, and under beds and tables—all without having to be pushed or plugged in. Resembling a king-size CD player, they are equipped with wheels, rotating brushes and sweepers, bumpers, and a small vacuum system. Their computerized "intelligent" navigation systems and infrared or ultrasonic sensors keep the vacuum cleaner away from obstacles, including pets, and from toppling off stairs. Powered by nickel-hydride batteries, robot cleaners can run about an hour before the batteries need recharging, and they can be confined to specific areas by magnetic strips placed near open doorways, or by electronic devices that set up invisible "walls."

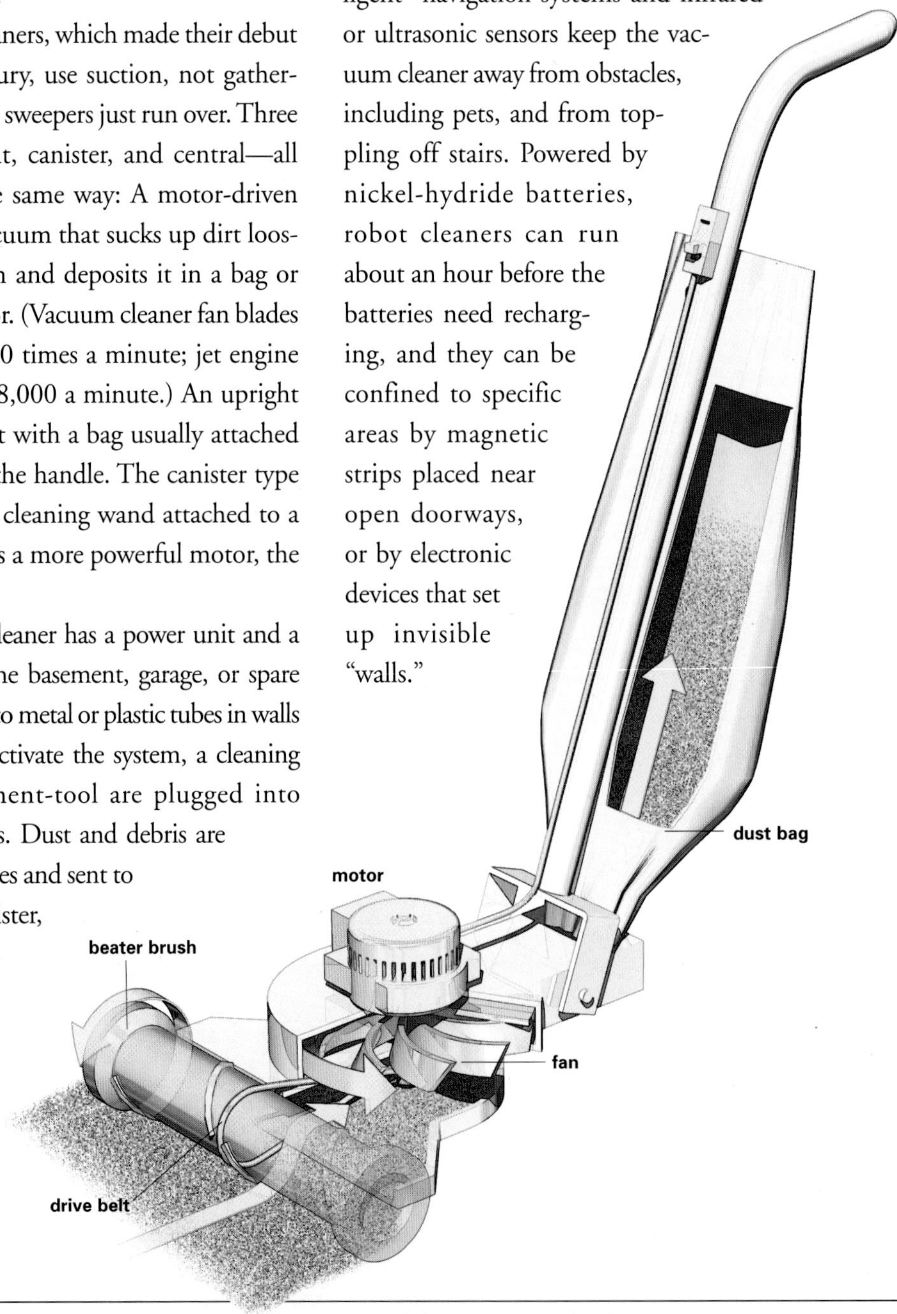

motor
beater brush
fan
dust bag
drive belt

HOUSEHOLD DUST

(Right) A vacuum cleaner may remove it from sight, but in an electron micrograph dust appears as a formidable opponent. Loaded with an assortment of substances, simple household grit can cause asthma and other allergic reactions. This batch has cat fur, twisted synthetic and woolen fibers, serrated insect scales, a pollen grain, and plant remains.

IT'S IN THE BAG

(Left) In the scientific sense, a perfect vacuum means a complete absence of matter within a space—a rare, if not impossible, phenomenon. In a more familiar sense, a perfect vacuum describes a powerful cleaning machine that removes a great deal of matter from our rugs. An upright unit uses a partial vacuum that fills a bag with so much dirt that it requires frequent emptying. Air blown through the unit by a motor-driven fan escapes through the bag's porous walls, but dirt remains trapped inside.

SEE ALSO
Small Appliances & Hand Tools · 24
Home Heating & Cooling · 32
Batteries · 56
Jet Engines & Propellers · 104

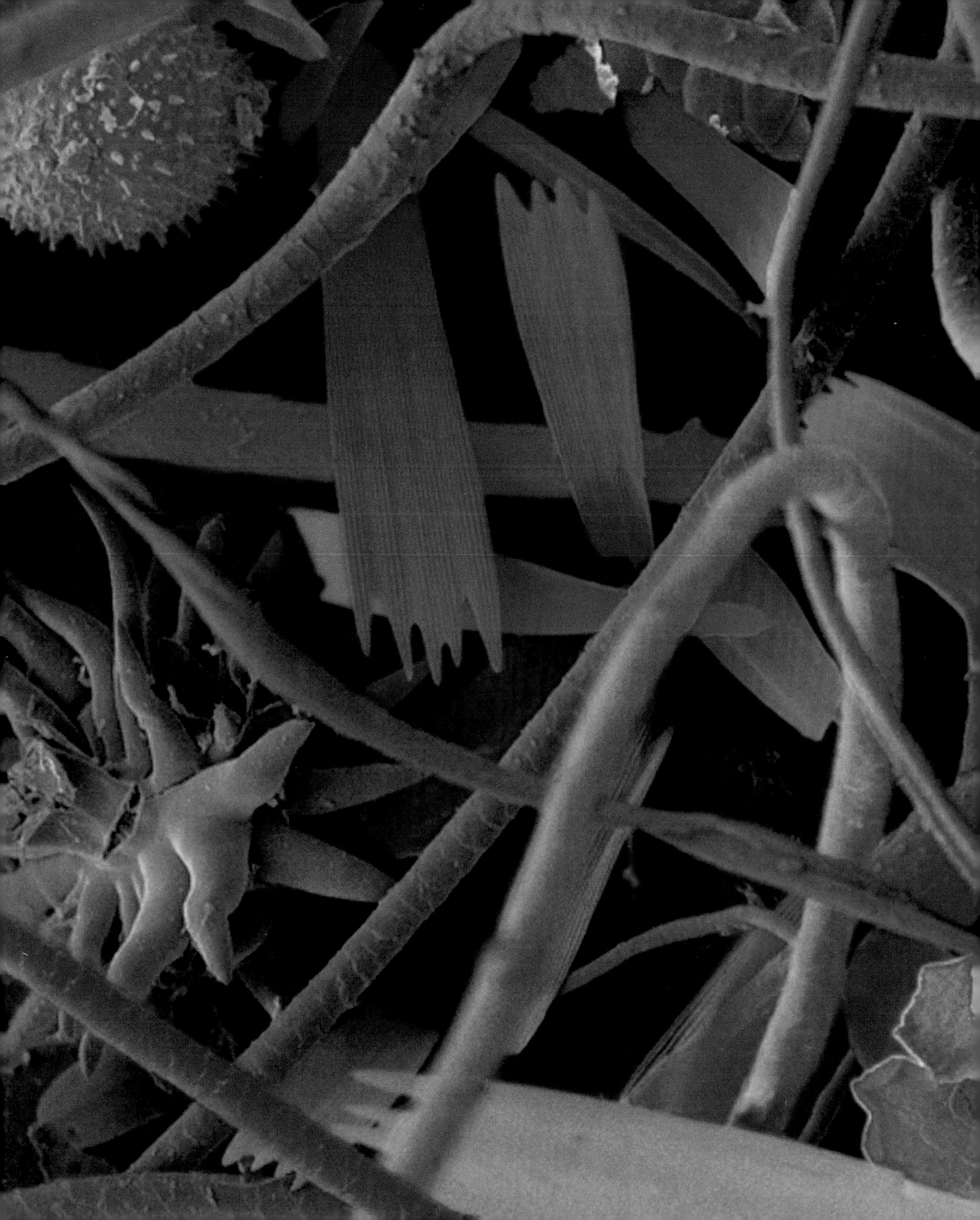

WASHERS & DRYERS

BEFORE THE VERSATILE WASHING MACHINE WAS invented, people used washboards to scrub, or they carried their laundry to riverbanks and streams, where they beat and rubbed it against rocks. Such backbreaking labor is still commonplace in parts of the world, but for most homeowners the work is now done by a machine that automatically regulates water temperature, fills, measures out the detergents and bleach, washes, rinses, spin-dries, and empties. With its intricate electrical, mechanical, and plumbing system, the washing machine is one of the most technologically advanced examples of a large household appliance. It not only cleans clothes, but it does so with far less water, detergent, and energy than washing by hand requires. Compared with the old wringer-type washers that squeezed out excess water by feeding clothes through rollers—and emptied wastewater into the kitchen sink through a hose—modern washers are indeed an electrical-mechanical phenomenon.

Manufacturers produce two types of washing machines: top-loaders and front-loaders. The top-loading model, which is generally more popular, relies on a motor-driven agitator unit that wraps and twists clothes around during the wash and rinse cycles. Pumps circulate, recirculate, and discharge water, while timers, switches, and sensors regulate the flow and temperature of the water, as well as the spin process. Some top-loaders can use an average of 40 gallons of water for each load of wash.

Front-loading machines are built without agitators. Instead of dragging and shoving clothes through a cycle, these machines churn the laundry about in high-speed, rotating drums. According to their manufacturers, front-loaders use 20 to 25 gallons of water for each washload.

The washer's companion is the dryer—electric or gas-powered—and, like electric and gas furnaces, this machine circulates the heat from coils or burners to do its job. A fan draws air into the dryer, where it is heated, passed through spin-damp clothes, and forced through a lint trap. The air is then ejected through an exhaust vent.

TAKE-OUT LAUNDRY
Laundromats, like this one, are the welcome answer for travelers and those with limited household space. Coin-operated and self-service, today's Laundromat traces its ancestry to the nation's first: "The Washeteria," which opened in Fort Worth, Texas, in 1936.

SEE ALSO
Home Heating & Cooling · 32
Pipes, Pumps, & Plumbing · 34

DETERGENT

A detergent molecule has a "nonpolar" and a "polar" end. The nonpolar end attaches to oily dirt while the polar end dissolves in the surrounding water, ionizing to form positive charges. These charges repel the ones on other soap-encrusted dirt particles, keeping the dirt suspended and allowing it to be rinsed away.

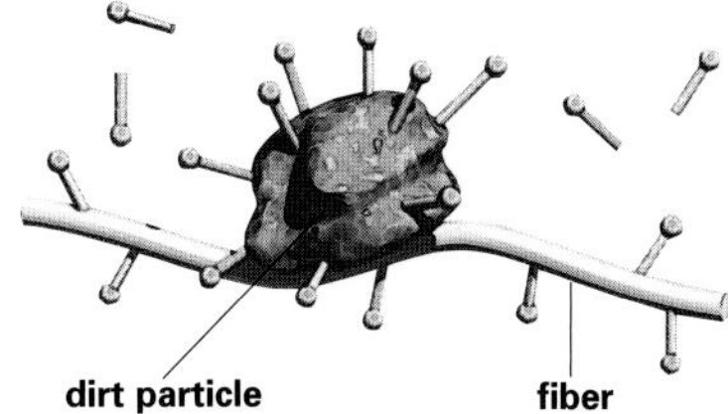

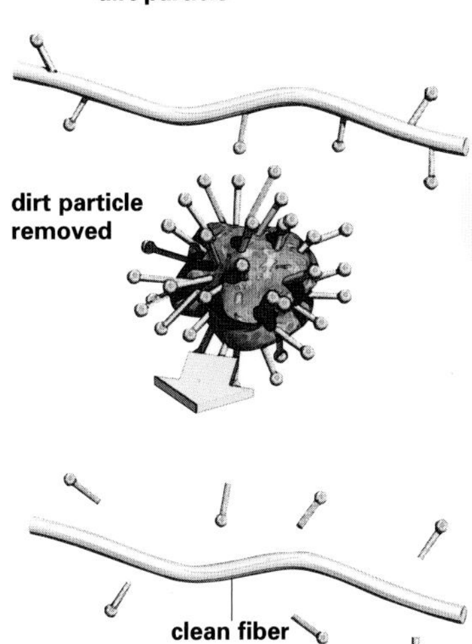

TAKING IT FROM THE TOP

Cone-shaped, motor-driven agitator in a top-loading washer moves clothes back and forth in a tub filled with soapy water. When the wash cycle ends, a timer directs the motor to free the agitator and spin a perforated inner basket. Centrifugal force spins water through the holes into the outer tub and presses the clothes against the sides of the basket. After the water drains out, valves open and refill the tub for the rinse cycle. A final spin damp-dries the laundry. Springs attached to the tub and the unit's frame keep the washer stable during the unavoidable vibration.

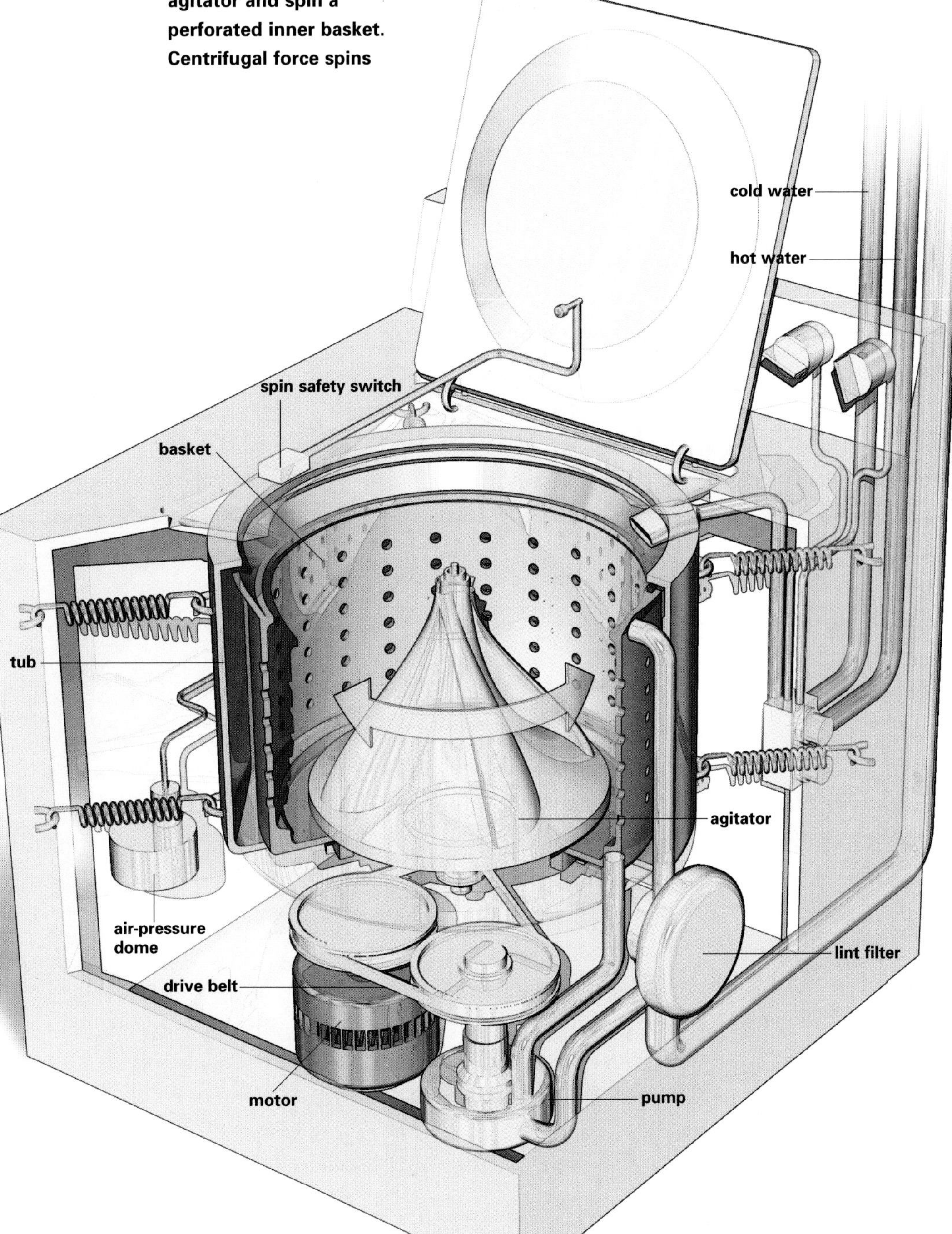

SMALL APPLIANCES & HAND TOOLS

Small household appliances and power tools require fairly strong motors that are capable of varying speeds. Many rely on universal motors, which are power units that operate on either direct or alternating current. To fulfill its role, an appliance might have a unique combination of gears, cams, drive belts, pivots, shafts, rotors, and pinions. These connect the motor itself, directly or indirectly, to the various attachments that do things like dice an onion or turn a screwdriver.

Appliances often contain gears, which transmit motion and power from one part of a machine to another and may be called upon when more twisting and turning force than usual is needed. Reduction gears slow an appliance when too much speed is undesirable. Motors may also be used to drive cams or other mechanisms that translate rotary motion into linear motion.

Many small appliances do not need gearing because their attachments work best when they are drawing full power directly from the motor. For example, while an electric mixer cannot be a direct extension of the motor's shaft because the speed would make a mess of the mix (a mixer employs a worm gear to slow down the action), a fan or the cutting blade in a food processor works directly off the motor, spinning at the same high speed.

When a homeowner turns a screw, he or she applies force to what is essentially a lever and, in so doing, produces torque, a rotational force. The effectiveness of the torque-producing effort depends on both the force exerted and the length of the lever's "arm." Thus, the handle of a screwdriver becomes more than a grip: It magnifies the force the hand uses to turn the blade and drive in a screw. The same thing is true of a pair of pliers twisting a nut. Both amplify the manual strength applied.

Power tools, such as drills and screwdrivers, draw extra strength from a nicely meshed system of gears attached to a chuck, a jawlike fixture that holds the drill bits and the screwdriver blades. The tools also enjoy variable speed because of regulators that control the flow of electricity through the brushes to the motor's rotating armature.

Like cordless toothbrushes and shavers, cordless shop tools depend on small, efficient motors and batteries, those ever present devices that convert stored chemical energy to direct-current electricity. Nickel-cadmium batteries that deliver about 1.3 volts have been partly responsible for the popularity of cordless appliances. Also, such tools can easily be recharged by being snapped into stands linked to household current. The current—its voltage reduced by a small transformer in each unit—is applied in the direction opposite the one in which it flows when power is dispensed, a restoration process taking several hours. Some battery-driven appliances can be plugged directly into a wall outlet, a feature that allows them to receive constant charging when not in use.

ESPRESSO MACHINE

An espresso-maker brews a strong cup of coffee by forcing steam through finely ground, dark beans. While coffee-drinking dates from 15th century Arabia, the modern espresso machine was created only in 1938 by Italian Achilles Gaggia.

SEE ALSO
Saws & Lathes · 26
Electrical Wiring · 36
Power Stations · 44
Batteries · 56

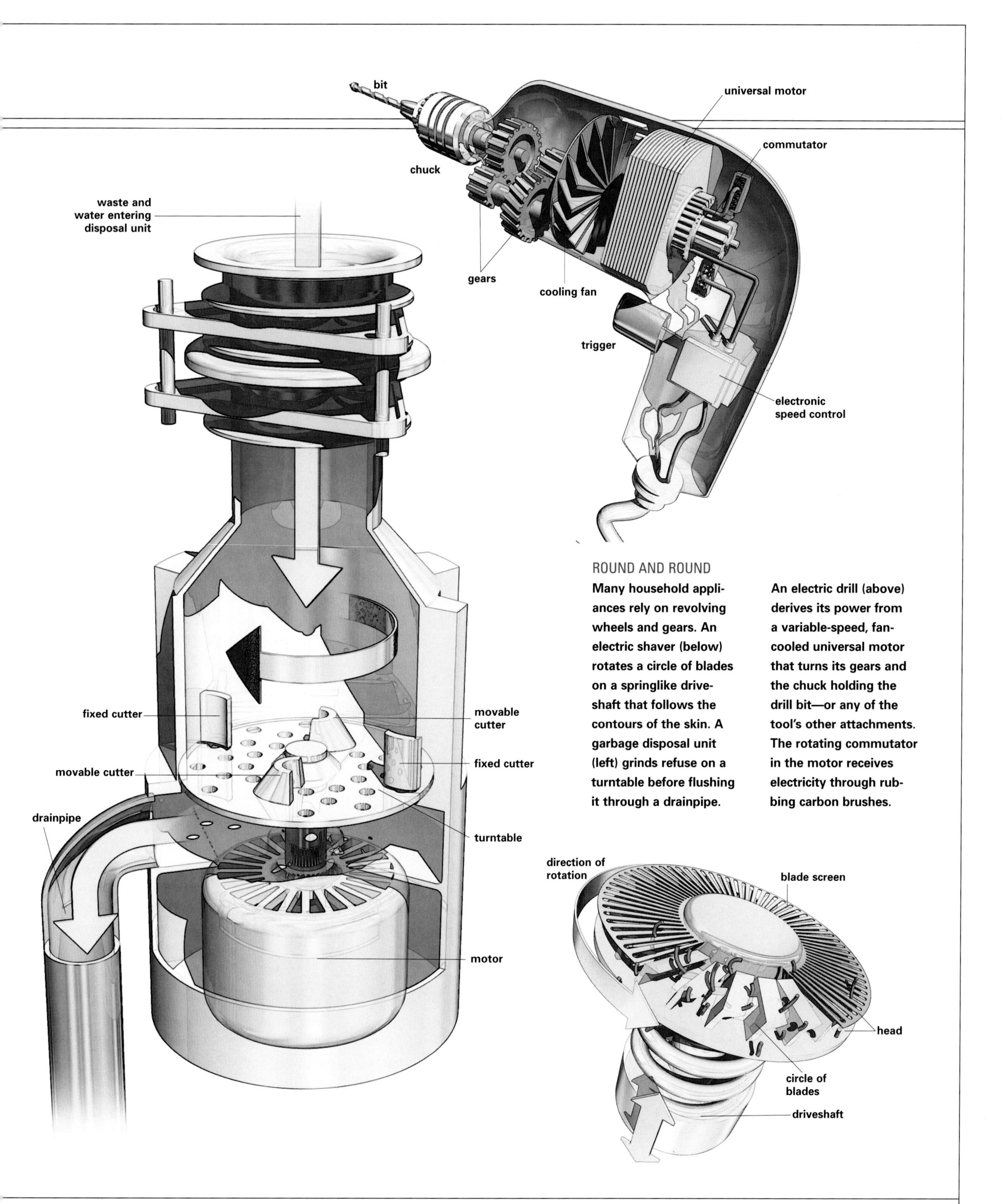

ROUND AND ROUND

Many household appliances rely on revolving wheels and gears. An electric shaver (below) rotates a circle of blades on a springlike driveshaft that follows the contours of the skin. A garbage disposal unit (left) grinds refuse on a turntable before flushing it through a drainpipe.

An electric drill (above) derives its power from a variable-speed, fan-cooled universal motor that turns its gears and the chuck holding the drill bit—or any of the tool's other attachments. The rotating commutator in the motor receives electricity through rubbing carbon brushes.

SAWS & LATHES

ROTARY MOTION—THE TURNING OF AN OBJECT around a center point, or axis—may be more important to machines than parts that move in linear fashion. Without it, there would be no wheels, no gears, no belts, no motors, and no enormously useful tools such as electric saws and lathes.

A handsaw, as everyone knows, is a simple cutting instrument whose blade has sharp, pointed teeth that abrade wood or other materials. Saber saws and other electric versions mimic the back-and-forth motion of a handsaw, while requiring circular gears and rotating motors to do so. Thin-bladed jigsaws move up and down to cut curves and to crosscut, bevel, and begin a cut in the middle of a wood panel. In some saws, the blade itself is circular and rotated by an electric motor. These power saws, which are useful in the construction industry, have a variety of teeth, from coarse to fine, and can quickly saw large quantities of wood. Chain saws are portable saws with blade-studded steel links that form an endless chain. Powered by a gasoline engine or an electric motor, a chain saw runs its cutting edge on an oval guide bar at a high rate of speed.

A circular saw has a motor that rotates a sharp, round blade at high speed to cut a straight line. The blade adjusts for depth and angle of a cut, while a blade guard springs into safety position when the saw finishes its work.

Another form of cutting tool, the lathe, uses rotating motion to turn wood, metal, and other materials into cylindrical, tapered, and conical objects. The material to be shaped is held at each end of the machine and rotated against a cutting tool.

LATHE

(Right) Chips fly as a carpenter shapes wood fixed to a lathe. This basic turning tool works by rotating an object, such as wood or metal, about a horizontal axis. A cutting tool then moves across or parallel to the rotational direction, shaping the object as it turns.

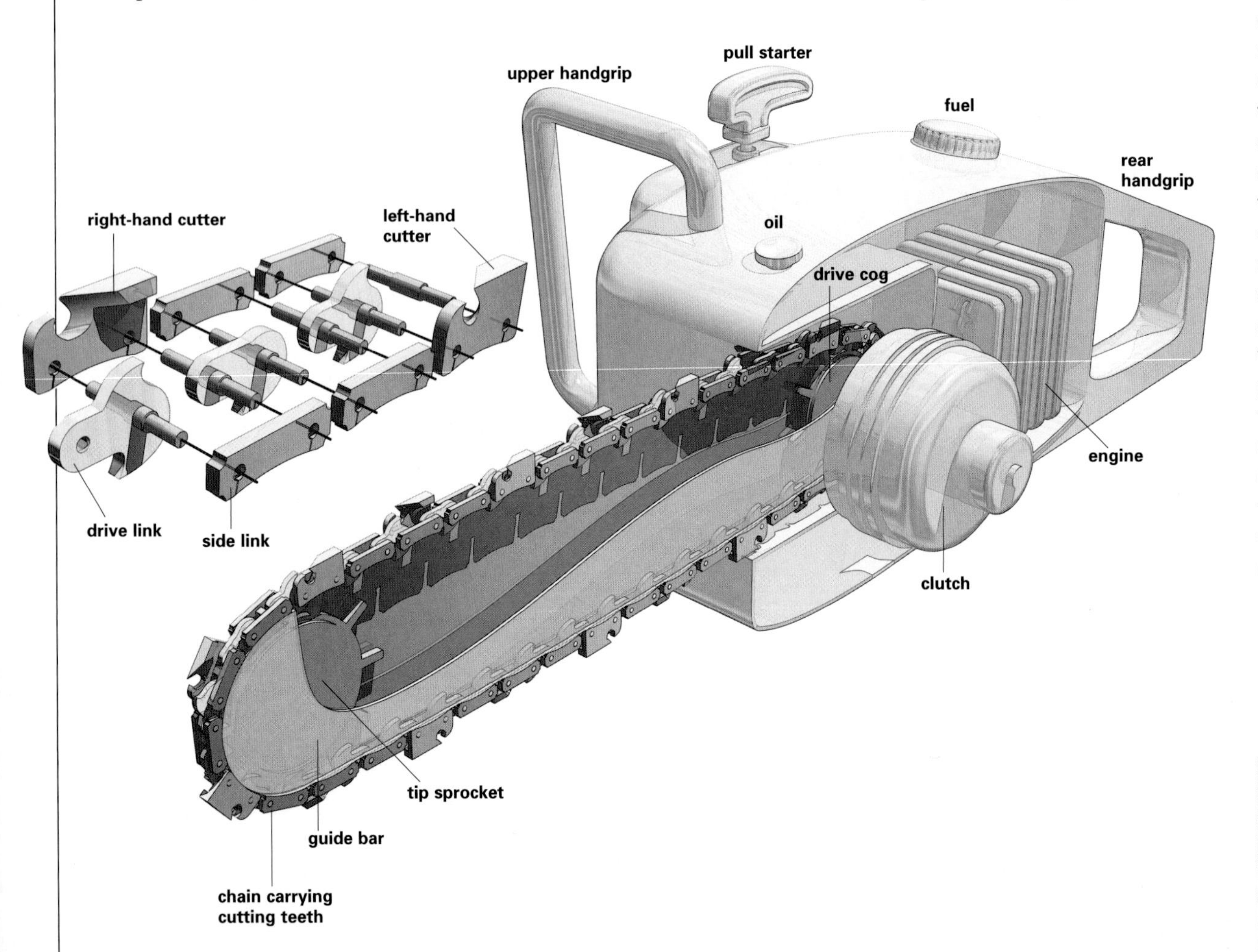

CHAIN SAW

(Left) A favorite of lumberjacks and farmers, the chain saw has no match when it comes to mechanical brawn and mobile cutting power. Driven by a gasoline engine or an electric motor, the chain saw relies on its dominant feature: the blade-studded, bicycle-like chain (see close-up at left) that hurtles around an oval guide bar. Like lawn mowers with internal combustion engines, the gasoline versions require fuel and oil tanks, spark plugs, mufflers, carburetors, and clutches. Electric models run on universal motors; shafts turn gears that drive sprocket-and-clutch arrangements to move their chains.

SEE ALSO
Small Appliances & Hand Tools · 24
Construction Equipment · 62

SECURITY SYSTEMS

Our natural gifts of sight, smell, hearing, touch, and taste are not necessarily enough when it comes to accurately measuring, and protecting us from, some of the often hazardous things that transpire around us. Neither is merely putting our trust in a door to provide security just because it is shut. So, engineers and scientists have come up with a spate of devices that help us ensure the safety and good health of our families and ourselves. Such devices include barometers, thermometers, pressure gauges, scales, Geiger counters, water and gas meters, seismographs, spirometers (for measuring air inhaled and exhaled during respiration), sphygmomanometers (for measuring blood pressure), and, of course, locks.

Smoke alarms and security systems exemplify the world of artificial senses. Smoke alarms "smell" fumes in different ways, mostly by interacting on a molecular level with the particles that make up the smoke. An ionizing alarm is armed with a bit of radioactive material in a chamber, and this material gives off atomic particles that form ions, which are charged atoms or molecules. A current is created in the ionized air, and when smoke particles attach themselves to the ions they disrupt the current and set off the alarm.

A photoelectric smoke detector has a photosensitive element, a cell popularly referred to as an electric eye. It has electrical properties that vary with the light that falls on it. Interrupting a light beam in the system opens an electrical circuit that then powers a work-performing mechanism, a process that enables photocells to open and close garage doors and trip burglar alarms. In a smoke detector, a steady light is beamed against a dark surface inside; if smoke enters the chamber, it scatters particles of light to a photocell that triggers the alarm.

Other security systems use photocells to measure light levels so that they can turn lights on and off. They also detect infrared radiation given off by the body of an intruder. To detect an object's density and motion, other devices rely on reflected sound waves that are above the range of human hearing.

FLAME MONITOR

A sensor relays information from a contained fire to a computer for analysis of its erratic behavior. New detection systems use a fire's fluctuations to separate dangerous ones from simple flashing lights and candles.

SEE ALSO
Electrical Wiring • 36
Nuclear Power • 52
Computerized Buildings • 66
Managing Money Electronically • 244

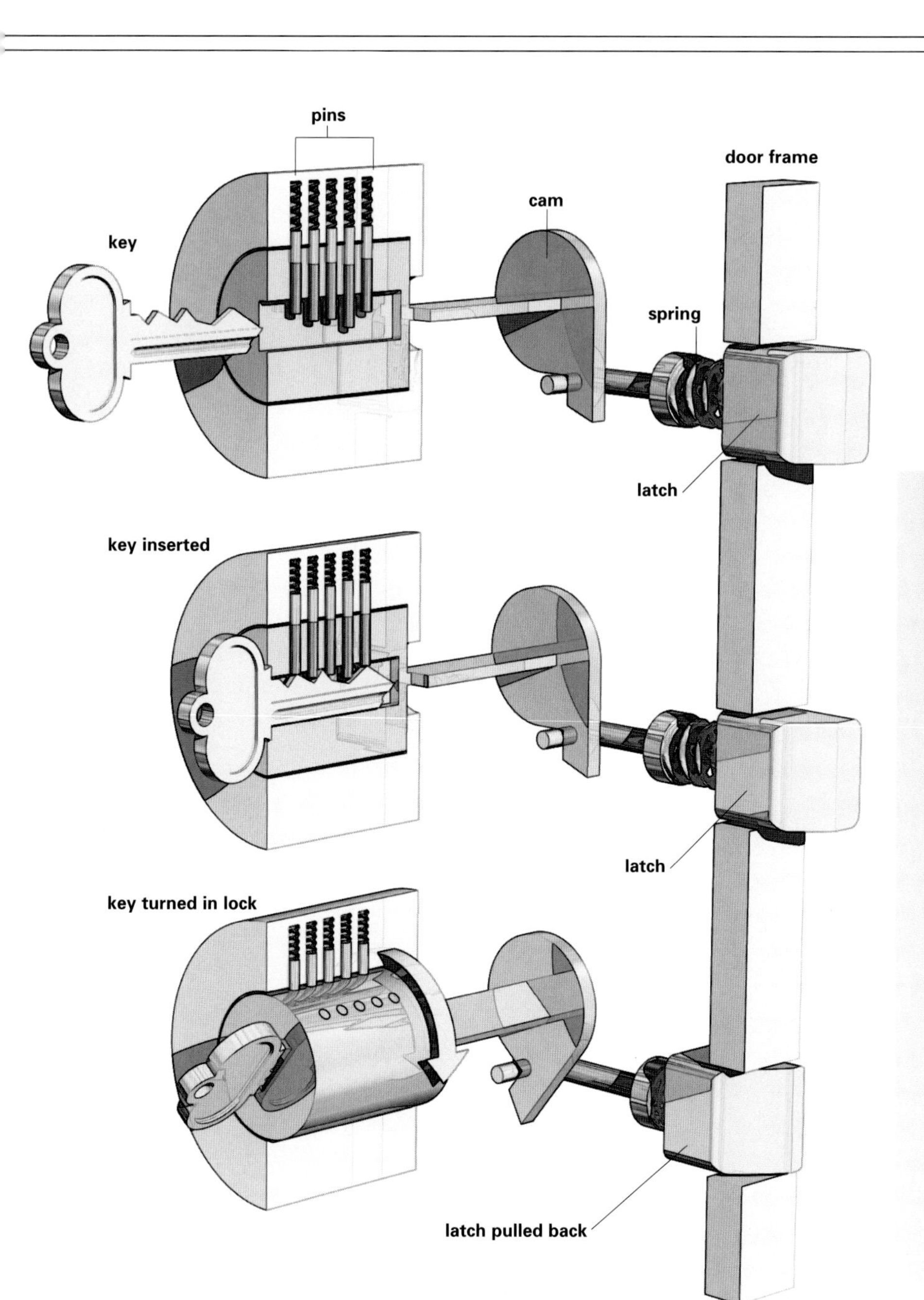

PIN-TUMBLER
CYLINDER LOCK

By blocking the shear point between a lock's plug and shell, the drivers and pins keep a plug locked in place. If the right key enters the lock, the pins settle into its notches, clearing the shear point and letting the key rotate the plug.

COMBINATION LOCKS

"Open, locks, Whoever knocks!" intoned one of the witches in Macbeth —not especially comforting to those trying to protect themselves and their property from intruders. But we try, at least, with locks of all kinds, shapes, and methods of operation. Among the most secure are combination locks. One form, like that outside a hotel room, uses an electronic code in a door slot, and a card to open it. Similar to an ATM's operation, when the card with an attached magnetic strip carrying the code is inserted in the slot, the lock reads the card and the door opens. A check-in computer sets the card's coded combination, and when a guest leaves, the computer erases the code, installs a new one, and sends the information to the room's lock.

Perhaps more familiar is the combination lock that secures safes and padlocks doors and bikes. The lock may be activated by twisting a dial, pushing buttons, or manipulating a row of movable numbers, as on a suitcase. Whatever the form, combination locks of this type open with the entering of a correct sequence of code numbers, which have either been factory set or changed by the owner. Inside the dial-type lock, a series of notched disks replaces the pin-tumbler mechanism in a cylinder lock. The disks catch onto protrusions on the interior part of the dial; when the dial is turned, the disks turn, and when the proper sequence of numbers is selected to align the notches, the bolt moves into a space created by the notches, and the lock opens.

CLOCKS

How did people tell time before they had clocks? They observed stars at night, the shadow of a sundial's gnomon, marks made on a candle, water that trickled through a clepsydra (water clock), or the grains of sand that sifted through an hourglass.

The oldest surviving mechanical clocks date from late 14th-century France. While the principles behind early clocks still apply to modern timepieces, such innovations as atomic clocks, illuminated dials, jeweled bearings, quartz movements, and digital alarms with snooze settings make the early versions seem closer to ancient water clocks than to our fancy hex bezels.

Frills and variations aside, clocks are essentially boxes containing several toothed wheels, pins, and springs, all precisely arranged to measure that impalpable continuum called time. In the basic clock, power is supplied by electrical impulses or by a falling weight or an unwinding spring that can be rewound when necessary. Movement of the hands is regulated by a train of toothed wheels, pinions, and spindles of different sizes, each taking its own appointed time to make a full revolution. In some clocks, a pendulum controls the wheel train's rate of rotation. In other timepieces, a mainspring turns a driving wheel that moves the minute and hour wheels. These wheels are connected to hands on the face of the watch or clock. The minute wheel turns so that the minute hand takes one hour to go once around the face. In 12 hours, it turns 12 times. The hour wheel turns so that the hour hand takes 12 hours to complete one revolution.

In the quartz versions, a small electrical charge is applied to a quartz crystal; the crystal begins to vibrate and give off pulses of current in a precise, predictable manner. These pulses, in turn, can be used to control the motor turning a clock's hands or to advance the numerals displayed by the liquid crystals in a digital display. (In a digital clock, alternating current powers the timepiece, and its precise cycles mark time's passage. Seven light-emitting diodes make up each digit and form numbers when energized in the correct combination.)

Quartz vibrations are known as a piezoelectric reaction, and when they are put to work in a clock or a watch, they make for a timepiece that is far more accurate than a mechanical one. Couple the pulses of current from the oscillating quartz with a microchip that reduces their frequency to a usable rate—one pulse a second—and you have a clock or a watch that relies on minuscule electronic circuitry instead of toothed wheels and a mainspring.

While a quartz-crystal clock has an error rate of less than one-thousandth of a second per day, the latest generation of atomic clocks can be accurate to plus or minus one second in ten million years. Atomic clocks are faceless and handless wonders that trade on the constancy of the frequency of a molecular or atomic process. Put another way, atomic clocks use the extremely fast vibrations of molecules or atomic nuclei to take the measure of time. Because the vibrations remain constant, these clocks measure short intervals of time with much more precision than a mechanical clock.

FORMIDABLE TIMEPIECE

(Left) A classic alarm clock stands ready to rouse a sleeper with a nerve-shattering jangle. A powerful mainspring runs the alarm mechanism and the rest of the clock, and an alarm "hammer" does the waking by swinging back and forth against metal gongs.

CLOCKWORK

(Right) Powered by an unwinding mainspring, a clock also depends on the linkage of precisely machined gears and on a hairspring that winds and unwinds to control the movement of the balance wheel, which regulates the escape wheel. A main driving wheel meshes with the teeth of a gear wheel and turns as the mainspring unwinds, controlling the movements of the minute and hour wheels. The number of their teeth determines the number of revolutions that the hands will make.

SEE ALSO
Small Appliances & Hand Tools · 24
Computer Chips & Transistors · 232

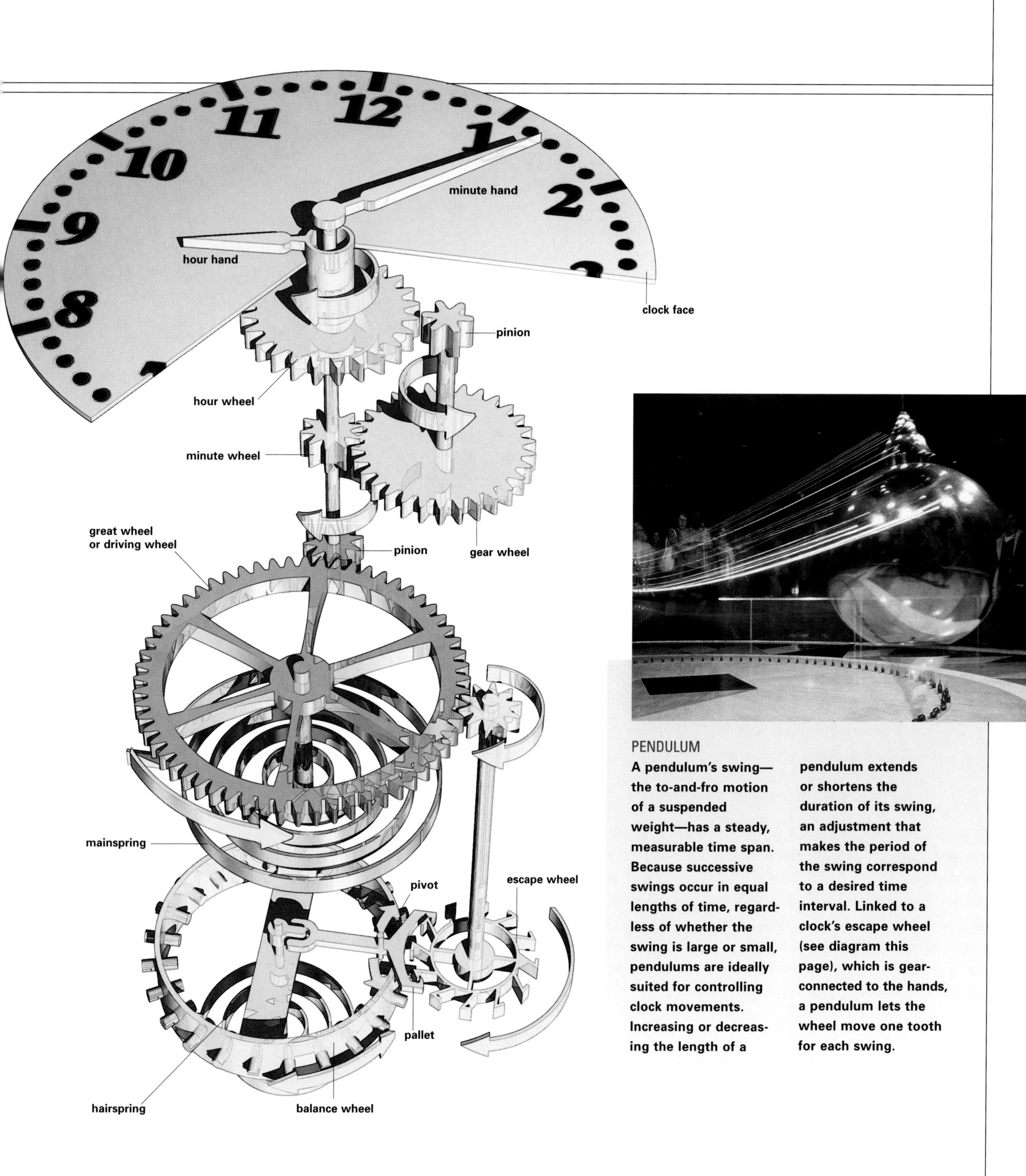

PENDULUM

A pendulum's swing—the to-and-fro motion of a suspended weight—has a steady, measurable time span. Because successive swings occur in equal lengths of time, regardless of whether the swing is large or small, pendulums are ideally suited for controlling clock movements. Increasing or decreasing the length of a pendulum extends or shortens the duration of its swing, an adjustment that makes the period of the swing correspond to a desired time interval. Linked to a clock's escape wheel (see diagram this page), which is gear-connected to the hands, a pendulum lets the wheel move one tooth for each swing.

HOME HEATING & COOLING

To warm their homes, the Romans devised a heating system that used wood- or charcoal-burning furnaces to supply hot air to rooms through tile pipes or channels under floors and inside walls. Today, gas and oil burners are at the heart of the home heating systems. They can also be modified to cool through a central air-conditioning unit. Beyond that, "intelligent" systems armed with battery-assisted computer-thermostats can monitor a home and calculate when it should turn on the air conditioning or the heating.

In the basic heating system, whether gas or oil, both must be burned, and both need to be mixed with air to do so. In a gas furnace, the gas-and-air mixture is ignited, and the heat produces hot water, steam, or hot air. A forced-air or hot-water circulating system spreads the heat throughout the home as waste gases go up a flue. An oil burner also sets fire to a mixture of fuel and air; in this system the fuel is first pumped from a tank and converted into a fine spray. It is then burned in a combustion chamber. Convector units then disperse the heat.

For cooling, the simplest device is the electric fan. It simply moves air around; or it pushes hot air out of a warm room. When fans are used with refrigerant, the coolants dissipate unwanted heat.

Window units and central air conditioners remove heat and humidity from enclosed spaces and eject it outdoors, reducing the air temperature inside with the aid of a closed system containing a refrigerant. All such systems are based on these principles: When a liquid vaporizes, it draws heat from its surroundings; when a gas condenses back to a liquid under high pressure, heat is released.

In central air-conditioning, known as a "split system," an evaporator coil is mounted on the furnace, and a condensing unit rests on a concrete slab just outside the home. A small tube carries liquid coolant in from the condensing unit, and a larger tube returns the coolant in a gaseous state to the unit outside. The furnace blower then draws in warm air from the house and sends it over the evaporator. As coolant in the evaporator changes from a liquid to a gas, it absorbs heat, thereby cooling the air, and causing it to lose moisture. The blower forces the cooled, dehumidified air into the supply ducts, and from there it circulates through the house.

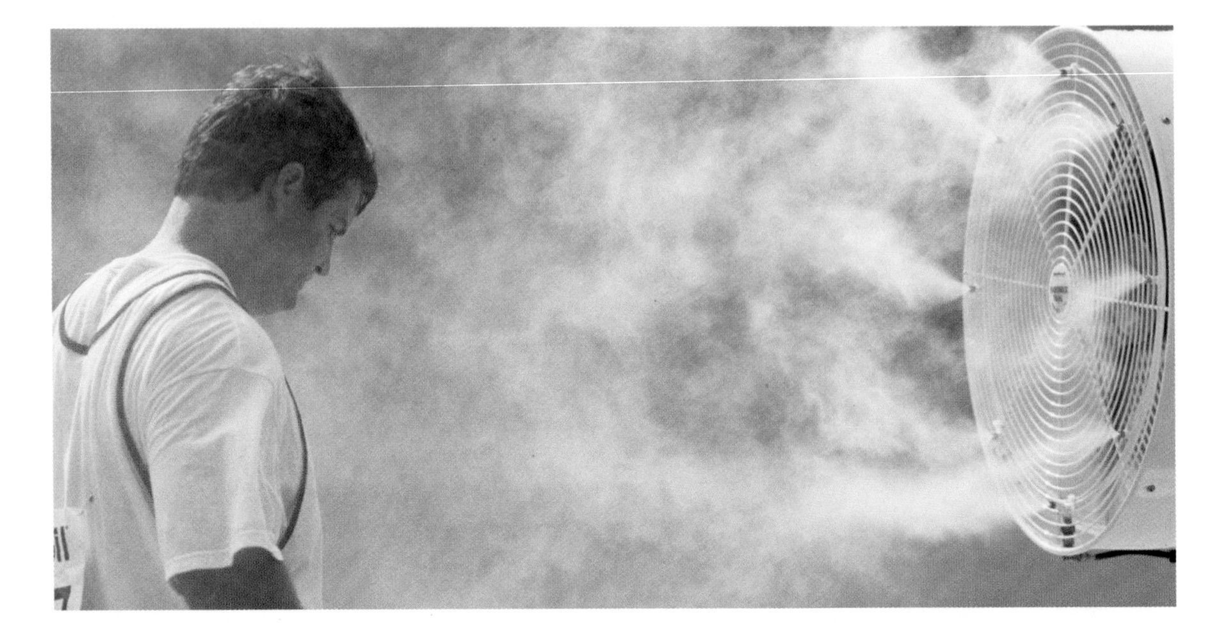

COOL DOWN

Seeking comfort, an Olympic hopeful at the 1996 trials in Athens cools off in front of a fan/mister in the hot, Georgia summer heat.

SEE ALSO
Refrigerators · 18
Solar Power · 50
Computerized Buildings · 66

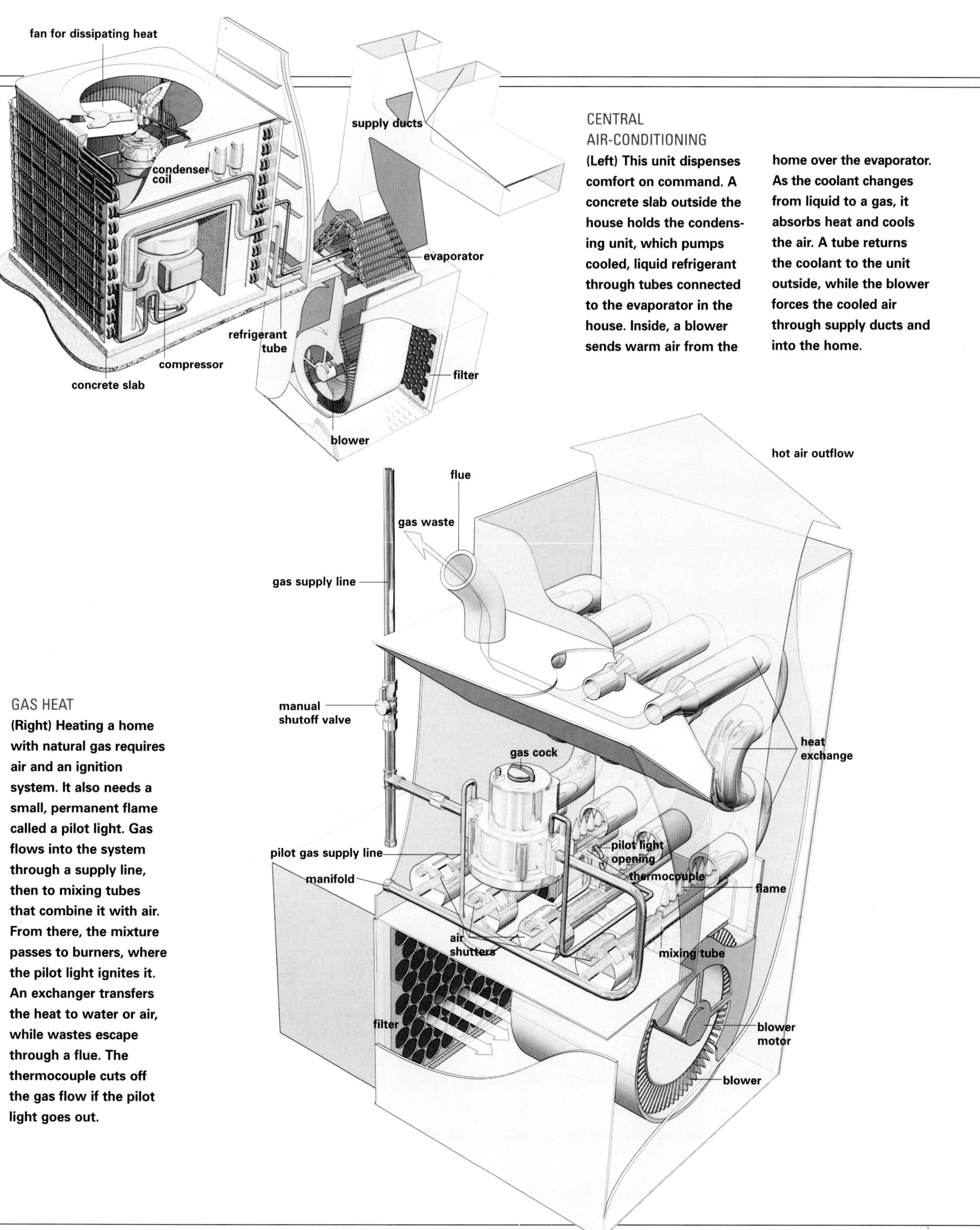

CENTRAL AIR-CONDITIONING

(Left) This unit dispenses comfort on command. A concrete slab outside the house holds the condensing unit, which pumps cooled, liquid refrigerant through tubes connected to the evaporator in the house. Inside, a blower sends warm air from the home over the evaporator. As the coolant changes from liquid to a gas, it absorbs heat and cools the air. A tube returns the coolant to the unit outside, while the blower forces the cooled air through supply ducts and into the home.

GAS HEAT

(Right) Heating a home with natural gas requires air and an ignition system. It also needs a small, permanent flame called a pilot light. Gas flows into the system through a supply line, then to mixing tubes that combine it with air. From there, the mixture passes to burners, where the pilot light ignites it. An exchanger transfers the heat to water or air, while wastes escape through a flue. The thermocouple cuts off the gas flow if the pilot light goes out.

PIPES, PUMPS, & PLUMBING

Historian Will Durant tells us that Romans carried plumbing "to an excellence unmatched before the twentieth century." Pipes made of lead brought water from aqueducts and mains into homes; fittings and stopcocks were of ornamented bronze; and leaders and gutters carried rainwater from rooftops.

A modern domestic plumbing system, more of a maze than anything the Romans ever constructed, is essentially an inner skeleton of main and secondary pipelines broken in places by shutoff and bypass valves, faucets, traps, toilets, showerheads, tanks, clean-out plugs, water heaters, and drains. Receiving and discharging water are, of course, the reasons for plumbing. Water from a municipal supply or private well usually enters a home under pressure and is routed through hot- and cold-water lines to the fixtures. When the water must be hot—for the bathroom shower or the automatic dishwashing machine, for example—it enters from the cold-water main and flows through a cold-water inlet into a gas-fired or electric water heater. In a thermostat-controlled tank, it is heated to the proper temperature and then sent to the hot-water faucet via a separate outlet pipe. Cold-water taps, along with the toilet, are linked directly to the cold-water drain.

Unlike the intake system, the drainage system is not pressurized; it relies on gravity to dispose of water that was flushed from a toilet or let out of a sink. Fixtures are connected to a large drainpipe by curved traps filled with water; the water acts as a seal to keep potentially dangerous sewer gases from flowing into the home. At the bottom of the system, a main drain transports waste to a town sewer line or to a septic tank on the property.

Today's plumbing systems are far more efficient, with automatic sensors, timing and flow-control mechanisms, and devices that monitor and deal with changing water pressures.

THE BATHTUB
(Above) A tub can be a repository for more than bathwater. Once just wood or metal barrels, today's tubs are constructed of cast iron, enameled steel, and fiberglass, and may have ergonomically designed arm and back rests, with perhaps a whirlpool system built in.

HOME NETWORK
(Right) Risers and supply-branches (shown in red for hot water, blue for cold) distribute water from a main line. Wastewater flows out through drainpipes (green) connected to the fixtures by water-filled traps. From the traps, water goes into a vertical soil stack and out into sewers or septic tanks. Vent pipes link drainpipes to rooftop stack vents and rid the system of gases.

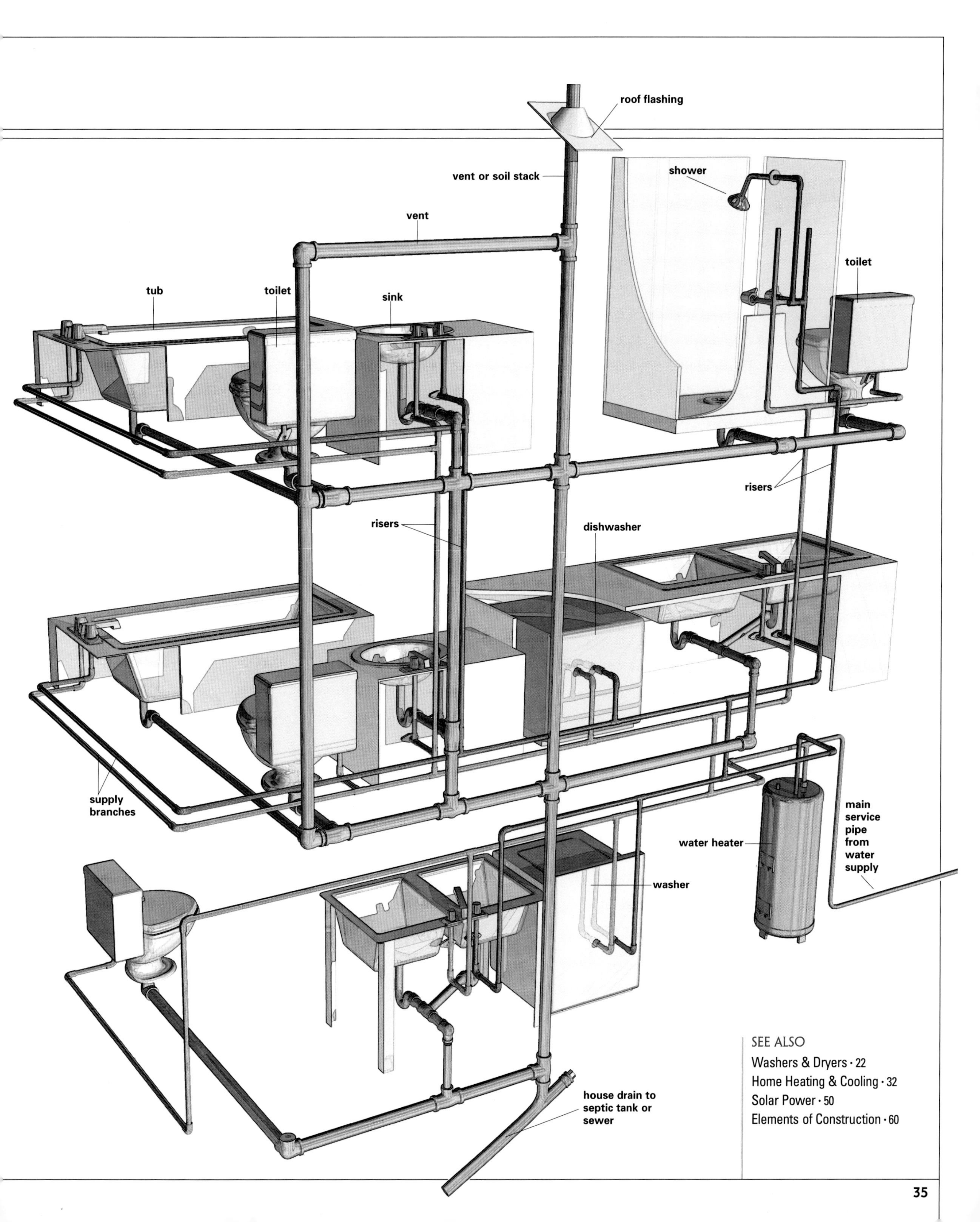

SEE ALSO

Washers & Dryers • 22
Home Heating & Cooling • 32
Solar Power • 50
Elements of Construction • 60

ELECTRICAL WIRING

IF A HOME COULD BE STRIPPED OF ITS WALLS AND framework, leaving the electrical system intact and upright, the remains would resemble a tangled and surreal sculpture of cables, wires, switch boxes, outlets, and dangling light fixtures. Invisible, yet flashing through it all, would be charged particles called electricity, which Benjamin Franklin referred to as a kind of fluid existing in all matter and whose myriad effects could be explained by too much or too little of it.

Probably our most versatile form of energy, electricity—with its whizzing electrons—is the lifeblood of a home, powering appliances and systems that are plugged into wall outlets or run on batteries. When supplied by a power company—which may generate its electricity from coal, water, wind, the sun, or nuclear fission—the electric current enters a home through overhead or underground cables. It goes through a meter that measures how much electricity is used and then flows to a service panel that distributes it over a wire network strung throughout the house.

Order is maintained by the service panel and the circuits—the continuous path that electricity follows through copper and aluminum conductors and other components. While the main power lines usually enter a house at one place, the power they carry is split and shunted out of the service panel over branch circuits, which are monitored by fuses and circuit breakers that are designed to prevent overloading, fire, and shock. The service panel generally contains three wires. Two of them are 120-volt wires that allow a household to run 120-volt and 240-volt appliances. (Voltage is actually the electromotive force with which a source of electricity sends electrons along a circuit to form a current.) The third wire is a zero-volt neutral wire; it connects to a ground wire attached to a metal pipe that goes into the ground. The cables forming branch circuits may hold two 120-volt wires, a neutral wire to complete the circuit, and a safety ground wire to protect the system when overvoltage occurs.

While the basics of wiring a home are relatively unchanged, new techniques and equipment permit wires to be run without ripping walls and ceilings apart. So-called structured wiring has also become a science; essentially, this innovation enables the homeowner to share a variety of electrical and digital services, all produced from a single source.

HIGH-WIRE ACT

(Right) Safety-belted and helmeted, a utility worker repairs a damaged power line. Such work is not without its risks. According to U.S. government statistics, four construction workers die on the job every day, with electrocution the second leading cause of death (after falls).

receptacles
junction box
switch
main service panel

OVERSEER OF CURRENTS

The main service panel in a home wiring system controls the flow of electricity. Its switches serve as circuit breakers, automatically flicking off when a circuit overloads. Cables and wires lead to junction boxes, protective enclosures that house receptacles; grounded, the boxes send dangerous accumulated charges directly into the Earth. Outside the panel, switches turn lights on, off, or down, or they delay breaking a circuit, letting people leave a room before it darkens.

SEE ALSO
Power Stations · 44
Batteries · 56
Computerized Buildings · 66

LIGHTBULBS

Before Thomas Edison's incandescent electric lamp, people depended on brushwood torches, candles, and oil lamps. Today, we brighten our homes with lightbulbs and fluorescent lamps. We find our way in the dark with flashlights, and we use streetlights and automobile headlights to get around our neighborhoods. New sources of light filter out unwanted colors and enhance vibrant tones in furnishings. We can change lighting to create ambiance. Lights also help houseplants grow, readjust our internal body clocks after long flights, and soothe seasonal affective disorders.

Lighting has evolved in astonishing ways since Edison invented his lamp. In 1879, he passed electricity through a strand of carbonized cotton sewing thread, causing the rudimentary filament to glow for more than 13 hours in a glass vacuum tube. His feat was marred, though, by the loss of power through heat and by the short bulb life.

The coiled tungsten filament, a metal with a high melting point that easily accommodates the heat needed for the best light, made a brighter, longer-lasting bulb. The life of a bulb got even longer when an inactive gas—nitrogen, argon, or krypton—was placed inside to slow the filament's evaporation. A variation on the theme is the halogen lamp. Inside its bulb are molecules of bromine or iodine, halogen elements that combine with tungsten given off by the filament to form a gas. When this gas comes in contact with the hot filament, the tungsten atoms separate from the halogen and adhere to the filament, essentially rebuilding it.

But it is the energy-efficient, long-lasting LED (light-emitting diode) that has begun to outshine other lightbulbs. LEDs are increasingly being used in flashlights, indicator lights in electronic gadgets, the "Walk" and "Don't Walk" in traffic signals, Christmas tree lights, and more. LEDs have no filaments, which makes them burn cooler and last longer (many thousands of hours). They are made of semi-conductor material that produces light when subjected to an electric current.

WELCOME BEACONS

(Left) Paris rooftops and windows frame the comforting glow of interior lighting at dusk. But even before electricity, the flickering gleam of a candle, a lamp, or a fireplace offered the same relief against darkness.

ELECTRIC LIGHTBULB

(Near right) An incandescent bulb works on the principle that a body—in this case, a thin tungsten wire filament—gives off visible light when heated to high temperature. Current passing through the filament heats the wire to more than 4000°F, resulting in the radiation of electromagnetic energy. An inert gas that fills the bulb reduces the possibility of the filament's oxidizing.

SEE ALSO
Electrical Wiring · 36
Power Stations · 44

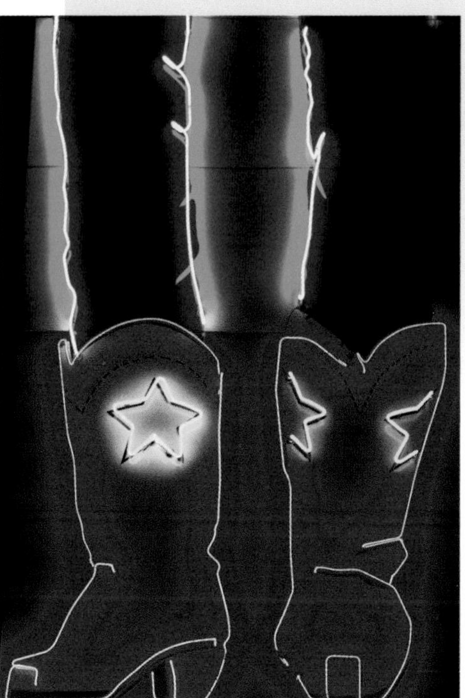

NEON

Cowboy boots glow with the light of rare gases. As electricity streaks through the tubular construction, its flow takes on various hues during encounters with neon, which glow orange-red if sparked. Other gases cause other colors: Argon glows blue and red; krypton, bluish-white; and xenon, blue. Tinted tubes or tubes with coatings of luminescent powder enhance the effect.

BOTTLED LIGHT

Fluorescent light comes from the interaction of electrons and mercury vapor atoms in the bulb. Pins pass current to an electrode, which emits electrons that strike atoms of vapor. This bombardment produces ultraviolet radiation that bounces off the phosphor coating, a fluorescent powder on the inside of the tube, energizing electrons in the phosphor's atoms. These atoms, in turn, radiate white light. Although far more complex than ordinary filament lamps, fluorescent lamps last longer, perform more efficiently, and produce much more light per watt of power consumed.

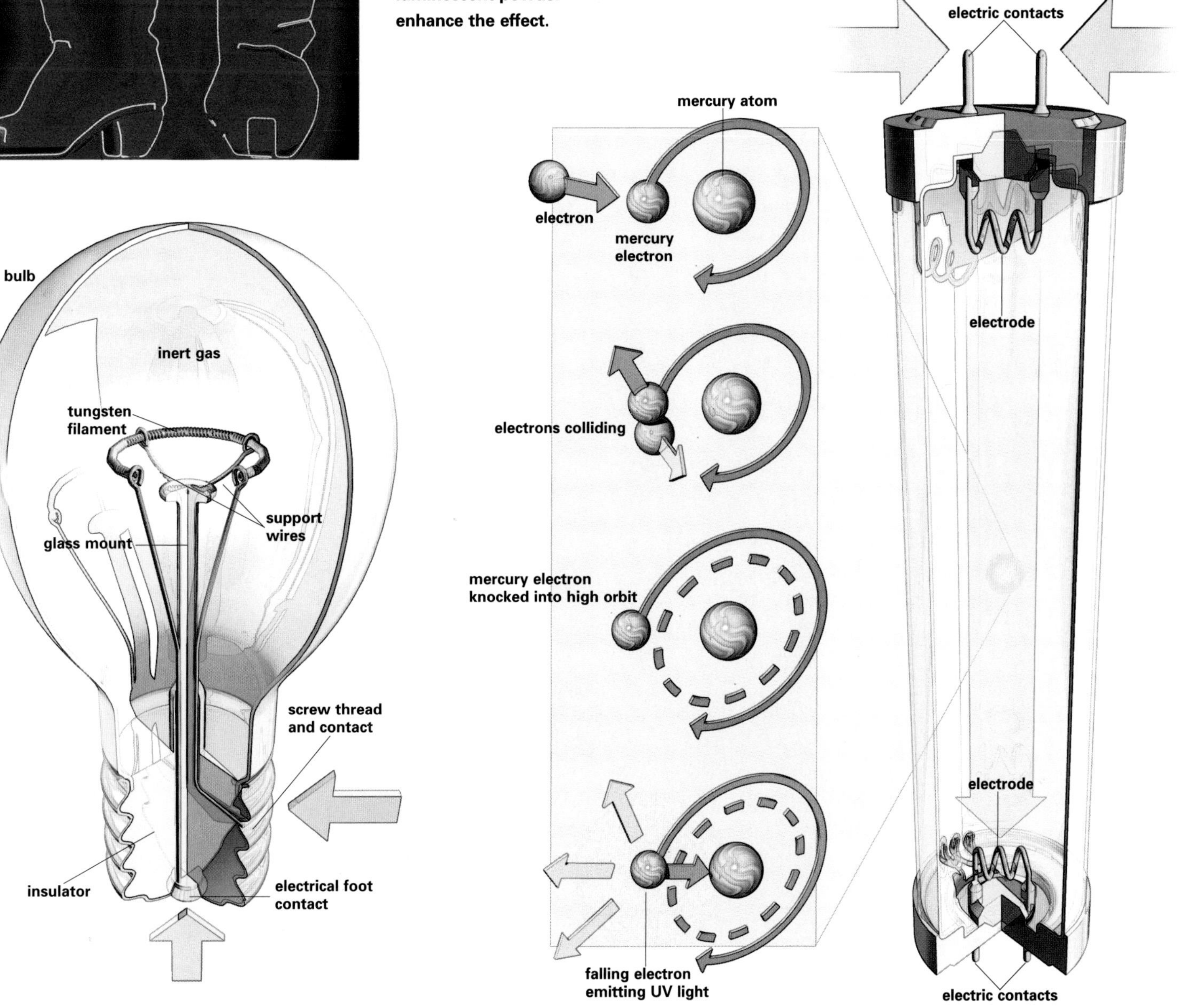

GARDEN MAINTENANCE

Because a well-manicured lawn and a weed-free garden can be a status symbol, a great deal of effort goes into doing what nature often did not intend. Once, hand-swung scythes and grazing farm animals kept the grass and native plants at reasonable height, while simple hoeing disposed of some of the weeds. Nowadays, home supply outlets offer a gardener's dream of electrical or gas-operated equipment—along with a variety of sprinklers and miles of hose—designed to achieve almost instant beautification.

Unless a homeowner relishes the exercise provided by an engineless reel mower, a power machine that spins a single blade can do cutting and even mulching. The blade runs parallel to the ground, while the mower's raised rear edge sets up a draft that lifts blades of grass for cutting, then blows the clippings out of a chute. In a mulcher-mower, grass clippings and leaves remain pressed against the blade and are continuously pulverized before being blown back onto the lawn as nutrients. In a reel mower, pushing revolves an arrangement of cutting blades that slice the grass against a stationary blade. Most gasoline-driven mowers have a two- or four-stroke, one-cylinder engine with a connecting rod that converts the piston's to-and-fro motion into the crankshaft's rotary motion; a very basic carburetor is mounted outside. Handheld grass and hedge trimmers have blades that move side to side and back and forth respectively, their engines and cutting blades powered by household current or batteries.

Since water is essential to growth, commercial irrigation systems send it flowing through fields in furrows, or the fields may be flooded, as in rice field irrigation; it may come from sprinklers or through a variety of tubes on and under the ground. A homeowner tending a lawn or a garden usually relies on an old hose with a hand-controlled nozzle, a network of plastic tubes that drip small quantities of water to the plants' roots, or an automatic sprinkler. Each option is based on the simple principle that as a fluid moves through a pipe, it exerts pressure on the walls that restrain it.

Some sprinklers are merely doughnuts of compact metal or plastic tubing with tiny holes punched into them; attached to a hose and moved manually from place to place, they spray out their fine arcs and curtains of water. Others spin or swing the water out of a punctured tube over a large area; a system of gears and a miniature turbine essentially converts the kinetic energy from water coming through an attached hose to mechanical rotary energy, thus requiring no other source of power to move the spray tube back and forth.

WATER SPRITE

(Right) An adult delights in water that keeps a lawn green. Sprinkler irrigation saves water and lets the gardener select where the water should go.

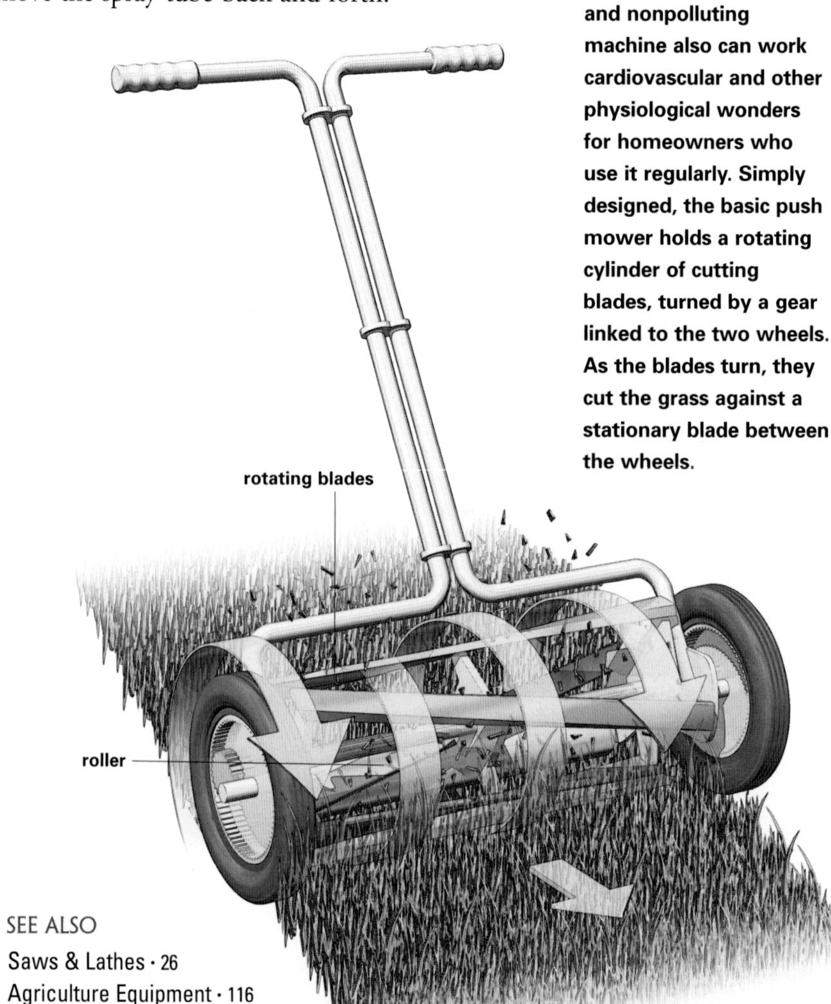

THE REEL THING

(Left) Its simple appearance notwithstanding, a push mower—also known as a reel mower—can admirably tackle almost any lawn. This low-maintenance and nonpolluting machine also can work cardiovascular and other physiological wonders for homeowners who use it regularly. Simply designed, the basic push mower holds a rotating cylinder of cutting blades, turned by a gear linked to the two wheels. As the blades turn, they cut the grass against a stationary blade between the wheels.

SEE ALSO
Saws & Lathes · 26
Agriculture Equipment · 116

CHAPTER 2

POWER & ENERGY

DEFINED AS THE CAPACITY FOR DOING WORK, ENERGY takes many forms: It can be mechanical, electrical, physical, or thermal. When it is expended or dissipated by machines or human activity, we have power, which performs work. Sources of energy include water, wind, coal, oil, gas, wood, radioactive rocks, the moon's tidal pulls, and the sun's rays. To a significant extent, many people rely on biomass energy, employing waste-to-energy incinerators and other means to retrieve it. Coal and nuclear energy generate most of the electricity we use to light our buildings and power our machines. Waterpower supplies about a fifth of the world's electricity, while wind and geothermal energy are important sources of power for several countries. In the future, we may even look to exploding neutron stars for energy and, perhaps more immediately, to fusion—the same process that powers the sun.

Monuments to energy, cooling towers belch steam at a nuclear power plant in France.

POWER STATIONS

In 1886, William Stanley, a pioneer in the generation and transmission of electric power, fired up what was probably the earliest central power station. Erected in Great Barrington, Massachusetts, his station used a 25-horsepower boiler and engine to turn a 500-volt generator and supply electric lighting to 25 businesses. Stanley's creation and other early power stations relied on the same basic principles and components we use today, most notably fuel, boilers, turbines, and generators.

Coal is the fuel of choice at power stations. Before the 1973 oil embargo, the United States depended on oil to generate much of its electricity, but from 1973 to the 1990s, the use of oil dropped from 17 percent to 3 percent. Coal arrives at most stations ground into dust, which is then mixed with air to make a highly explosive fuel. Fired, it heats water in a boiler, thus producing high-pressure steam that pushes against the propeller-like blades of a giant turbine attached to the shaft of a huge generator. As the shaft spins, coils of wire interact with a circular array of magnets to create electricity.

Electricity then goes through a set of transformers, which are devices that alter voltage. First, they step up the voltage for main distribution along tall, latticed towers. Later, at a substation, other transformers bring it back down to accommodate normal domestic voltage of 240 and 120 volts before it is distributed.

POWER PLANTS

(Right) Power plants help supply the world with electricity, its most versatile form of energy. While plants can use virtually any fuel to make electricity, about 55 percent of the electricity in the U.S. comes from coal-burning facilities.

CURRENT

A simple AC generator, which consists of magnets, a spinning coil with connecting slip rings, and carbon conductor brushes, uses electromagnetic induction to convert mechanical energy into electricity.

As the coil spins and the induced current reverses in the magnetic field, the current passes from the rings to the external circuit by way of brushes that press against the rings.

SEE ALSO

Electrical Wiring · 36
Mining Coal · 178
Drilling for Oil · 182

HYDROELECTRICITY

PLACE A WATERWHEEL BENEATH A SMALL waterfall. As it turns under the force of the water on its paddles, the energy can be shifted from its central, axle-like shaft to linked wheels, belts, and gears, and ultimately to useful tools. For centuries, people used waterpower to grind corn and wheat, to spin fiber, and even to fire up their charcoal-burning, metal-smelting furnaces. The Chinese, for example, are known to have used a water-powered bellows in A.D. 31. Waterpower, or hydroelectric power, is an efficient, economical way to harness the potential of falling or dammed water; obtaining power from waves and tides is also feasible, but the method has its drawbacks.

A hydroelectric power station requires a large head of water and a fall or gradient to take advantage of gravity and the water's momentum. For these reasons, such a power station is usually located in mountains, near a waterfall, or below a dam holding back the enormous force of a reservoir that results from blocking a river's flow. A hydroelectric plant is really a king-size version of the waterwheel, but its powerhouse contains generators and turbines whose blades spin under the carefully controlled force of a jet or flow of water.

Wave power, which has been used in Japan and Norway, takes advantage of the vast store of energy that is created in water when wind drives into it. Wave-to-energy devices called OWCs, for oscillating water columns, do the work. Fixed in the water, when waves strike them air inside is compressed and forced though turbines; other versions are floats that drive pumps as they bob in the waves. Wave-power devices must withstand the vagaries of wind and weather, and still need refinement if they are to be truly practical.

Harnessing the tides can be an expensive operation, but despite the cost and the potential stations have for damaging the habitats of fish and birds, tidal power facilities are in use in France and a handful of other countries. The U.S. government funded a tidal project in the 1930s in Maine's Passamaquoddy Bay area, which was to be dammed so that the incoming tide would be trapped by gates. The water would then be released through turbine generators, after which it would escape into another bay and kept there until low tide, finally flowing out into the Bay of Fundy. Dikes were built, but lack of funding and politics ended the project.

THE FORCE OF WATER
(Right) Outlet tubes in the Glen Canyon Dam direct water from the Colorado River. Completed in 1966, the 710-foot-high dam supplies power and regulates the river's flow.

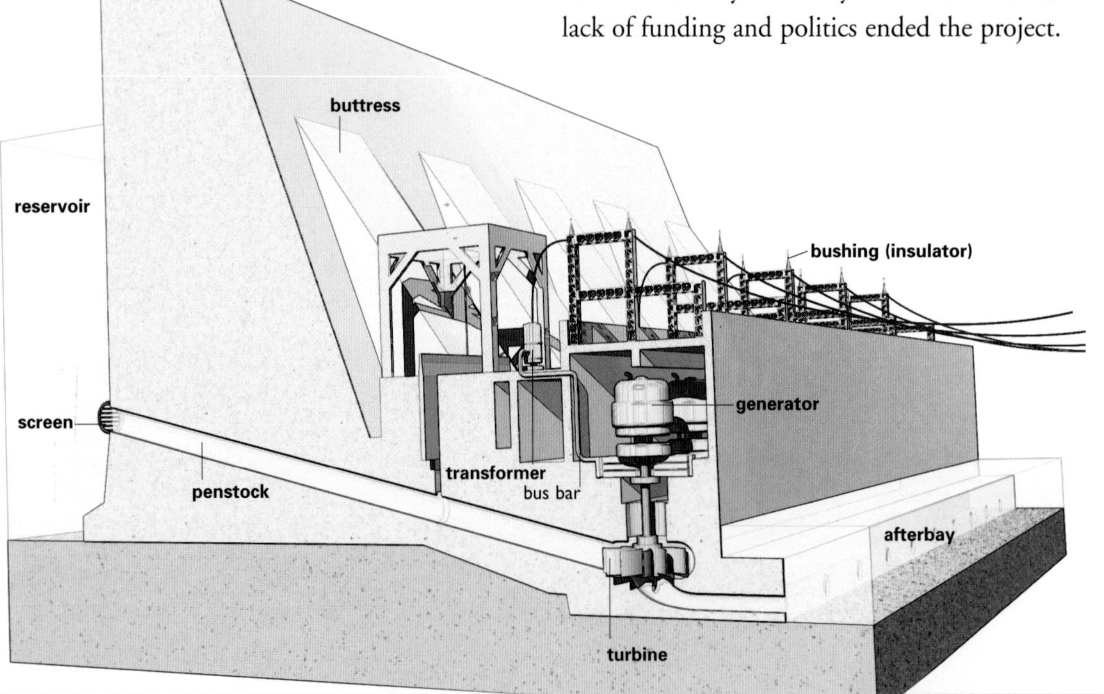

POWER STRUCTURE
(Left) Dams can produce electricity, store water for irrigation and public use, and control floodwater. This massive barrier controls water by regulating its flow through gates and within long tunnels and pipelines called penstocks, or sluices. In a hydroelectric plant, water feeds through penstocks into the turbine. Bushings provide electrical insulation, and bus bars—lengths of conductors—collect the electric current and distribute it.

SEE ALSO
Electrical Wiring · 36
Power Stations · 44

WIND & GEOTHERMAL POWER

HOT WATER

(Left) People relax in the bathlike waters near a geothermal power plant in Iceland, a country that, despite its name, has numerous volcanoes and hot springs. A tempting energy source, geothermal power cannot warm everyone: Only certain regions hold natural reservoirs of hot water, and even those sources run dry, may contain contaminants, and cost a great deal to develop.

WIND FARMS

(Right) In some parts of California the wind helps generate electric power. Computers can adjust the propeller-like blades on wind turbines to function in all wind conditions. To be effective, however, such windmills must stand where winds blow consistently.

WIND POWER CAPTURES THE AIR WE KNOW as the wind; geothermal power taps naturally occurring steam and hot water below the surface of the ground. From time to time, both forms of power are painted as rather fanciful sources of energy, unrealistic answers to dwindling supplies of oil, coal, and gas. But where they are feasible, they work very well.

Like the waterwheels they resemble in principle, windmills have been used for centuries, their wind shafts connected to a series of gears and attachments that milled grain, irrigated fields, and pumped seawater from low-lying land. A wind turbine, the modern version of the windmill, produces electricity, driving a generator whose shaft-spin can be increased by gears; a computer may also control the rotor's movement and the position of the huge propeller-like rotor blades. The world's largest concentration of such turbines is in California.

Geothermal energy, in the form of hot water and steam, is the product of decaying radioactive elements in the ground. When tapped, it can drive a turbine to produce electricity, or it can be piped into buildings and used for space heating. In the United States, geothermal power plants operate in California, Hawaii, and other states. Such plants are common in Iceland, where many homes are warmed by piped-in underground heat. Most underground heat is not exploitable, however, and it is concentrated in reservoirs found only in certain parts of the world. Often, the sources manifest themselves quite dramatically as geysers, steam vents, fumaroles, and the hot springs that have been used therapeutically for centuries. But surface displays may be far from a reservoir, and some sources of geothermal energy give no sign of their presence, requiring prospectors to use seismic and geologic probes that are complex and costly.

SEE ALSO
Power Stations · 44
Solar Heating · 50

SOLAR POWER

Fossil fuel supplies will not last forever, a fact that underlines the importance of conserving resources and finding alternative energy sources to heat our homes. One option is solar power, the energy emitted by the sun as electromagnetic radiation. It is plentiful, nonpolluting, and free.

Today, water or air heated by the sun fulfills a variety of household needs, providing hot water for kitchen sinks and showers or warm air to take the chill out of the house itself. Capturing solar energy to heat a home can be done in two ways: passively or actively. Passive heating depends on a structure's design, materials, and location to maximize the warming effect of sunlight. The method works best in a well-insulated home with large, south-facing windows (in the Northern Hemisphere), an interior of dark stone or tile, and a location in a sunny region. Such a home receives lots of sunlight, which warms the air inside; the dark interior absorbs sunlight and reradiates the heat during the night.

The active method requires equipment that collects, stores, and distributes solar energy. A flat-plate collector fixed to the roof of a house serves either an air- or a water-based system. A cover made from a clear material, usually glass, allows sunlight to pass through and strike a dark, heat-absorbing metal plate. Warmed air between the cover and the plate helps to insulate the collector. As the plate soaks up heat, it warms air or water contained in pipes or tubes inside the collector. If the system is air-based, a fan blows the warmed air into a large, rock-filled storage area under the house. A water-based system pumps the warm water into an insulated storage tank for household use or through tubes placed under floors and in ceilings to provide space heating.

SUN WORSHIP
(Left) Resembling windows in a cathedral, the solar panels on this low-energy house collect the power of the sun on bright, clear days and feed a generator. Greater use of nonpolluting solar power could reduce the energy drain caused by buildings, which account for a third of all U.S. energy consumption.

MAKING ENERGY
(Right) Sunlight is converted to electricity, a so-called photovoltaic process (PV), in this solar energy plant. A quarter of a million American homes get their electrical power in this way.

SEE ALSO
Home Heating & Cooling · 32
Pipes, Pumps, & Plumbing · 34
Alternative Fuels for Cars · 80

NUCLEAR POWER

POWER PLANT

(Left) The cooling stacks of a nuclear plant spew nonradioactive steam, not smoke. Steam spins the electricity generator, and changes back to water to cool the system.

CONTROL RODS

(Opposite, top right) Twelve-foot-long rods initiate and control a chain reaction. Shown in yellow, uranium fuel in ceramic pellets packs rods made of zirconium, a metal that resists heat, radiation, and corrosion. Control rods, shown in red, absorb neutrons, preventing them from hitting and splitting uranium atoms; inserting these rods stops the chain reaction, and withdrawing them speeds it up.

NUCLEAR REACTOR

(Opposite, bottom) In the reactor process, fission heat turns water into steam that spins the turbine and the generator. Water also acts as a moderator; it removes heat from the chain reaction and slows down neutrons, increasing the probability of fission. Loss of water can slow or stop the reaction.

A SINGLE OUNCE OF THE URANIUM-235 ISOTOPE can generate an immense amount of energy, a fact that makes it easy to understand why more than 400 nuclear plants in 30 countries—a quarter of them in the United States—rely on it to produce electricity. Nuclear energy now provides about 20 percent of U.S. electricity, and the distinctive domes and towers of nuclear plants have become sights as familiar as gas stations and shopping malls.

Each plant generates electricity with the same steam-turbine-generator arrangement used by a coal-fired plant. The fuel, however, consists of solid ceramic pellets containing isotopes of uranium atoms that are split apart in a process known as fission. The pellets are packed in long metal tubes, which are bundled together and installed in a heavily shielded and water-cooled reactor where fission occurs. When uncharged subatomic particles called neutrons are released, the particles collide with uranium atoms and split them. As the nuclei burst, the atoms release their own neutrons, which then strike other atoms to create a chain reaction; the immense heat generated by the process vaporizes water, creating steam that turns the turbine and generator. The power plant controls the reaction by inserting and removing rods made of neutron-absorbing material. Inserting these rods prevents neutrons from hitting atoms, and withdrawing them speeds up the reaction.

SEE ALSO
Electrical Wiring • 36
Power Stations • 44
Fusion • 54
Submersibles • 100

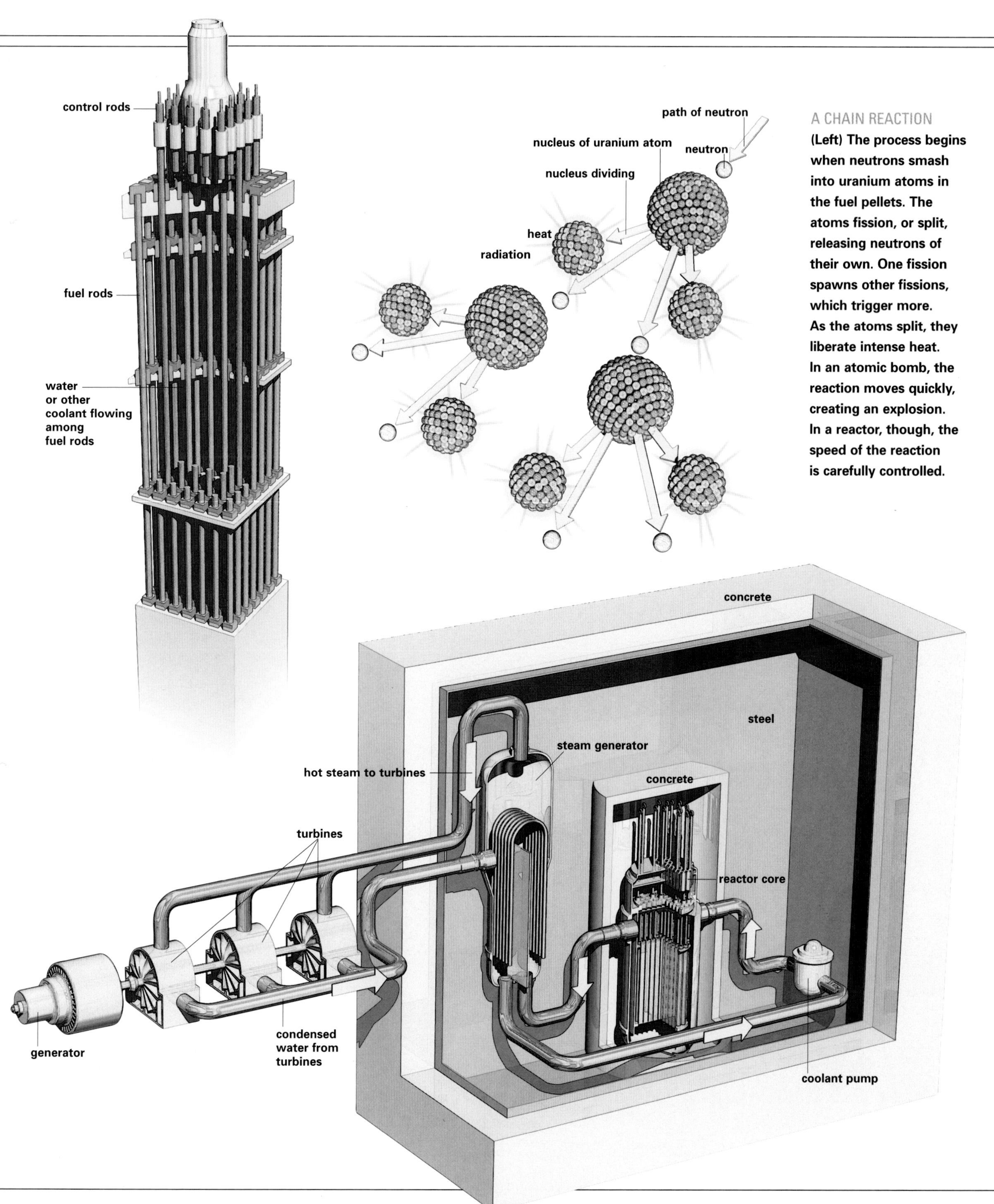

A CHAIN REACTION

(Left) The process begins when neutrons smash into uranium atoms in the fuel pellets. The atoms fission, or split, releasing neutrons of their own. One fission spawns other fissions, which trigger more. As the atoms split, they liberate intense heat. In an atomic bomb, the reaction moves quickly, creating an explosion. In a reactor, though, the speed of the reaction is carefully controlled.

FUSION

FUSION IS THE NUCLEAR REACTION THAT energizes not only the sun and stars, but also, in an equally uncontrolled fashion, the hydrogen bomb. Unlike fission, which produces energy by splitting large, heavy atoms into smaller ones, fusion occurs when the nuclei of small light atoms are compressed under intense heat to form larger and heavier nuclei, a reaction that also creates enormous bursts of energy.

In theory, this energy could be captured and converted into electricity, a process that appears simple on paper. It is also appealing, because the fuel to start the process is plentiful: Deuterium and tritium, the heavy isotopes of hydrogen, can be extracted from ordinary seawater. Deuterium is a likely fuel candidate because only 1/250th of an ounce—the amount in a gallon of seawater—contains about the same energy as 300 gallons of gasoline.

The problem is how to control such titanic energy. Reactors would be required to withstand heat from plasma, the seething gas of charged particles resulting from a fusion reaction. Scientists have tested experimental containers that have no resemblance to a fission reactor's vault; they are instead made of coils of wire—"magnetic bottles" that create magnetic fields powerful enough to confine the manufactured "hell." The doughnut-shaped Tokamak Fusion Test Reactor at the U.S. Department of Energy's Princeton Plasma Physics Laboratory—shut down in 1997 after 15 years of operation—was one with 587 tons of coils of superconducting materials in a riblike arrangement; 24 feet tall, with a diameter of 38 feet, and an 80-ton vacuum chamber, it generated a horizontal field that forced charged plasma particles to circle inside the container. The particles collided, fused, and released energy without touching the chamber walls. During its experimental life, the Princeton tokamak set records for fusion performance, among them a world-record temperature of 510 million degrees Celsius, more than 25 times that at the sun's center.

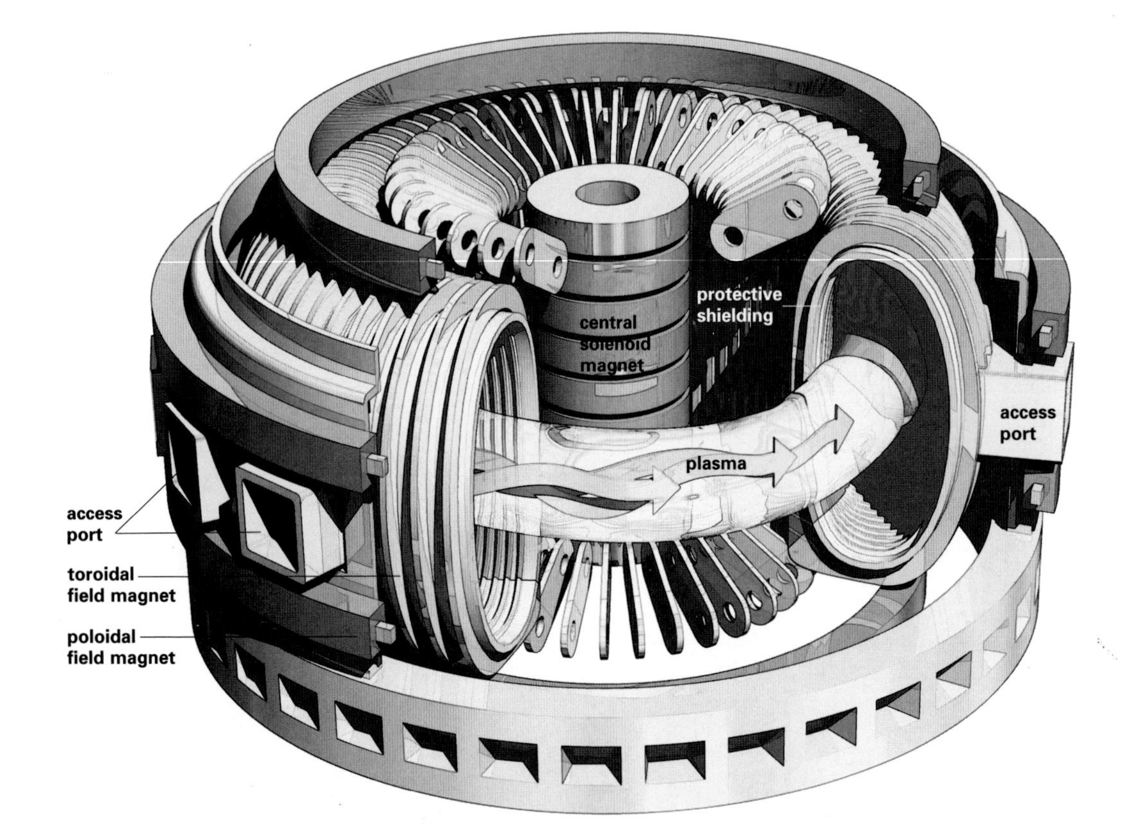

TOKAMAK

(Left) In 1951 Soviet physicists Andrei Sakharov and Igor Tamm first proposed the tokamak, a device that contains hot plasma, fusion's charged gas. Inside the tokamak, which stands for "toroidal chamber with an axial magnetic field," superconducting materials create a magnetic field that causes plasma to flow within a doughnut-shaped chamber.

BELLY OF THE BEAST

(Right) Inside the tokamak, Princeton University researchers toil in an attempt to achieve the elusive break-even point at which energy released by fusion equals that required to produce it. Building plasma containers has proved a daunting task, given the furious heat they must hold. But with superconducting materials that contain heat more efficiently than copper, fusion power may live up to expectations as a source of safe and limitless energy.

SEE ALSO
Nuclear Power • 52
Alternative Fuels for Cars • 80
High-Speed Trains • 90

BATTERIES

WHERE WOULD WE BE WITHOUT THEM, these multishaped and multisized powerhouses of the electrical world that run our watches and clocks, computers, cell and cordless phones, flashlights, automobiles, boats, hearing aids, toys, cameras, pacemakers, power tools, toothbrushes, radios, and dog collars? Probably hopelessly entwined in the countless miles of cable and wire that would be required to tether so many gadgets and pieces of equipment to wall outlets.

Ever since 1800 when Italian physicist Alessandro Volta (1745-1827) invented the first battery, known as the voltaic pile (and got an electrical unit, the volt, named after him), batteries have become commonplace. Volta's bulky battery—literally an alternating pile of disks made of zinc and silver (or copper) separated by brine-soaked pads—hardly resembled any of today's. But the principle is essentially the same: the conversion of stored chemical energy, through a reactive process between chemicals, into the flow of electrons we know as electricity.

Three major components inside the battery are responsible for the reaction: a negative and a positive electrode (electrodes are conductors through which electricity is brought in or out), and an electrolyte (a paste or liquid which carries the current). To start an electrochemical reaction, a number of materials chosen for their ability to attract or relinquish electrons are used to make up the three components. These materials and their functions are key, since a current won't move through a circuit unless there are more electrons at one terminal than the other. When the chemicals and other materials are precisely arranged inside the battery, and an outside circuit is attached—a bulb, motor, or any other device—electrons collect on and are released from the negative electrode and flow through an outside circuit to the positive one. Since the outside connection is what starts and draws the current, if a battery is unused it will retain power for some time. On the other hand, because powering a device uses up the fixed amount of reactive materials inside the battery, the battery will eventually die.

A battery's operating voltage (which must match that of the device), performance, durability, and ability to be recharged depend, among other factors, on the chemicals and combinations used and on the number of battery cells connected in a series. A standard lead-acid battery in automobiles operates at 12 volts; the AA alkaline battery provides 1.5 volts; a tiny, round zinc-mercury-oxide battery, often used in hearing aids and watches, 1.4 volts.

Many batteries can be recharged by placing the battery in an electric "cradle" that reverses the process of what goes on during discharge; electrical energy is now stored as chemical energy when electrons are made to flow in the opposite direction of the discharge. Nickel-cadmium batteries, used in everything from calculators to cordless power tools, are among those that can be reused. So are car batteries, the lithium-ion batteries in cell phones, and lightweight zinc-air batteries used in hearing aids.

EVER READY?

(Right) A heap of spent and discarded batteries is sorted for rejuvenating and reuse, or disposal. Lead-acid automobile batteries are most often recycled, while many of the household types end up by the billions in solid waste facilities.

BATTERY MAKER

(Above) Robotic equipment turns out batteries at a plant. Industrial battery making is complex, but a simple battery can be made at home by basically hooking up a lemon, a galvanized nail, and some copper wire to a voltmeter.

SEE ALSO
Small Appliances & Hand Tools · 24
Automobiles · 76

DURACELL
ALKALINE BATTERY
ENERGIZER
RAY-O-VAC
HEAVY DUTY
Alkaline
General Purpose
EXTRA LIFE
EVEREADY
SUPER
D SIZE
1.5V

CHAPTER 3

BUILDING & BUILDINGS

TODAY'S CONSTRUCTION PROJECTS OFTEN USE wood, brick, stone, and concrete—the same materials ancient builders relied upon. To erect buildings of great size, complexity, and beauty, many projects also rely on iron, steel, aluminum, glass, and synthetic materials that were never dreamed of by early architects. Simplicity of design and function remain, but building today is an exact science, a major shift from the days when construction was a practical craft based often on experience and observation alone. Modern buildings, along with older ones that have been modified, are engineered to withstand winds, earthquakes, traffic, hurricanes, and fire, and most are probably far more comfortable, safe, and efficient than anything that was built in the past. The equipment used to erect these structures is also impressive. Occasionally, it is even more dazzling than the building it shapes.

Stark against the sky, an ironworker, girders, and a

ELEMENTS OF CONSTRUCTION

The keystone is a simple, wedge-shaped block of stone on which much depends. Fitted last into the top of an arch, it locks the other pieces in place, providing support for the curved structure that, in turn, supports weight above it. Keystones and all of the other basic elements of construction work as a team to ensure stability and distribute weight. In constructing an arch, for example, wedge-shaped stones called voussoirs are placed on each side of the keystone. As gravity pulls the keystone downward, the thrust is carried on either side by the voussoirs immediately flanking it. The total thrust of the voussoirs is then distributed through the semicircle until it reaches the vertical part of the wall, which carries it directly to the structure's foundation. The result can last for centuries: Among the arch's exemplars of form and function are the aqueducts built by the ancient Romans which curved across valleys and cities, carrying water from distant sources to baths and fountains.

The forces at work in an arch, or in any building—bending, compression, tension, and torsion (twisting)—create internal stresses. Yet, structural forms abound, each with its own function, to deal with these stresses. Natural base or prepared foundations support buildings ranging from sheds to skyscrapers. Columns support beams and lintels; horizontal spans carry the load of a roof. Piers, arches, rib-vaulting, and buttresses, as well as flying buttresses and semidomes, absorb thrust and support huge domes.

Most structures adhere to these principles that preserve structural integrity, as well as to stringent building codes that enforce them. Homes and buildings referred to as modular are also governed by the elements of construction. Contrary to what some believe, modular homes are not "mobile" homes, which may be built on a steel frame with wheels. Modular buildings—which include homes, hospitals, correctional facilities, military housing, churches, and restaurants—are true to their name. Their structural components, called modules, are manufactured in a controlled factory environment, and then assembled with the help of a crane into a complete structure on a site foundation. Constructed of the same materials that go into build-on-site "stick homes" (the term for a conventionally built dwelling), a modular building can be finished in a few weeks after it is set on the foundation.

MEN AT WORK

(Right) The geometrically precise framework of buildings under construction in California's Orange County has support, as well as design, as its goal. Horizontal sills of timber rest on the foundations; joists form horizontal frames for flooring; vertical studs provide support for walls; horizontal collar ties connect rafters on either side of roof ridgepoles.

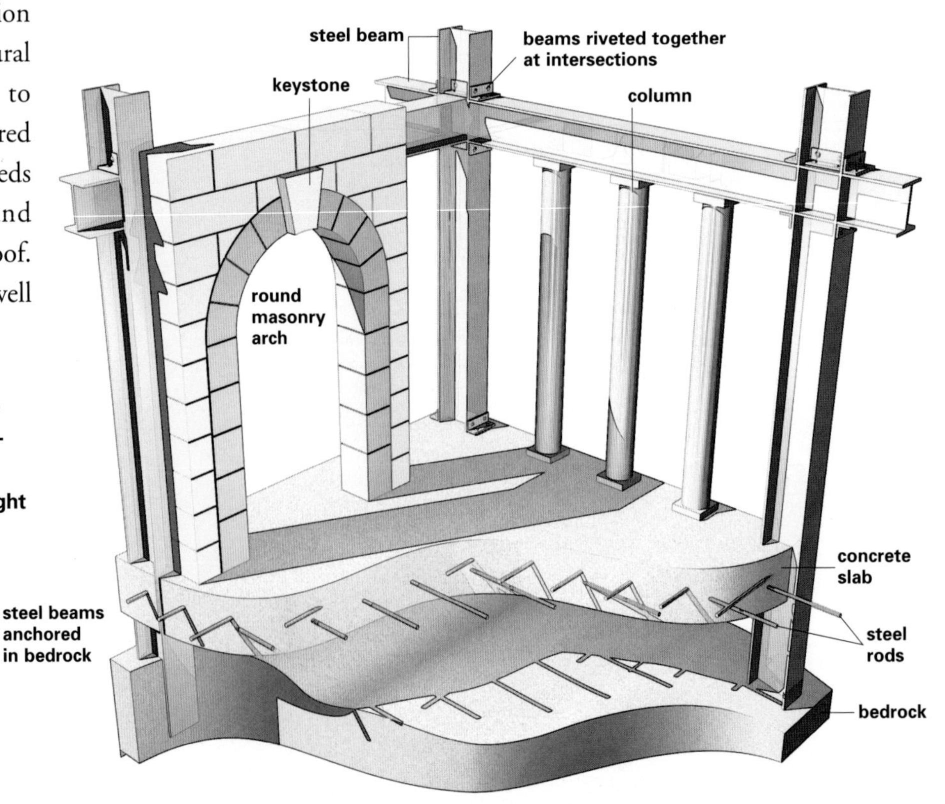

BEARING THE WEIGHT

Columns, beams, an arch, and a foundation of steel-reinforced concrete provide interior and exterior support for a building. Carefully connected to one another, these elements distribute and support the great weight of a large structure.

SEE ALSO

Construction Equipment · 62

Skyscrapers · 64

Lumber · 172

CONSTRUCTION EQUIPMENT

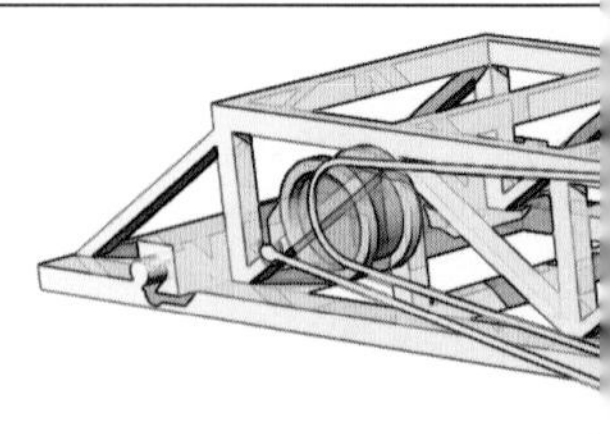

ONCE, THE MAIN THINGS THAT CONCERNED a home builder were slope, drainage, and proximity to a water supply. Site preparation amounted to clearing away trees, underbrush, and rocks. Excavation, if any, was minimal, and a foundation was laid rather quickly with rough stones gathered from nearby fields. Using wood, bricks, mortar, and hand tools, the builder soon had the bearing walls up and the roof raised. Today's builders take more into consideration: Faults can trigger earthquakes; soil can heave or sink structures when changes in weather or soil composition occur; high water tables can flood cellars; radon may be present in rocks. Tools are now nail guns, power saws, routers, jackhammers, and electronic surveying devices. On a larger scale, they are dynamite, water pumps, and earth-moving bulldozers, graders, backhoes, and power shovels. A variety of cranes can work on the ground or atop skyscrapers.

Design, site preparation, and foundation laying must consider the strength of the supporting soil and rock. They must consider the "dead load" of the building—the total weight of the structure and of all its fixed equipment—and the changing "live load," which includes seismic forces, wind, and vibrations caused by equipment, people, and even moving furniture. If subsurface conditions are adequate, foundation concrete is poured into an excavation. Other times, piles made of wood, concrete, or steel are driven into firm soil or to solid rock. Caissons, which are watertight cylinders filled with concrete to form columns, may also be sunk into weight-bearing soil that lies beneath fill or mud.

CONSTRUCTION SITE
The site at first gives little hint of the splendid edifices that will rise above it. But even before the buildings take shape, work sites invariably attract passers-by to the noisy drama in the belowground arena.

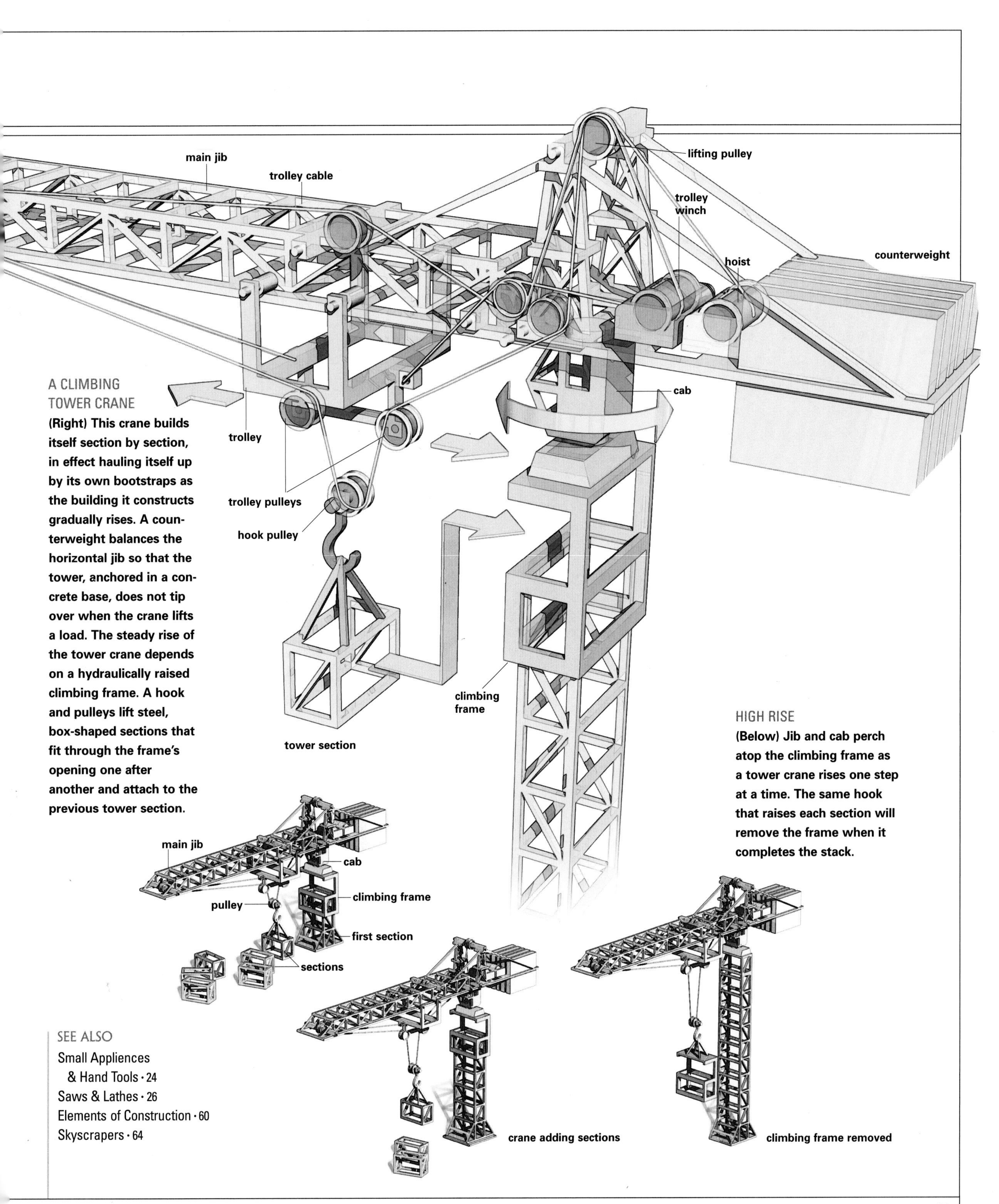

A CLIMBING TOWER CRANE

(Right) This crane builds itself section by section, in effect hauling itself up by its own bootstraps as the building it constructs gradually rises. A counterweight balances the horizontal jib so that the tower, anchored in a concrete base, does not tip over when the crane lifts a load. The steady rise of the tower crane depends on a hydraulically raised climbing frame. A hook and pulleys lift steel, box-shaped sections that fit through the frame's opening one after another and attach to the previous tower section.

HIGH RISE

(Below) Jib and cab perch atop the climbing frame as a tower crane rises one step at a time. The same hook that raises each section will remove the frame when it completes the stack.

SEE ALSO

Small Appliences & Hand Tools · 24
Saws & Lathes · 26
Elements of Construction · 60
Skyscrapers · 64

SKYSCRAPERS

An American original, the skyscraper began its gravity- and wind-defying rise in the late 19th century. It was a soaring testimonial to economic boom times, mass production, and the burgeoning technology of structural engineering. Developed first in Chicago, Carl Sandburg's "city of the big shoulders," the skyscraper relied on its internal steel skeleton of columns and beams, rather than conventional heavy masonry walls, to support most of its weight. A "curtain wall" made of nonbearing materials, such as glass and thin marble sheets, sheathed the framework, giving engineers and architects a relatively lightweight building with excellent tensile strength.

Innovations such as I-shaped steel beams, concrete reinforced with embedded steel, and tubular concrete designs allowing lighter and stronger walls to bear loads enabled skyscrapers to rise even higher. New designs provided more internal room by reducing the need for diagonal wind-bracing; prefabricated materials speeded construction; and improvements in concrete gave architects more freedom of expression.

Beyond aesthetics, efficiency, and the comfort of its occupants, a skyscraper is built to withstand high winds and other weather assaults, as well as the shaking of earthquakes. To protect against the lateral forces generated by earthquakes or winds, engineers today are constructing "nonrigid" buildings that, in effect, go with the flow or sway with the forces. One technique places layers of rubber and steel plates between a building's base and foundation. Another relies on sliding bearings under load-bearing columns, an arrangement that helps dissipate a quake's energy through friction.

But even with new techniques and materials, skyscrapers must still rely on the engineering requirements of solid bedrock and well-compacted soil, and deep concrete foundations to bear the building's enormous weight. The popular tubular concrete design, for example, actually transforms a building's exterior walls into a rigid tube that effectively carries gravity and wind loads.

The kind of glass that wraps a modern skyscraper, both to define it and serve as its windows, is also an important consideration in the construction. Made in multiple layers that are bonded with various plasticizers, skyscraper glass is built to prevent shattering and separation from its framing.

But whatever the technology used to build a skyscraper, the fact remains that erecting one can take years of proper planning.

ON THE JOB

(Left) Building a high-rise requires the vision of architects, acres of urban land, tons of material, and a king's ransom. It also needs the skill of steelworkers, stonemasons, glaziers, welders, electricians, plumbers, and countless other workers seemingly oblivious to doing their jobs at great heights, often exposed to the elements.

EMPIRE STATE BUILDING

(Right) No longer the world's tallest building, the Empire State Building still draws tourists. Completed in 1931, the building features setbacks, steplike creations that allow sunlight to reach lower floors and the street.

SEE ALSO
Elements of Construction · 60
Computerized Buildings · 66
Elevators & Escalators · 68

COMPUTERIZED BUILDINGS

SKYSCRAPERS AND OTHER LARGE BUILDINGS are more than walls, partitions, and foundations. From air-conditioning and heating ducts to mail chutes that drop letters to a central pickup point, buildings are laced with familiar systems for environmental control, transportation, communication, power, water, and waste disposal.

With new, computerized technology, "intelligent" buildings can respond quickly to the needs of occupants. Electronically enhanced and regulated by central computers, buildings are wired to respond to many contingencies. Fire alarms and security systems can be linked so that doors can unlock automatically to provide the quickest evacuation route, while dampers regulate airflow to inhibit the spread of flames. Sensors in sheets of "smart" composite materials detect leaking gas and alert engineers when repair is needed. Whole walls studded with actuators are transformed into speakers, and some systems monitor and correct other systems. From a central control station—not necessarily in the same building but perhaps in another structure miles away—a building manager can turn out lights, limit power consumption, run boilers more efficiently, and stop elevators in an emergency.

Workplace automation that relies on shared equipment and systems to integrate the data stream has contributed greatly to the efficiency of companies and offices. Among a smart building's many tools are centralized data and word processing, teleconferencing and electronic mail systems, and message centers that answer the telephones automatically.

TELLTALE FINGERS

(Right) Fingerprints get an electronic scan for the right match to a prescanned picture in order to unlock a secure system. The familiar ridged whorls give each of us a distinctive characteristic and, thus, a means of identification. When the hand is placed on the scanner, sensors pick up an image of the fingertip's ridges and valleys that are analyzed by a processor; if the prints connect with those on file, access to whatever the system guards is approved.

EARTHQUAKES

Earthquakes can devastate rigidly constructed buildings, but damage-control technology largely attributed to Japanese and U.S. engineers has made structures less vulnerable to nature's assaults. Using machines that simulate earthquakes, scientists study the effects of quakes on tall buildings, bridges, nuclear power plants, gas tanks, oil pipelines, and even household equipment such as refrigerators and ranges. What they learn helps engineers build structures that absorb and dissipate a quake's destructive energy. The Osaka World Trade Center Building in Tokyo (left), for example, uses a computer-directed sliding weight to shift the structure's center of gravity when the Earth trembles or the wind rises. To dampen shock, another method uses alternating layers of steel and rubber between a building's base and foundation.

SEE ALSO
Security Systems • 28
Skyscrapers • 64
Sending & Receiving Messages • 216
E-mail • 240

ELEVATORS & ESCALATORS

PERHAPS THE MOST ESSENTIAL PIECE OF technology in a high-rise building is the people mover. Indeed, elevators and escalators make skyscrapers possible. They have had an extraordinary economic impact on businesses and public facilities, and they are factors to be reckoned with when architects design new buildings or transportation centers.

The first elevator was developed in the middle of the 19th century by Elisha Otis, an American inventor. Powered by steam, his device ran up and down between guide rails and had an automatic safety device that clamped on the rails to prevent the car from falling if the hoisting rope broke. Hydraulic systems powered later elevators. Some small buildings still have them, but most of today's models operate with steel hoist cables drawn by a motor over a grooved pulley wheel, called a sheave. The cables are attached to a counterweight that goes up when the car descends.

Escalators, on the other hand, are moving stairs mounted on a continuous chain that is drawn over a drive wheel run by an electric motor. The stairs form a sort of level treadmill at the beginning and end of the ascent, becoming an arrangement of treads and risers during the incline. While elevators are swifter and can fit into small wells inside a building or run up its sides, escalators do not stop, and can carry more people.

This conveyor-belt technology—the same that has moved coal, sand, and grain for years—is also used in another so-called people mover. A horizontal version of the escalator, it is a moving sidewalk complete with handrails and other safety devices. One early model, a multibelt affair built by a coal company, was unveiled at the 1893 World's Columbian Exposition in Chicago, and used to transport visitors past exhibits in an orderly fashion. Moving walkways are fixtures in airports. Airports also use automated driverless vehicles that run over steel or concrete guideways. Powered by electricity, they are guided by a computer that starts and stops the cars at the right stations at the right times and operates the doors.

PATHS CROSS

(Right) Shoppers enjoy the ease of an escalator network in a department store in Osaka, Japan.

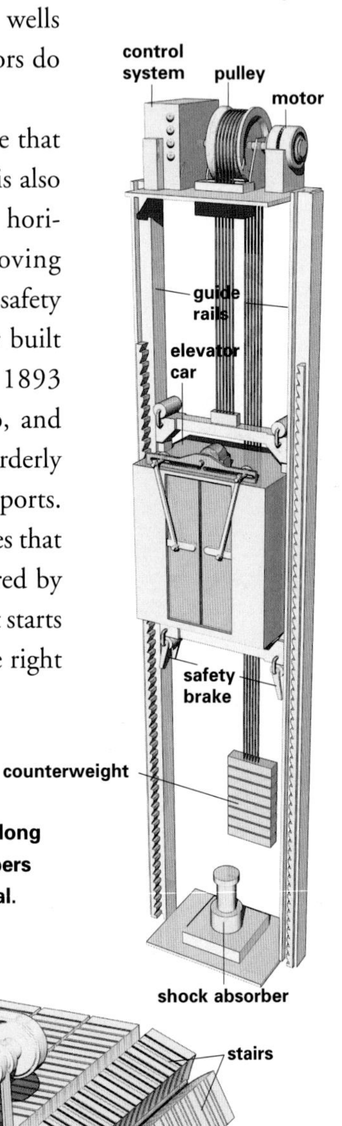

THE LIFT

(Right) What goes up goes down, and vice versa. The elevator, or "lift," follows this simple principle. Without elevators, people would have long walks, and skyscrapers would be impractical.

INCLINED ROUTE

(Below) Up stairs become down stairs, and handrails appear and disappear as an escalator moves effortlessly over its inclined path. Built on the endless chain principle, an escalator relies on an electric motor at the top. A chain—drawn around a drive wheel at the top of the landing and a return wheel at the bottom—moves the treadmill-like stairs attached to it. As the stairs move upward, they form treads and risers that flatten at the end of the ascent and again at the beginning.

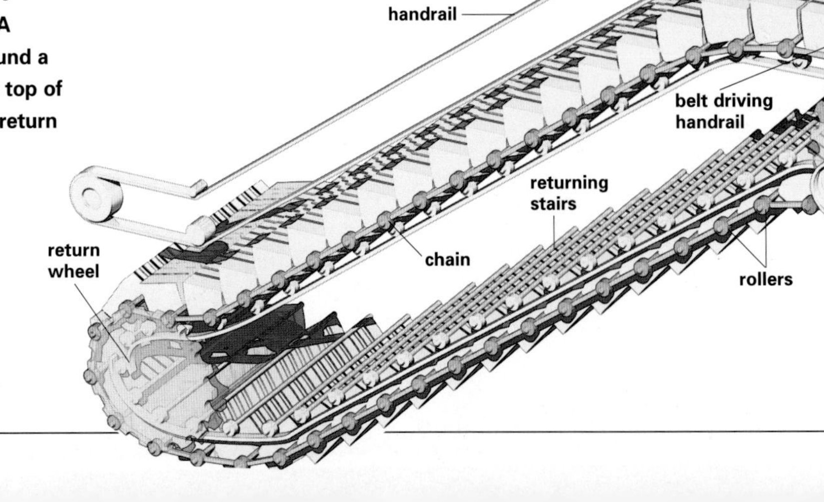

SEE ALSO
Skyscrapers · 64
Computerized Buildings · 66

db
Dress

TUNNELS

Digging a tunnel involves a slow, seemingly blind, passage of equipment and workers underground—through a mountain, under streets, or beneath an expanse of water—with the crews and machines excavating, blasting, and reinforcing their way from the two opposite ends until they meet.

Tunneling requires heavy cutting equipment, such as huge tunnel-boring machines (TBMs) that have rotating drill heads. Moved slowly along by gigantic hydraulic rams and guided by computer-linked lasers and many other electronic distance-measuring contrivances, these engineering marvels simultaneously cut through and eject tons of soil and rock. They also line the tunnel with prefabricated concrete or cast-iron segments to keep it from caving in.

For at least two and a half centuries, people in England and France dreamed of digging a tunnel under the English Channel—and more than a few thought that the dream was a nightmare. To them, a tunnel linking the two shores would serve more than peaceful pursuits. It might also provide an avenue of invasion. But in the most expensive privately financed engineering project in history, TBMs completed the 31-mile-long, three-tunnel "Chunnel" linking the two countries in 1994. The tunnel's 24-mile underwater section, burrowed into a chalk layer well beneath the Channel, is the longest in the world. More than just a tunnel, it has two-way rail tubes for shuttles, freight trains, and high-speed passenger trains. The Chunnel's central service tunnel provides fresh air and allows for emergency evacuation and maintenance.

Designed to bore smooth walls to the exact size, TBMs eliminate blasting accidents and deafening noises. They are very expensive, though. An alternative is to lay precast tunnel sections into trenches where they are joined and backfilled. Despite the high-tech safety precautions of either method, tunneling remains one of the most dangerous engineering projects.

TUNNEL-BORING MACHINE

(Right) One of the 11 mammoth boring machines on the Channel Tunnel project dwarfs workers. Each machine's rotating drill head cut through rock 200 feet below the floor of the English Channel.

CHANNEL TUNNEL

Embedded in a chalk layer, the "Chunnel" contains two one-way rail tunnels for trains and shuttles and a central service tunnel that provides ventilation, room for maintenance, and an evacuation route. Crossover caverns allow trains to change tracks, while piston relief ducts balance pressure waves caused by fast-moving traffic. Cross passages carry air from the service tunnel to main tunnels.

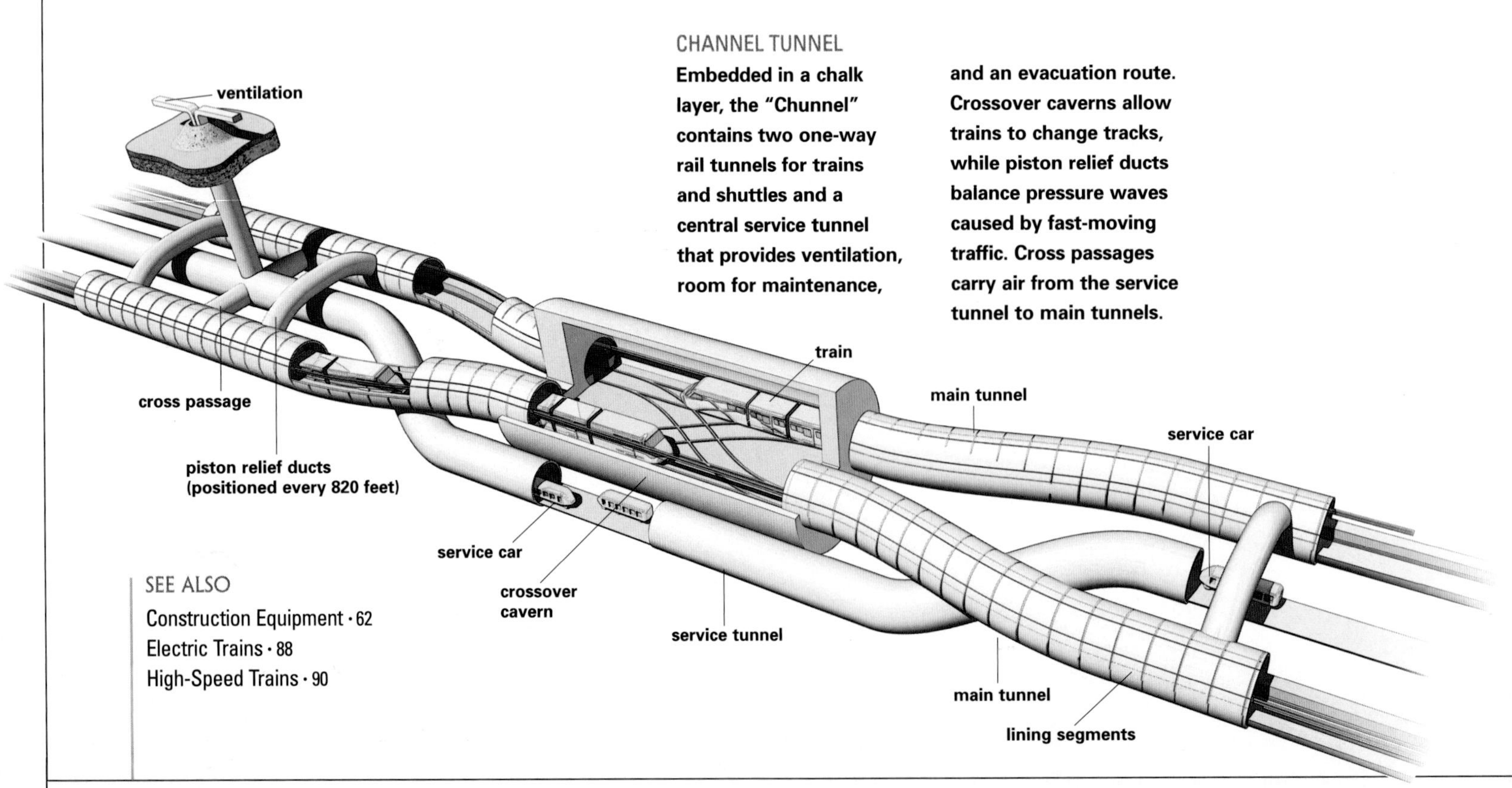

SEE ALSO
Construction Equipment • 62
Electric Trains • 88
High-Speed Trains • 90

BRIDGES

WHETHER THEY ARE THE QUAINT, WOODEN, covered structures that still span some of our rivers and streams, or magnificent stretches of steel and heavy cable that loom high over deep bays and valleys, all bridges have the same function: to provide easy passage over natural and artificial obstacles.

There are five general types of bridges: beam, arch, suspension, cantilever, and cable-stayed. Reinforced and prestressed concrete and steel are the major materials used. Which type of bridge to build depends on many factors—span, winds, vulnerability to earthquakes and tides, clearances for ships and boats, and the weight of traffic. How to build bridges requires a keen understanding of loads and stresses, metal fatigue, strength of materials, and soil and sediment science. Because a bridge is in a sense a landmark and a monument, aesthetic form is also a major consideration in the construction, just as it is in that of a skyscraper.

In taking into account both the bridge's own weight and the weight of the traffic, engineers must calculate the total effect on the foundation, and how best to build a structure that will resist stresses and strains. Builders know that when any force is imposed on a structure it tends to deform, and that materials respond in different ways when a load is applied. Thus, the choice of materials and style of the bridge are vital considerations.

A beam bridge (a span supported at either end, like a log across a ditch) is the most common form of bridge design. However, when resting merely on supports at either end, its length is limited due to the bending effect of a downward load on the beam. So, for longer spans, additional supports, or piers, and spans are added. The arch bridge puts more stress on the end supports than does an ordinary beam, but it can accommodate longer spans than beams. However, since its shape makes it virtually impossible for ordinary wheeled traffic to pass over it, a true arch bridge has a flat deck hung above or below it.

Suspension bridges, because of the efficient way they handle and distribute loads, have no equal in the distance they can cover. These bridges have several key parts: towers set on piers, steel cables that arc between them, steel hangers that connect the traffic-carrying deck to the structure and carry the loads up to the cables, and steel and concrete anchorages that hold the cables at either end. Simply put, when the decks' loads are transferred from the cables to the towers, the weight is sent to the bridge's foundations. These bridges are not always practical because they can be expensive to build.

Cantilever bridges, often used over roads and to carry railways, are stiffer than suspension bridges and do not have attachment sites at opposite ends, like beam bridges and arches. Instead, they have two rigid, deep beams extending from opposite piers, like brackets. Fixed at one end only, they may join in the middle to form a strong, heavyweight-bearing surface, or their gap may be closed with a concrete slab. The cable-stayed bridge is characterized by its diagonal steel cables, called stays, which attach the bridge's deck to tall, mastlike towers that sit atop the weight-absorbing foundation piers.

GOLDEN GATE BRIDGE

(Right) Inching his way on a high cable on San Francisco's Golden Gate Bridge, a worker heads up through a fog bank to do some painting.

AGUAS LIVRES

(Above) Not a bridge in the true sense, the Aguas Livres aqueduct once carried water from distant sources to baths and fountains, as did similar structures built by the ancient Romans, including bridges. The arch is still used in bridge design today.

SEE ALSO
Elements of Construction • 60

CHAPTER 4

TRANSPORTATION

WITHOUT FUEL AND ENGINES, WIND AND SAILS, or arms and oars, all our vehicles would remain as motionless as bowling balls waiting to be bowled. We know this is true, but most of us cannot explain why. It took Sir Isaac Newton (1642-1727) to shed light on it all: "Every body perseveres in its state of rest or of uniform motion in a straight line, unless it is compelled to change that state by forces impressed thereon." His other laws of motion tell us that acceleration depends on the amount of force exerted and that every action has an equal and opposite reaction. The forces instigating change may be physical, mechanical, electrical, magnetic, or gravitational, but no matter which furnishes the drive, the goal is the same: a change of position. While the concept is simple, it and all of its ramifications created the foundation for modern science and for all of our means of transportation.

Rolling leisurely past hedgerows in Wales, father, son, and canine companion

AUTOMOBILES

THE MODERN AUTOMOBILE IS A FAR CRY FROM Leonardo da Vinci's 15th-century concept of a steam-propelled vehicle, and it's a long way from the three-wheeled, three-mile-an-hour steam carriage that was first built in 1769 by Nicolas-Joseph Cugnot, a French Army officer. While the standard car's basic operation has changed little over the years—the gasoline-fueled internal combustion engine is still its source of power—the modern automobile is a paradigm of technological progress. Electronic ignition greatly extends the life of spark plugs; front-wheel and four-wheel drives help provide more traction; and a fuel-injection system helps control the air-fuel mixture to the engine. It also has superchargers for more power, antilock disc brakes that self-adjust for skids, and emission controls that are designed to reduce the amount of pollutants released into the environment.

Early cars, some electrically driven, were produced in small numbers just before the start of the 20th century. Many of the models were made in Europe, and only a wealthy few owned them. The Duryea Motor Wagon Company, founded in 1895 in Springfield, Massachusetts, built 13 cars that year. Today, many millions of passenger cars are in operation in the United States, a testimonial not only to the overwhelming demand for comfortable, reliable, and economical transportation, but also to mass production and a revolutionary manufacturing process: the assembly line.

Two Americans share credit for supplying the demand. Beginning in 1901, Ransom Eli Olds made the first commercially successful U.S. car, the Oldsmobile. But the car that captured the popular imagination was Henry Ford's lightweight yet strongly built Model T. First manufactured in 1908, it was the first automobile built by modern mass-production methods. Engine and chassis assemblies were tested on one level and driven to the bottom of a chute; there, bodies were slid down and bolted on, and the cars were driven off.

Today's modern automobile line is a round-the-clock combination of human assemblers and computerized operations that may reach ten miles in length. Along the way, giant presses shape and stamp out parts, and robots weld the automobile frame. The body is dipped into a rustproofing bath, and coats of paint are baked on.

One new type of robot is called a cobot. Many industrial robots are dangerous to work with, and assemblers don't always relish being too close to them. In the case of the cobot, which was developed by researchers a few years ago at Northwestern University, it provides only guidance while workers provide the power. As "intelligent assist devices," cobots help workers guide heavy and unwieldy parts into cars—doors, windshields, and seats—without damaging the body. Cobots work by providing computer-directed "virtual surfaces," invisible boundaries that guard, say, a doorframe and the interior of the car. This guidance system either deflects or redirects the way an auto part is pushed by a worker.

AUTOWORKERS
Unhindered by doors and windows that special machines will add later, workmen install essential parts and systems in the suspended space frame of a nearly completed car.

SEE ALSO
Automotive Electronics • 78
Alternative Fuels for Cars • 80

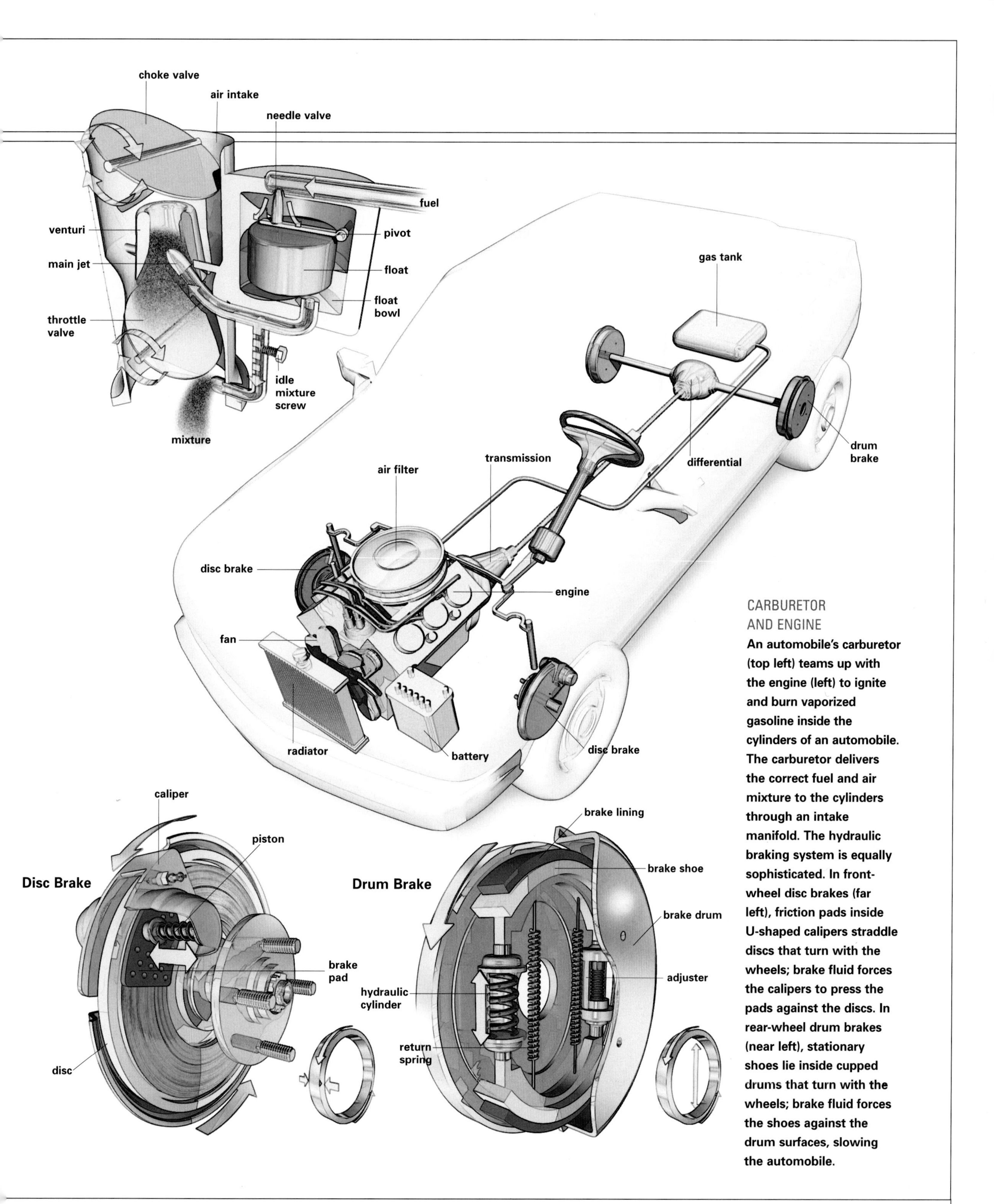

CARBURETOR AND ENGINE

An automobile's carburetor (top left) teams up with the engine (left) to ignite and burn vaporized gasoline inside the cylinders of an automobile. The carburetor delivers the correct fuel and air mixture to the cylinders through an intake manifold. The hydraulic braking system is equally sophisticated. In front-wheel disc brakes (far left), friction pads inside U-shaped calipers straddle discs that turn with the wheels; brake fluid forces the calipers to press the pads against the discs. In rear-wheel drum brakes (near left), stationary shoes lie inside cupped drums that turn with the wheels; brake fluid forces the shoes against the drum surfaces, slowing the automobile.

AUTOMOTIVE ELECTRONICS

NOT LONG AGO, WHEN A MECHANIC PEERED under the hood of a car or truck to perform a tune-up, it was all pretty straightforward. Now vehicles are equipped with almost the same technology found in computers and other electronic gear and made with new lightweight and resilient materials (ceramic-electronic devices that can deal with drastic temperature changes under the hood, and a variety of polyester and nylon resins for sensors and connectors). The tune-up, which used to mean an engine overhaul every 10,000 miles or less, has been stretched to 100,000 miles due to lead-free gasoline, fuel injection systems, and platinum spark plugs that resist erosion. Moreover, much of what used to be adjusted by hand is now handled automatically.

According to engineers and analysts, more than 80 percent of automotive innovation is now based on increasingly complex electronics. In the modern car, so-called data networks—which enable the engine to communicate with other parts of the vehicle—rely on over 80 microprocessors and even more "power" semiconductors to direct all the mechanical and safety systems, such as airbag and antilock brakes, steering, and antitheft devices. The transformation has also given technology some new words: telematics, an application that covers navigation, communication, and entertainment devices, and mechantronics, which covers examining systems in their entirety rather than on a part-by-part basis.

Perhaps one of the most valuable systems is the OBD, for on-board diagnostics. Computer based, it monitors the performance of an engine's major parts, including emission controls. For car owners, it is an early-warning dashboard light that alerts the driver to "Check Oil," or "Check Engine." The system works on a "diagnostic trouble code" that is stored in the computer memory; newer cars all use the same computer language to monitor an engine's components. To fix the problem, mechanic-technicians plug the car's system into special diagnostic equipment that allows them access to far more information than in the past. This is done by using a computerized "scan tool" that enables the technician to speedily retrieve the stored diagnostic trouble codes from the memory system.

Other monitoring systems—which work somewhat like the "black boxes" used to record flight data—may now be installed by car owners themselves. These can be plugged into the car's OBD connector and can chart trip speed, fuel pressure, braking technique, RPMs, and various driving habits. At home, the device can be linked to a personal computer to retrieve stored data.

GPS IN THE CAR

(Right) Computerized navigation comes to the aid of a baffled traveler on the road. Installed in a vehicle, an electronic navigation system maps a car's precise position on a highway, identifies routes, and selects the best way to the driver's destination.

SAFETY FEATURES

(Left) An air bag inflates with explosive force against a test dummy when a strong impact ignites a detonator cap. Air bags save lives, but they also can cause serious injury, even death, in rare instances. New models that inflate with less force reduce this risk.

SEE ALSO
Automobiles · 76
Alternative Fuels for Cars · 80
Satellite Communications · 218

VISIONAUTE
TOTAL 0:15
SAGEM
PHILIPS

ALTERNATIVE FUELS FOR CARS

TWO OF TODAY'S MAJOR CONCERNS ARE AIR pollution caused by gasoline-powered automobiles and our limited fossil fuel supply. Engineers are looking to electricity for help.

The ordinary, gasoline-powered car already relies on electricity—up to a point. Its electrical system furnishes electricity to operate the starter, ignition system, and accessories, and it recharges the storage batteries. But this kind of automobile also needs pistons and a carburetor, water pump and muffler, none of which are found in an electric car. A simpler machine, an electric car sends electrical energy stored in the battery directly to the motor, where it is converted into mechanical energy. An onboard recharger plugs into an outside electrical outlet for "refueling."

While owners of electric cars do not have to deal with oil changes, their vehicles are relatively slow and their batteries generally take two or three hours to recharge after only about a hundred miles of road time. With improved batteries, or possibly batteries that store solar energy, the cars may one day be more popular.

Engineers are also testing hybrid automobiles that use fuel cells or electric motors in combination with gasoline engines. A fuel cell works like a battery except that it does not need recharging. It is basically a pair of electrodes wrapped around an electrolyte, an arrangement that generates heat and electricity by combining oxygen and hydrogen without combustion. The hydrogen fuel powering the cell is actually separated from the car's gasoline, but the fuel cell utilizes so much more of the gasoline's potential energy than a piston engine does that a car's range may be doubled on the same amount of gas, with less pollution. Even more environment friendly is hydrogen that does not have to be extracted from fossil fuel; it can be removed from water by sunlight, a feat already accomplished by scientists.

Some hybrid cars use the gasoline engine to charge batteries running the electric motor. Others switch between engine and motor according to driving conditions. A car developed in Japan starts on battery power and runs on a combination of gasoline and electric power when it speeds up.

The automotive industry won't be giving up gasoline and diesel fuel in the near future, given the enormous effort of revamping refineries, gas stations, parts manufacturers, and more. Yet technology and science are never ending quests—and scientists in the distant future are certain to make the gasoline engine a relic, like the horse and buggy it replaced.

SEE ALSO
Solar Power • 50
Batteries • 56
Automobiles • 76
Drilling for Oil • 182

THE ELECTRIC MOTOR
(Right) A mechanic tunes up an electric rally car. The motor has few moving parts and has electrical or solid state components but no complex emission-control systems.

HYBRID POWER
(Below, left to right) A gasoline engine splits power between wheels and generator, running the motor and recharging the battery. As the engine shuts off, energy flows from the wheels through the motor to be stored in the battery. When the car accelerates, the battery's energy assists the drive power.

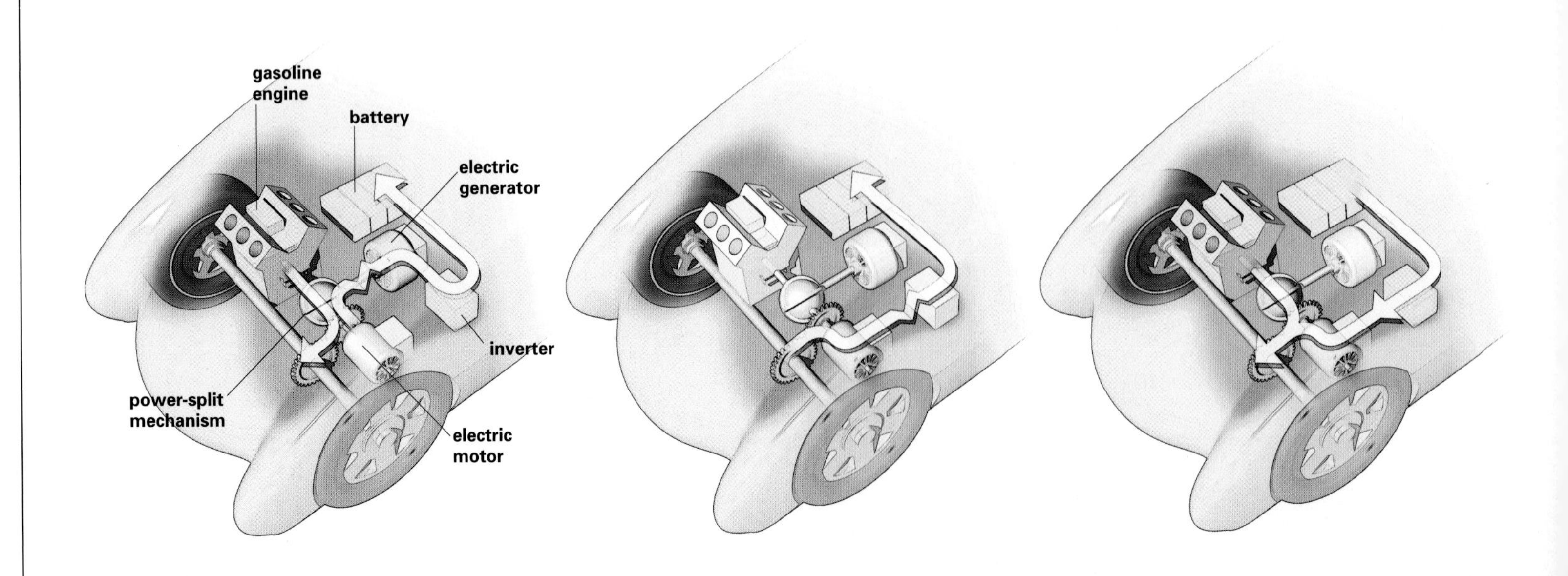

RENAULT

BICYCLES

RIDING A BICYCLE IS ONE OF THE SIMPLEST WAYS to travel: It generally requires a minimum of expertise beyond knowing how to shift gears and keep one's balance. But the bicycle, which dates back more than 200 years, is an invention whose relatively humble appearance belies its complexity.

The early bike had iron tires that were first propelled by the rider's feet, then by ropes wound around an axle and hitched to a lever for driving power. The seat was uncomfortable, to say the least, because it was usually no more than a wooden beam. Even with refinements, the first bikes were frequently ridiculed: One 18th-century model was dubbed a "dandy horse" by Englishmen who thought the frame resembled a pack animal. A modern bicycle, however, can be a high-tech, designer dream-machine equipped with a score of gears, an aerodynamic configuration, airfoil tubing, and a titanium cantilever frame.

In a sense, the rider is a bicycle's engine, supplying the physical energy to be converted to mechanical energy. The bike's chain serves to transmit force from one place to another, in this case from the pedal sprocket to the rear driving wheel. Gears on the front sprocket and the rear wheel serve the same purpose as other gears: They change one rate of rotation to another. This makes the bicycle very efficient, because the effort applied to the pedals can be geared up for high speed or geared down for hill-climbing power. A pair of devices called derailleurs transfers the chain from one sprocket to another. In highest gear, the rear wheel turns many times for each turn of the pedals, and the bike moves along swiftly; but in low gear, when a stronger forward shove is required to go, say, up a hill, the wheel turns fewer times in relation to the pedal, thereby trading speed for ease of pedaling.

THE HUMAN BODY
Muscle power is all that a bike needs to get moving, although new materials and designs help make bikes more aerodynamic, and cycling easier. Frames may be constructed of aluminum, lightweight metal alloys, or carbon fiber composites, with shock absorbers of various polymers.

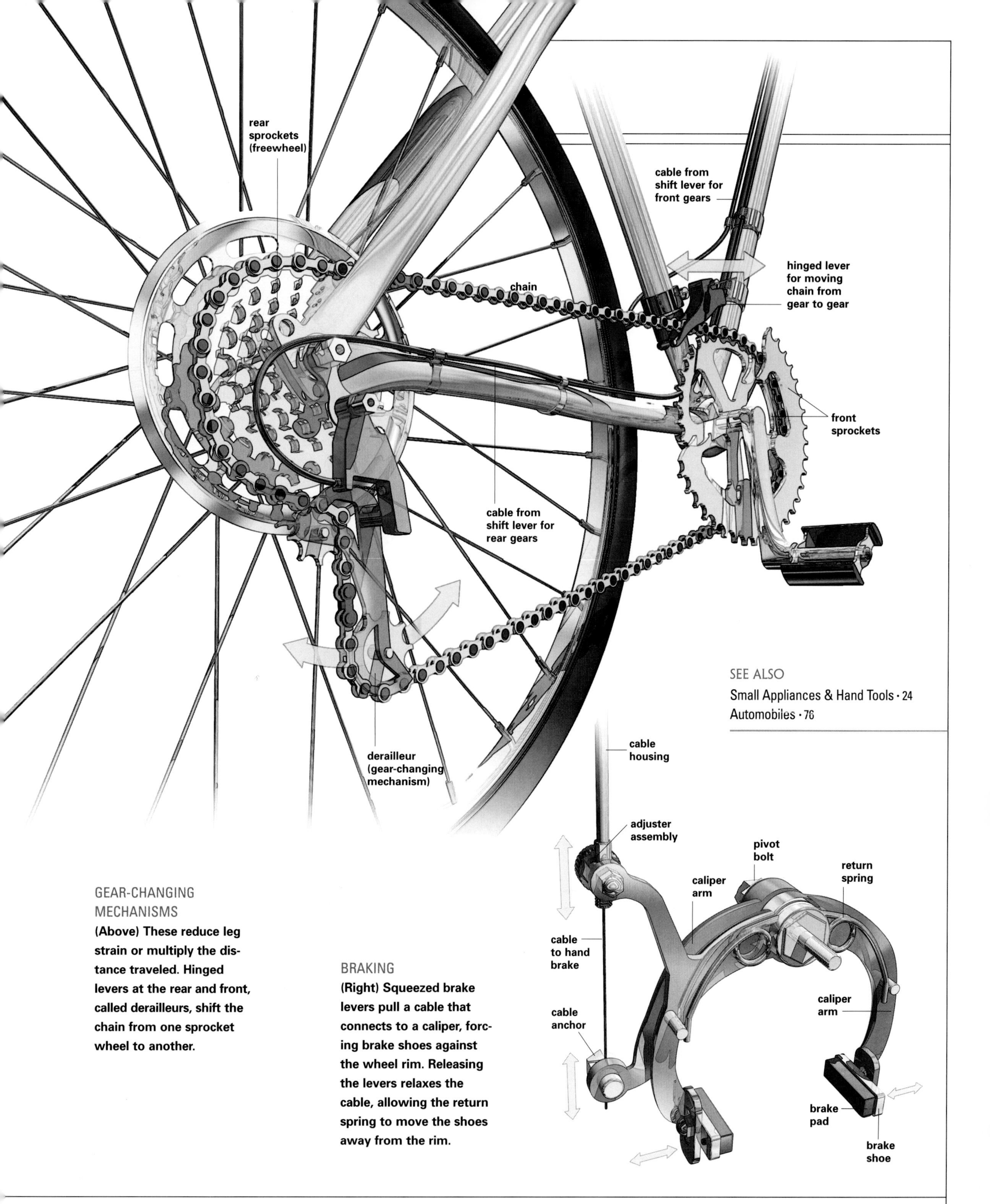

SEE ALSO
Small Appliances & Hand Tools • 24
Automobiles • 76

GEAR-CHANGING MECHANISMS

(Above) These reduce leg strain or multiply the distance traveled. Hinged levers at the rear and front, called derailleurs, shift the chain from one sprocket wheel to another.

BRAKING

(Right) Squeezed brake levers pull a cable that connects to a caliper, forcing brake shoes against the wheel rim. Releasing the levers relaxes the cable, allowing the return spring to move the shoes away from the rim.

STEAM ENGINE TRAINS

"TRAINS ARE WONDERFUL," THE MYSTERY writer Agatha Christie once observed. "To travel by train is to see nature and human beings, towns and churches and rivers, in fact, to see life." Steel wheels on steel tracks do the job of travel admirably, and the development of the steam locomotive in the early 19th century proved that even an inefficient piece of equipment can suffice.

Its iron horse image notwithstanding, a steam locomotive was a coal stoker's backbreaking nightmare, a noisy machine that wasted energy, dirtied the air with soot, and cost a great deal of money to operate. One such juggernaut, which ran at a mile a minute in Rhode Island and Connecticut in the early 1900s, burned three tons of coal before pulling out of the roundhouse, carried four tons on a tender, and required 4,000 gallons of water in its tank. Revered in ballad and folklore, and romanticized by model railroaders and preservation societies, the world's remaining steam locomotives puff along these days in only a few countries.

Admittedly, there is something nostalgic about a huffing steam locomotive hauling a line of coaches, boxcars, and a red caboose. But something more stirring may be found in the technology behind the thing: the enormous piston-pushing power of steam produced at high pressure. With its piston rods connected to driving wheels as much as 85 inches in diameter—an arrangement akin to a tricycle's pedal wheel—the steam locomotive is the perfect visible example of how a machine can convert one form of movement into another. As the driving rods move back and forth under the force of steam, their linear movement is shifted into rotary mode, turning the wheels. The legendary John Henry may have wanted to die with a hammer in his hand rather than let a steam drill beat him down, but it wasn't worth the effort, technologically speaking.

BELCHING SMOKE

(Right) Once rulers of America's rail lines, steam engines and their intimidating bulk gave way to more refined, and more efficient, diesels and electrics.

BOILER WORKS

(Below) Essentially a furnace and boiler on wheels, a steam locomotive burns fuel, usually coal, in a chamber called a firebox. Heat passes through tubes inside the huge water-filled boiler and generates steam, which is collected and sent through U-shaped "superheater" tubes to cylinders on each side of the locomotive's front end. When high-pressure steam enters the cylinders (see bottom inset), it moves the pistons and drives the train.

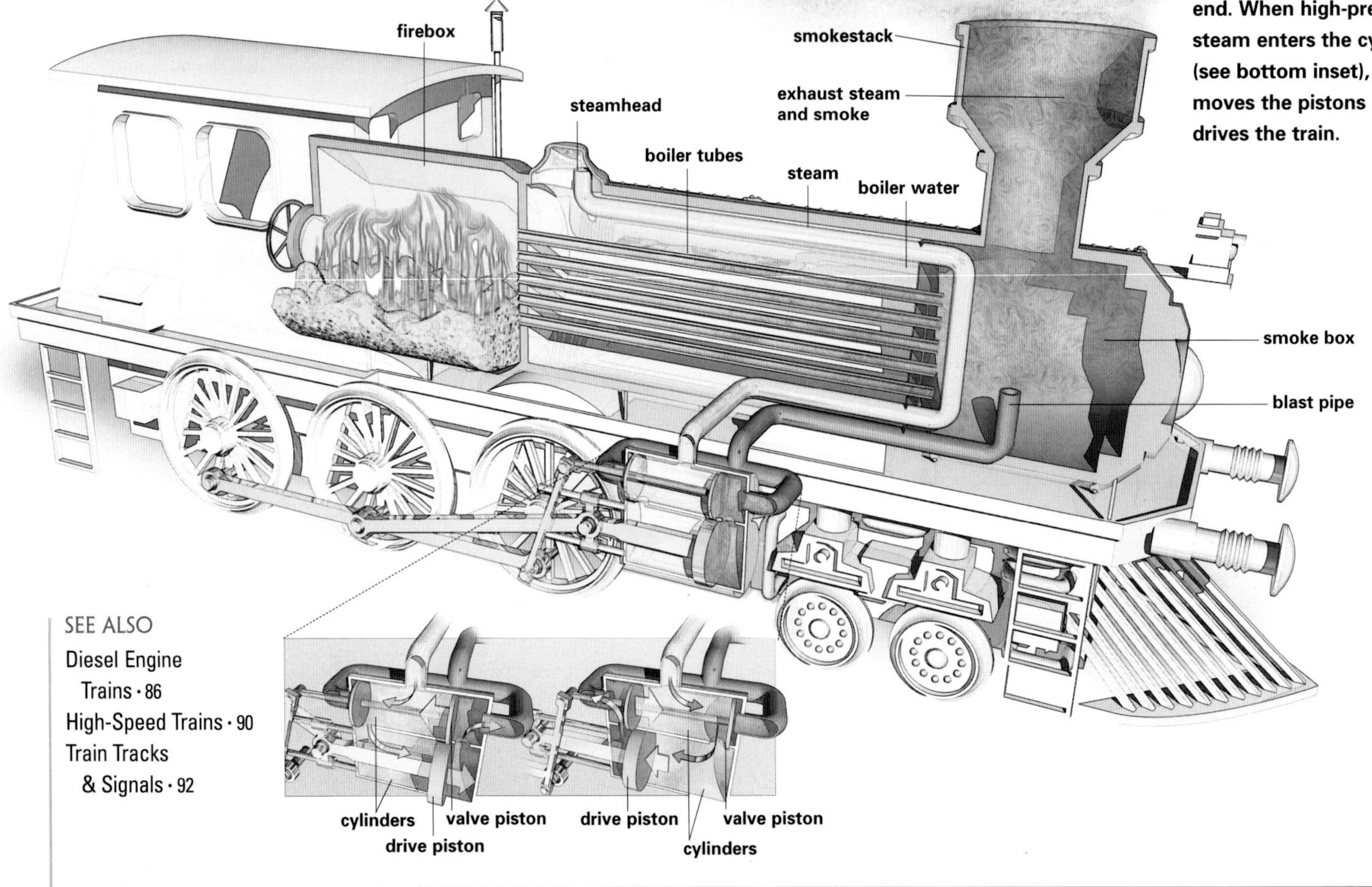

SEE ALSO
Diesel Engine Trains · 86
High-Speed Trains · 90
Train Tracks & Signals · 92

DIESEL ENGINE TRAINS

THE DIESEL ENGINE SPELLED THE END OF the line for the steam locomotive. Put on the market in 1898 by inventor Rudolf Diesel, a German mechanical engineer, it relies on the same piston strokes and basic moving parts as the gasoline engine. Its fuel, though, is a heavier, thicker, less expensive oil, and it does not need spark plugs for ignition. Instead, this brawny brother of the lighter gasoline engine compresses air inside a cylinder to temperatures that may reach 1000°F and, at the moment of maximum compression, ignites a charge of fuel oil sprayed into the heated air, producing the power stroke.

Because a diesel engine performs more work per gallon of fuel, it is a boon for long-hauling trucks, buses, ships, and trains, and for heavy agricultural and road-building equipment. It works one way in trucks and another in trains, however. In a road vehicle the engine's power is transmitted directly to the wheels, but in a train the engine is connected to an electrical generator. The current produced is stored in huge batteries, then fed to electric motors installed in so-called bogies, which are pivoting carriages that not only house the motors for the driving wheels but also enable the train to negotiate curves. A diesel locomotive is an enormously heavy machine that would be difficult to control if power went straight to the wheels. An electric motor's output can be regulated fairly easily and produces power at very low speeds.

The diesel engine may not evoke images of sleek sports cars and racing boats, but it nonetheless reminds us that function, especially when efficient, is as admirable a quality as form.

RIDING THE RAILS

(Left) The Amtrak Cascades races between Seattle and Portland. Diesel trains do away with concerns about overhead electric wires exposed to weather and accident, and a steam engine's smoke.

FIRING UP A DIESEL ENGINE

(Right) The piston in the cylinder squeezes and heats air trapped inside; at the top of the stroke, the system injects oil; the air and oil mixture burns and drives the piston down to turn the crankshaft connected to the generator.

SEE ALSO
Steam Engine Trains • 84
Train Tracks & Signals • 92

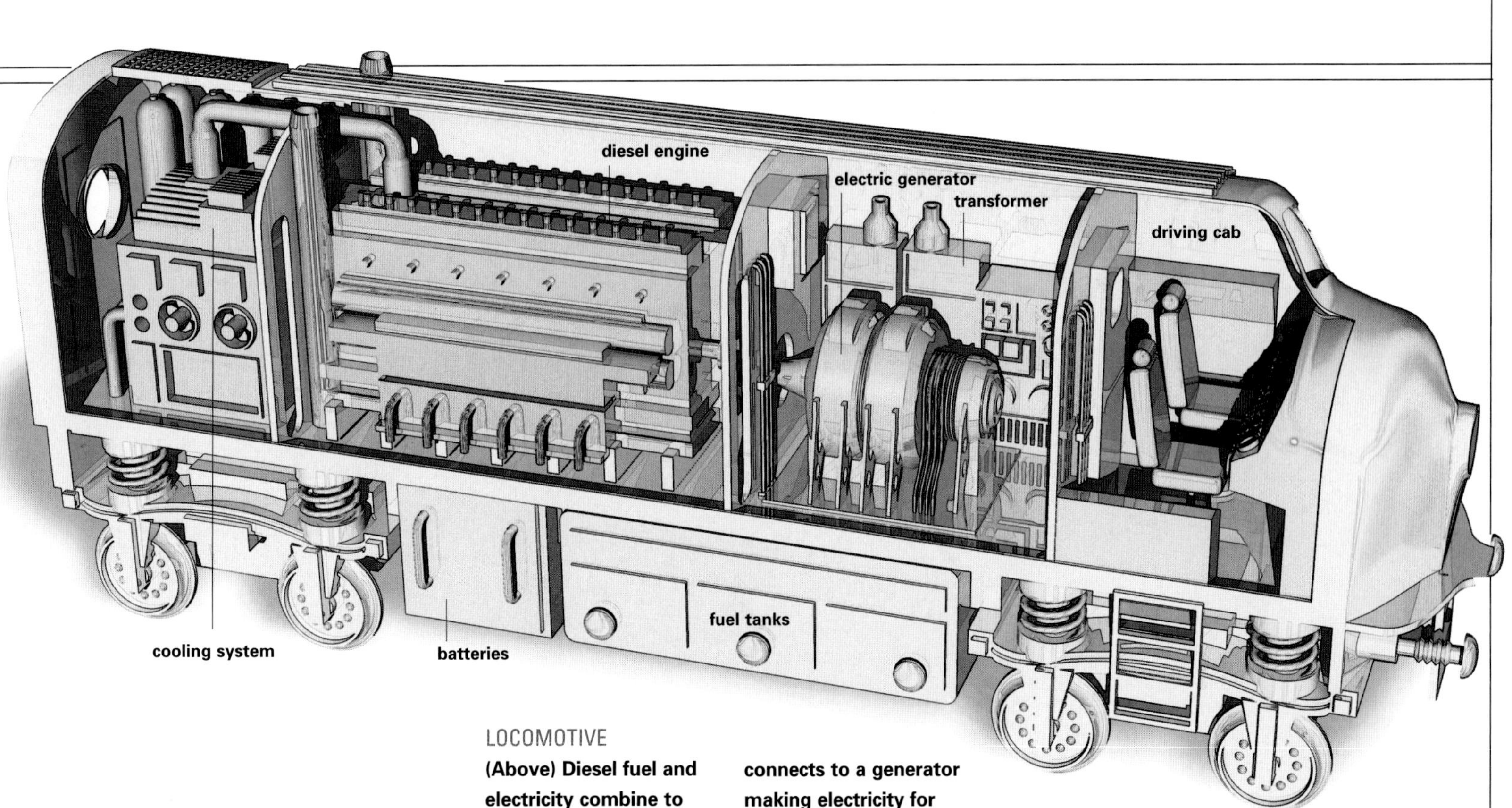

LOCOMOTIVE

(Above) Diesel fuel and electricity combine to provide power without the huff and puff of steam and without overhead wires. The engine connects to a generator making electricity for storage in large batteries alongside the wheels. Power for the driving wheels comes from motors that draw on the batteries.

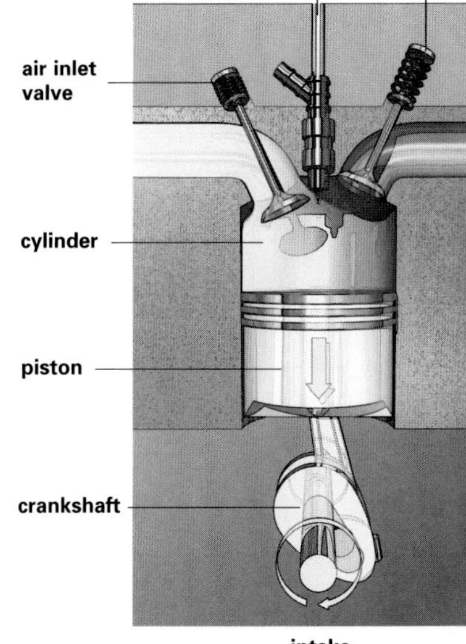

intake

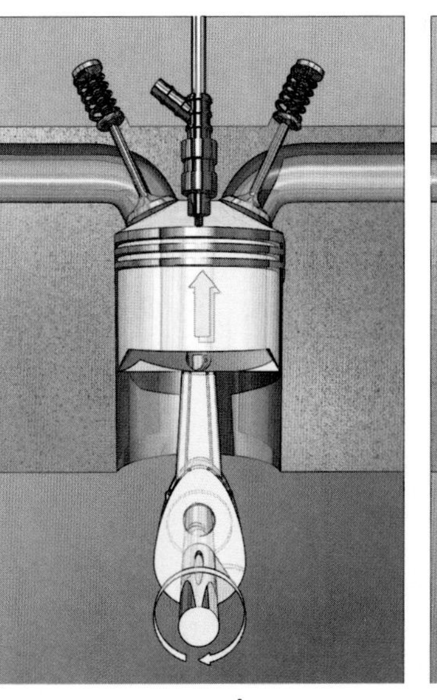

compression

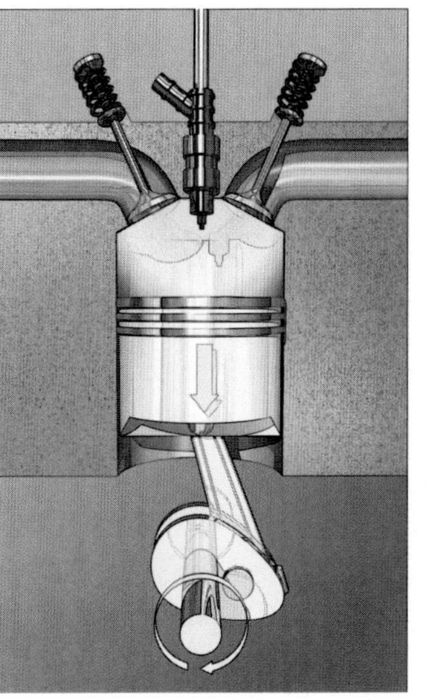

combustion

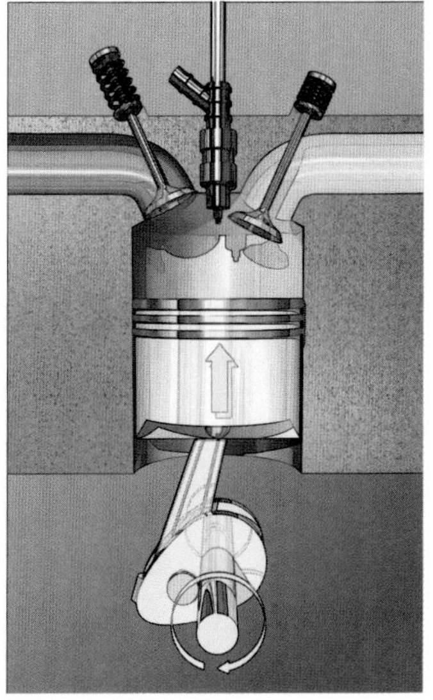

exhaust

ELECTRIC TRAINS

UNDERGROUND
(Left) An underground Mass Transportation Rail train rolls into a station in Hong Kong. Essential to public transport, electric railway systems also have miniature, and entertaining, counterparts in remote-controlled, low-voltage model systems, which move their cars about with small transformers that increase and reduce voltage.

SKY CAR
(Right) A combination of electrical and mechanical components, a resort's gondola hovers over a skier. A type of aerial lift, the gondola's passenger cabins are connected to a loop of steel cable suspended between two stations; the cable is driven by a wheel connected to an electric engine at a terminal.

IN 1873, ANDREW SMITH HALLIDIE, AN AMERICAN engineer and inventor, patented the first cable cars: rickety trolleys hauled up San Francisco's steep hills by an endless wire-rope cable running in a slot between the rails. Drawn by a steam-driven mechanism in a powerhouse, the cable cars—still operating today but with technical improvements—eventually spelled the end of horse power as a means of moving passenger coaches. Other cities, including Los Angeles, Washington, D.C., and Kansas City, followed suit, but traffic congestion was a major problem. As a popular joke in late 19th-century Boston had it, the city's streets were so clogged by trolleys that passengers could reach their destination more quickly by climbing onto their car's roof and walking across the tops of stalled vehicles.

The electrified subway changed all that. The world's first subway, run with steam locomotives, opened in England in 1863 and converted to electricity in 1893. That year, Budapest, Hungary, became the first European city to build an electric subway line. America's first electric subway system opened in Boston in 1897, followed in 1904 by the one in New York City.

Subway trains run on a pair of rails in systems that may include hundreds of miles of track, and they are generally powered by voltage running through a third rail. Electric surface trains work on the same principle, but they usually pick up voltage from overhead lines with a folding, scaffoldlike apparatus known as a pantograph. The same arrangement drives electric buses, sometimes known as trackless trolleys. Monorails, which are elevated trains built for relatively short distances, run on a single rail, either straddling it or hanging beneath it.

SEE ALSO
Power Stations • 44
High-Speed Trains • 90
Train Tracks & Signals • 92

snowbird
snowbird

HIGH-SPEED TRAINS

TOP SPEED FOR A SUBWAY TRAIN IS AROUND 75 miles an hour, faster certainly than an early steam locomotive's 25 mph. (In fairness, though, it should be noted that in 1893 a steam engine known as the 999 thundered between Batavia and Buffalo in New York, at 112.3 mph, the fastest humans had ever moved.)

Today, Amtrak's Acela—derived from the words excellence and acceleration—is America's high-speeder, capable of hitting about 150 miles an hour. Acela, which links Boston, New York, and Washington, D.C.—the Northeast Corridor—relies on a signal system that receives electrical information transmitted through the rails and displays it to the train crew in the cab. An automatic train control system supervises the engineer, who is also "watched" by a speed enforcement system that automatically adjusts speed.

France's electric streaker, the relatively lightweight TGV (Train à Grande Vitesse) is faster, making 185 mph—a pace made possible by a well-crafted combination of aerodynamic styling, specially spaced track centers, and heavily computerized signaling systems. But it is the system known as electromagnetic levitation that promises far more speed over the rails. Maglev technology relies, as its name suggests, on magnets, not wheels. Set on the underside of a train and in the track, they lift the train and propel and guide it with shifting magnetic fields. The idea is not new. In the 1960s, James Powell of Brookhaven National Laboratory was stuck in a New York traffic jam, wishing he could fly over it. A few years later, he and a colleague patented a magnetic levitation system. Eventually, MIT's Henry Kolm built a scale model of a contraption he called the magneplane, and though it moved only a fraction of an inch above a 400-foot aluminum track, it reached an astonishing 56 mph. Today, the record for a commercial maglev system belongs to the Shanghai Transrapid, a German-built vehicle that hit 311 mph in 2003.

Apart from their speed and low noise-emission, maglevs are perhaps the most energy-efficient systems ever developed. According to one mechanical engineer who has studied the Transrapid, the train uses 30 percent less energy compared with existing high-speed rail systems, three times less energy than an automobile, or a full five times less energy than an airplane for the same passenger mile.

SPEEDY FLEET
High-speed electric trains poise for a horizontal takeoff at Lyon Station in Paris. Sleek, and virtually noiseless, high-speed trains offer unparalleled comfort—and they spew out no pollutants.

SEE ALSO
Steam Engine Trains • 84
Diesel Engine Trains • 86
Electric Trains • 88
Train Tracks & Signals • 92
Hydrofoils & Maglev Ships • 98

SUPERCONDUCTIVITY

Maglev trains as well as many of the artifacts of computer and information technology, will be running faster if a phenomenon known as superconductivity goes into routine use. In 1979, Japanese scientists set a speed record of 321 miles an hour with a maglev train that had superconducting magnets on board. Superconductivity works this way: When a metal becomes superconducting—that is, it transports electricity indefinitely without loss from resistance when subjected to intense cold—it produces a magnetic field with almost magical properties. Here, a thallium-based superconductor bathed in supercold nitrogen vapor floats magnets above and below it through repellent magnetic force. The potential is great because superconductors have the ability to affect everything that uses electricity. Current applications include medical imaging systems, infrared sensors, microwave devices, and magnetic shielding apparatus.

THE MAGLEV TRAIN

Electromagnets on the train's underside and on the track float the train on a magnetic cushion with no ground contact. Fed with alternating current, propulsion magnets in the sides of a U-shaped aluminum guideway pull and push the train along and control its speed in a smooth sequence of events. Swift and comfortable, maglev trains will give new meaning to "ground transportation."

train magnets
track magnets
train magnets
track
track magnets
train magnets

TRAIN TRACKS & SIGNALS

A CENTURY AGO, SIGNALMEN POSITIONED ALONG rail routes controlled travel by telegraphing information to one another about their sections of track. Block-signal systems manually operated at stations along the route kept trains safely apart; to get trains from one track to another, a job also performed by hand, workers used a switch stand and lever. Modern train control still depends primarily on dividing the tracks into zones, or blocks, but each is now watched by computers that are fed information by an intricate network of transmitter-receivers, called transponders, on and under the tracks. This trail of electric circuitry activates signals and switching devices, sets speed limits, warns operators of trains in their path, and ensures a safe distance between trains. If a train operator is incapacitated, the train can automatically be slowed or braked to a full stop. At the heart of the rail system, in a control room holding large, lighted maps, a train's position and speed can be charted in real time by dispatchers, and track switches can be changed by a few strokes on a central-control keyboard.

Today, train conductors receive information from dispatch or electronic monitoring devices that relay any equipment problems on the train or the rail. Most rail systems have phased out locomotive firers and brake operators, and freight trains use only an engineer and a conductor because new visual instrumentation and other apparatus have eliminated the need for crew members located at the rear of a train.

TRAFFIC CONTROL

(Left) Casey Jones, with a hand on the throttle and his eyes on the tracks, could not have imagined today's train traffic control centers. A train-spotter in a Paris control room has safety in mind as he monitors high-speed TGVs. Highly computerized, train control centers know the location, speed, condition, and destination of all rolling stock.

RAILHEAD

(Right) A railroad yard worker takes advantage of a brief halt in incoming and outgoing train traffic to inspect a section of rail. But the absence of activity in what is normally a busy location is seen by some as proof that trains are not always profitable due to competition from discount airlines, and a desire to get to destinations faster.

SEE ALSO
Steam Engine Trains • 84
Diesel Engine Trains • 86
Electric Trains • 88
High-Speed Trains • 90
Computers • 230

215
216
217
218

SAILBOATS

WITHOUT THE WIND, BECALMED SAILBOATS are nearly as still as seascapes hanging preserved in time on museum walls. Only paddles, a motor, or the wind can get things going again. Long ago, primitive peoples moved their vessels across the water with crude paddles, a locomotive power carried to its finest extreme in sleek oar-propelled war galleys that were built in 16th-century Venice.

It is the sail, though, that conjures up images of graceful movement, of nature's own technology at work. "Of all man-made things," wrote a sailor-author, "there is nothing so lovely as a sailboat. It is a living thing with a soul and feelings, responsive as a saddle-horse, loyal as a dog, and thoroughly downright decent."

Whether a sailing vessel is a simple day-sailer or an elaborate, full-rigged merchantman or man-o'-war carrying acres of canvas, it is a wind-driven work of art, and because its sails act as airfoils, it is a nautical version of an airplane. Indeed, although the medium through which sailboats move is water, similar forces of lift, drag, and thrust are at work. Sails are cut and sewn to form a curve much like that of a plane's wings, and lift, or drive, results when wind exerts less pressure on the convex side of a sail. A boat sailing into the wind is pulled diagonally forward by lift generated as wind flows over the sails; running before the wind, it is shoved forward by wind pressure coming from behind.

The complex array of sails and rigging used by the classic square-riggers—20 or more enormous sails in a three-masted ship—provides more drive for the mainsails by compressing, funneling, and deflecting air. The auxiliary sails may also furnish drive themselves. But although sailing is still largely a hand-controlled endeavor, winches may be motorized, and automated and hydraulic systems can hoist, reef and furl sails, all of which enable one person to handle a sailboat of almost any size.

It is the sail itself, however, that must do the lion's share of the work. Today, Dacron is the cloth of choice for most sails; templates are used to achieve the sail's right shape (called belly) from triangular to quadrilateral, and size. Sewn with heavy waxed thread, the sails have metallic eyes through which rope can be passed set into the corners, and Dacron sail tape reinforces the sail's edge.

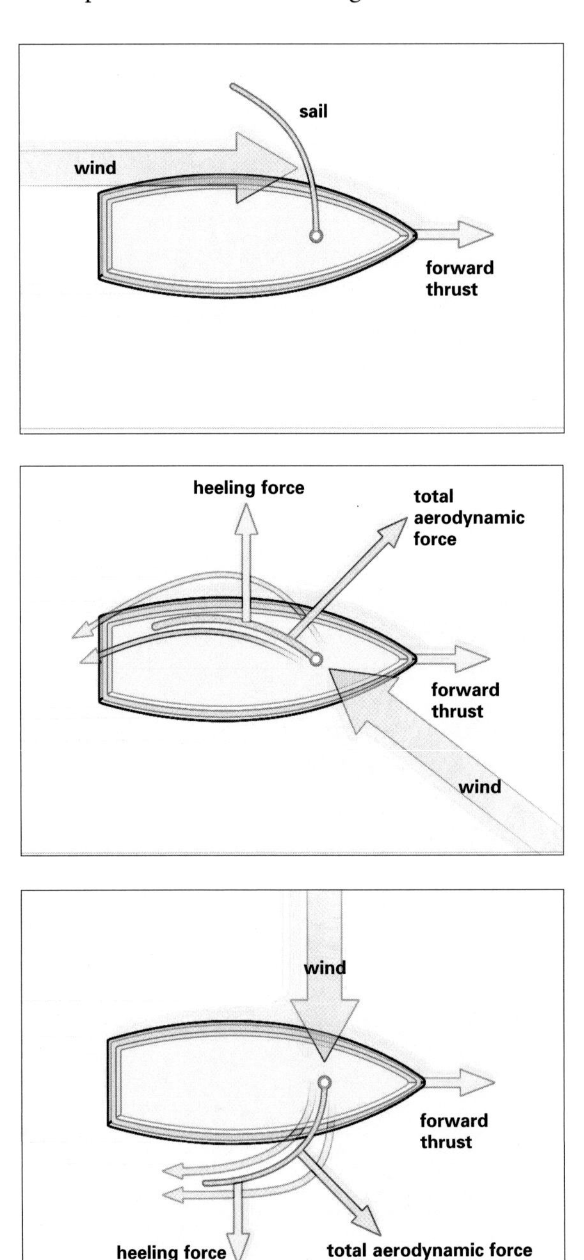

MANEUVER WITH WIND
(Left) Sailing before the wind is an easy push forward; sailing forward against the wind, or with wind blowing across, requires complex maneuvers, such as setting a zigzag course and changing the sail's angle.

MASTHEAD CLASSROOM
(Right) Crew members of the Coast Guard cutter *Eagle* furl sails in a training exercise. The largest tall ship flying the Stars and Stripes, and the only square-rigger in U.S. service, the *Eagle,* with its home port in New London, Connecticut, is a seagoing school for future officers of the Coast Guard. The ship has some 22,000 square feet of sail and five miles of rigging.

SEE ALSO
How Boats Stay Afloat · 96
Hydrofoils & Maglev Ships · 98
The Principles of Flight · 102

HOW BOATS STAY AFLOAT

According to legend, the Greek physicist and mathematician Archimedes (287-212 B.C.) was sitting in his bathtub, taking note of the amount of water his body had caused to overflow onto the floor, when he formulated the basic principle of buoyancy. Thus began the science of hydrostatics, which deals with the laws of fluids at rest and under pressure. Its main point is that a body immersed in fluid loses weight equal to the weight of the fluid displaced. Put another way, things that float—ships and swimmers, ducks and dugouts—do so when the weight of water they displace is exactly the same as their own. If, however, an object displaces a weight of water less than its own, it sinks.

But there's more to it than weight. A wooden or fiberglass platform will not float if too many heavy rocks are put on it, and neither will a bracelet or chunk of steel placed on the water's surface. They sink because their density is greater than that of water and because they do not displace enough water to create the upward force necessary for flotation. On the other hand, wooden boats and steel ships float because their hollow configuration makes their average density less than that of water. Their weight is spread over a larger volume and they displace more water than, say, a steel block. We float, too, when we inhale and rest on our backs in a lake, and we sink when we exhale and point our feet downward.

BUOYANCY

Oblivious to the rules of hydrostatics, floaters enjoy the moment. They do what they do because of the ability of water, or other fluid, to thrust upward when a body is placed on it. Salt water has greater buoyancy than fresh because it is heavier.

SEE ALSO
Sailboats · 94
Hydrofoils & Maglev Ships · 98
Submersibles · 100

STABILITY

(Below) The steadiness of a ship depends on the vessel's center of gravity. Keels and pontoons help maintain stability and minimize the risk of capsizing, but very large vessels may employ stabilizers, winglike protuberances that provide lift to counteract the sea's roll.

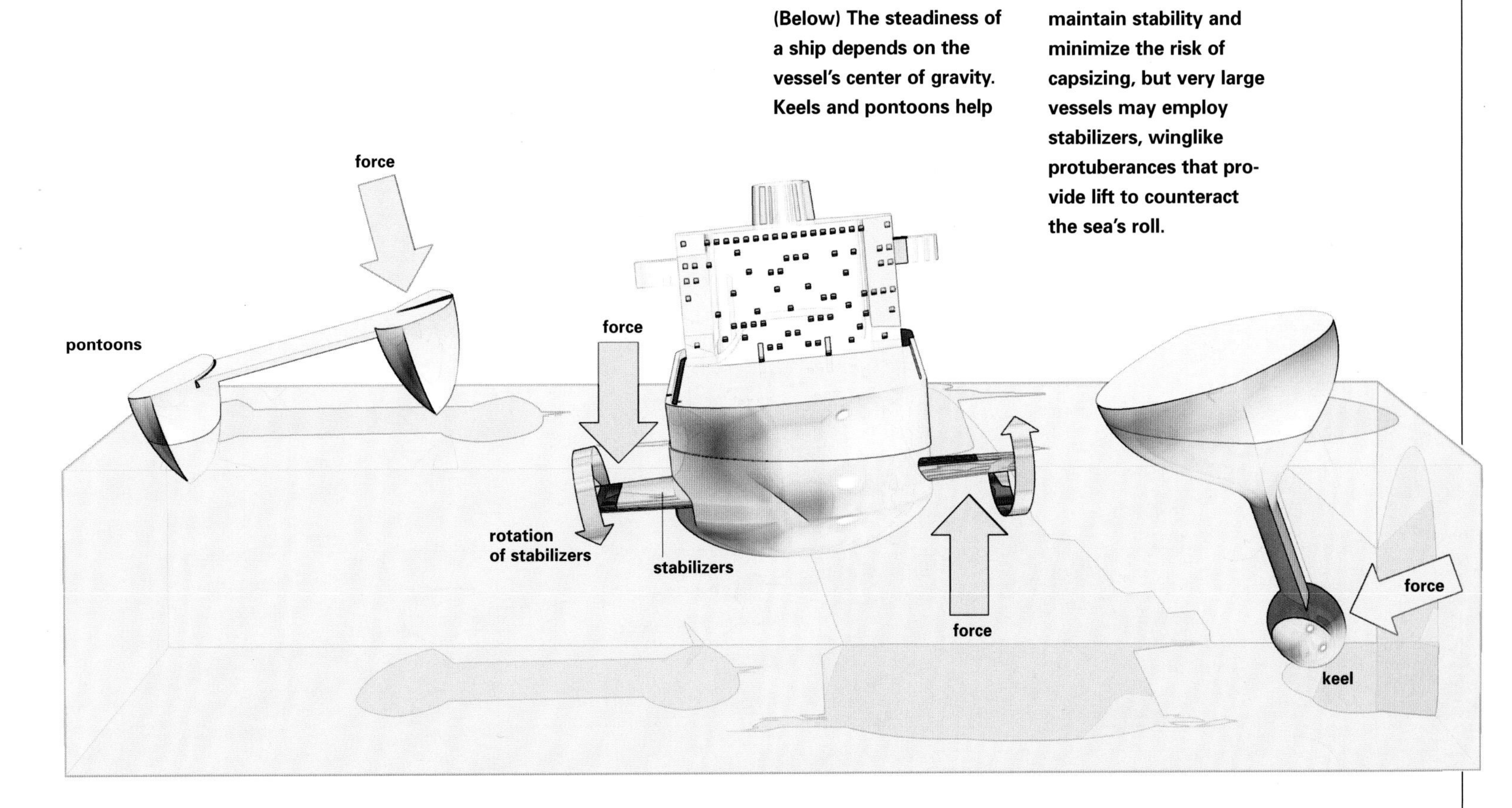

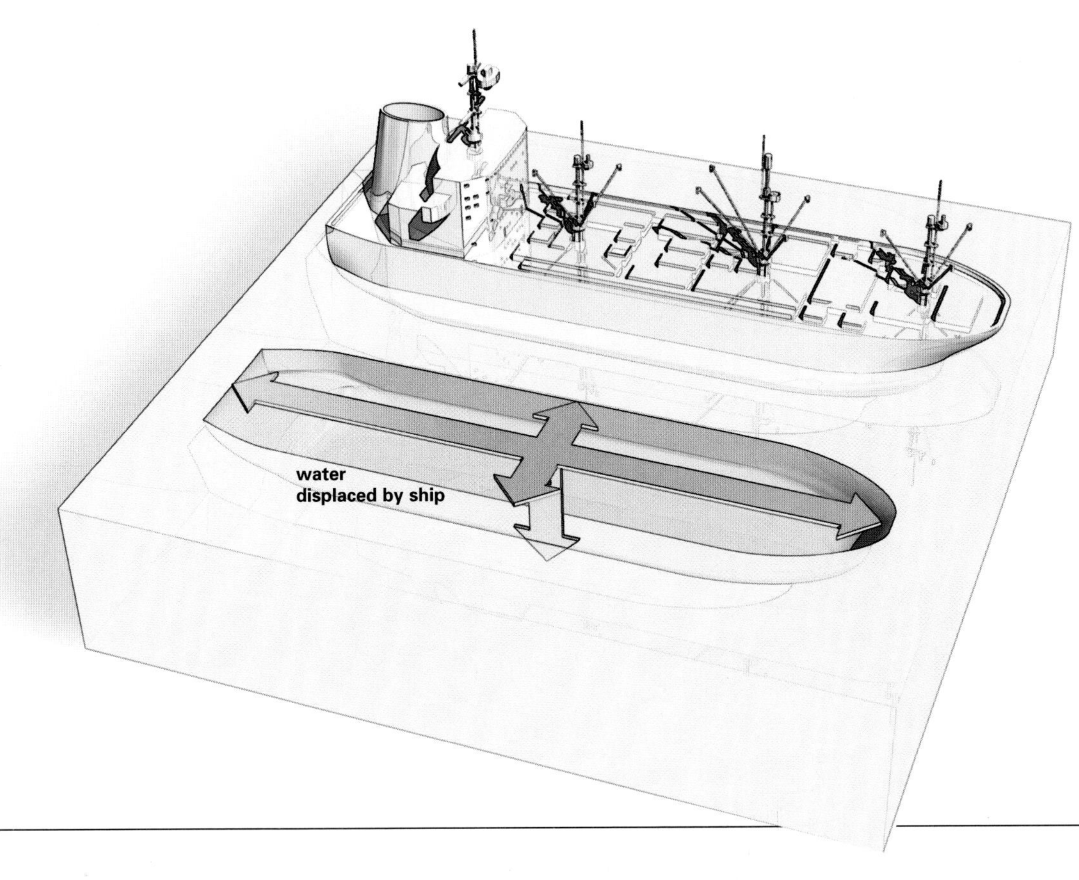

DISPLACEMENT

(Left) A ship floats because it creates an upthrust force from the water equaling the vessel's own weight, and it does so before the point of submersion. For example: A thick, flat slab of steel manufactured without a hollowed-out, bowl-shaped hull to distribute the slab's weight would quickly sink like a stone. Design can also affect speed; the deeper a boat sits in water, the slower it goes.

HYDROFOILS & MAGLEV SHIPS

All boats displace water when at rest and create bow waves when they move. Planing boats—powerboats, aquatic scooters, and hydrofoils, for example—climb over their bow waves and skim the surface as they pick up speed. In the case of hydrofoils, underwater "wings" lift them free of the water as they gain speed, enabling the vessels, driven by turbine-powered pumps and water-jets, to skim along above the surface. So-called hovercraft rise higher off the water because huge fans create cushions of air beneath them.

The most innovative craft that may be a reality one day is the magship, a vessel that works on the same principle that drives the maglev train. Conceived in the 1960s by Stewart Way, an American engineer, it had a test run as a 10-foot-long, submarine-like model that ran at 2 knots for about 12 minutes. Years later, Japanese physicist Yoshiro Saji constructed a 12-foot, wooden-hulled model that ran at 1.5 knots.

A magship is propelled by the same force that spins rotors in electric motors: electromagnetism. Under the dictates of a principle known as Fleming's Left-Hand Rule (a magnetic field plus electric current produces a linear force), the ship is an ingenious version of a DC motor, except that it has no moving parts. Instead of wire, the conductor of current is the ocean, a medium full of dissolved salts that are excellent carriers of electricity. An onboard generator sends current into the water between electrodes attached to the hull's underside. At the same time, superconducting magnets on the hull beam a powerful magnetic field into the water. Because the current flows at right angles to the field, force is exerted against the seawater, driving it backward and thrusting the ship forward. Like all visionary technology, the magship concept has its critics.

SHIP OF THE FUTURE?
(Right, top) A maglev ship substitutes superconducting magnets and a generator for conventional engine and propeller. The generator sends electric current between electrodes attached to the underside of the hull, while the magnets beam a field into the highly conductive surrounding seawater. The current is at right angles to the magnets' field, a phenomenon that exerts force against the water, driving it back and shoving the ship ahead. The magship, though, has a drawback: Its magnetic field might attract metallic debris—or even another ship.

SKIMMING THE WAVES
(Left) A hydrofoil carries passengers through the Wu Gorge in China. Because much of the hull does not touch the river surface and the foils have minimal contact with the waves, the hydrofoil's water resistance drops, and the boat achieves high speeds.

SEE ALSO
High-Speed Trains · 90
How Boats Stay Afloat · 96

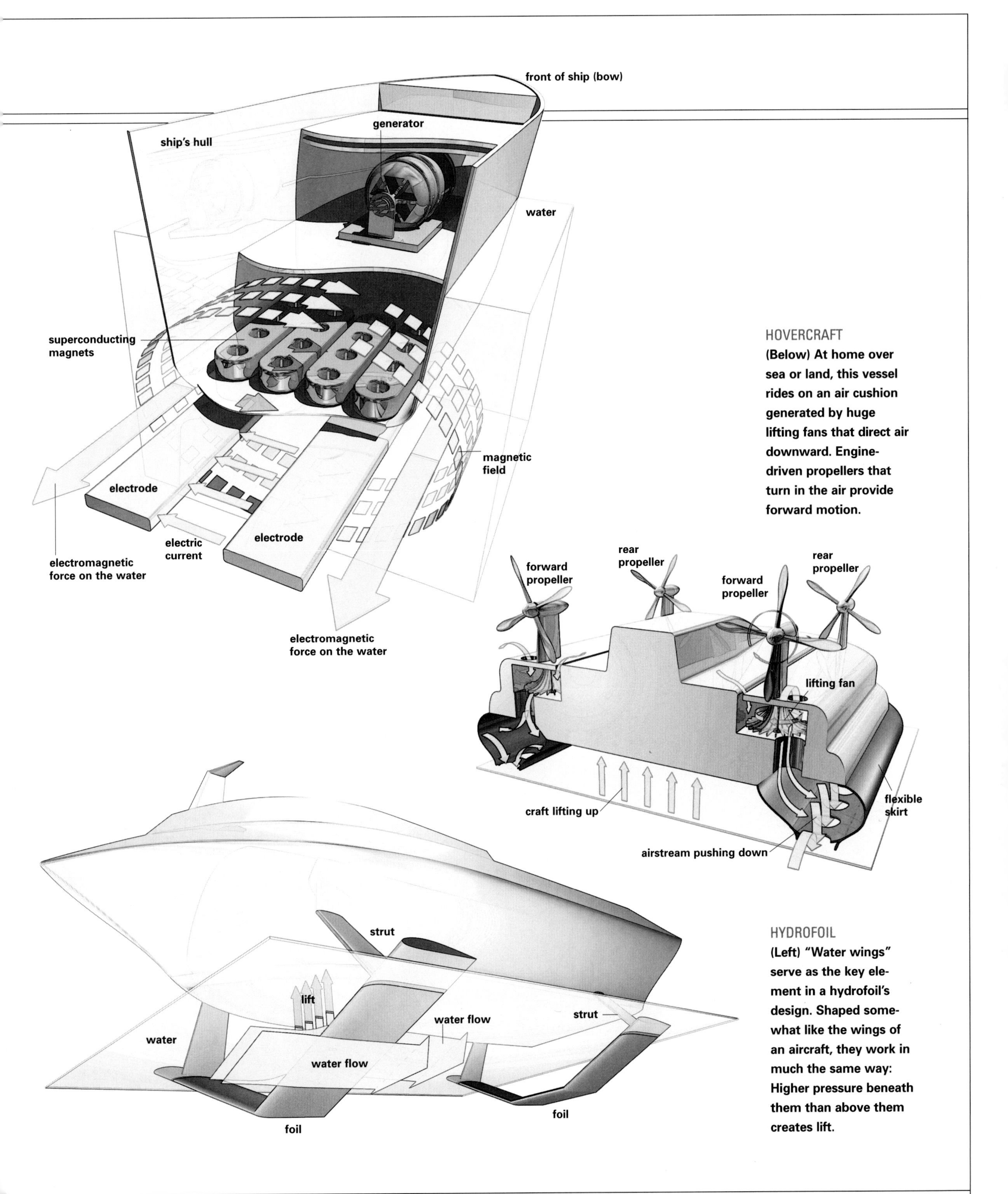

HOVERCRAFT

(Below) At home over sea or land, this vessel rides on an air cushion generated by huge lifting fans that direct air downward. Engine-driven propellers that turn in the air provide forward motion.

HYDROFOIL

(Left) "Water wings" serve as the key element in a hydrofoil's design. Shaped somewhat like the wings of an aircraft, they work in much the same way: Higher pressure beneath them than above them creates lift.

SUBMERSIBLES

NAVY SUBMARINES DIVE, RISE, AND FLOAT UNDER the water by adjusting the amount of water and air in their ballast tanks. On the surface, the tanks are full of air, making a ship weigh less than the volume of water it displaces. Flooding the tanks causes a submarine to sink because it then weighs more than the water it displaces. Rising requires a vessel to reduce its weight by forcing compressed air into its tanks to expel seawater. To float just beneath the surface, the amount of water in the tanks must equal the weight of the water displaced.

While the detailed technology behind nuclear propulsion is highly classified information, one can safely say that a submarine travels essentially on steam. Intense heat is produced from the fission of nuclear fuel in the ship's heavily shielded reactor. This heat generates steam, which drives not only the turbine generators that supply electricity to the ship but also the main propulsion turbines that turn the propeller. The electricity supplied by the generators also makes fresh water and oxygen from the ocean, letting crews live submerged for months.

Research submersibles that do bathymetry (bottom topography) and other ocean chores may be manned, carrying their own fuel and air supplies, or they can be remotely operated vessels (ROVs) tethered to a research vessel (which may also tow underwater platforms packed with instruments) and controlled by the mother ship's computer guidance system. Built to withstand the crushing pressures in the deep sea, they are very maneuverable and carry a wide array of tools for observation, retrieval, collection, and repair.

One unique device is the drifter, a waterproof tube of data-collection and transmission equipment. Extending from the tube are arms, which have vinyl or cloth "sails" stretched between them; floats suspend the sails beneath the surface where they catch a current. When the drifter is freed, its transmitter sends a signal to a polar satellite that calculates its position and relays the data to a receiving station. Sensors aboard the drifter also measure surface temperatures, ocean color, pressure, and salinity.

SUBMARINE DESIGN

(Below) A rounded configuration and double-walled hull resist crushing deepwater ocean pressures. A submarine can dive, travel underwater, rise, and run on the surface—all by adjusting the volume of water and air in ballast tanks. A nuclear reactor generates steam that drives turbines turning the propeller.

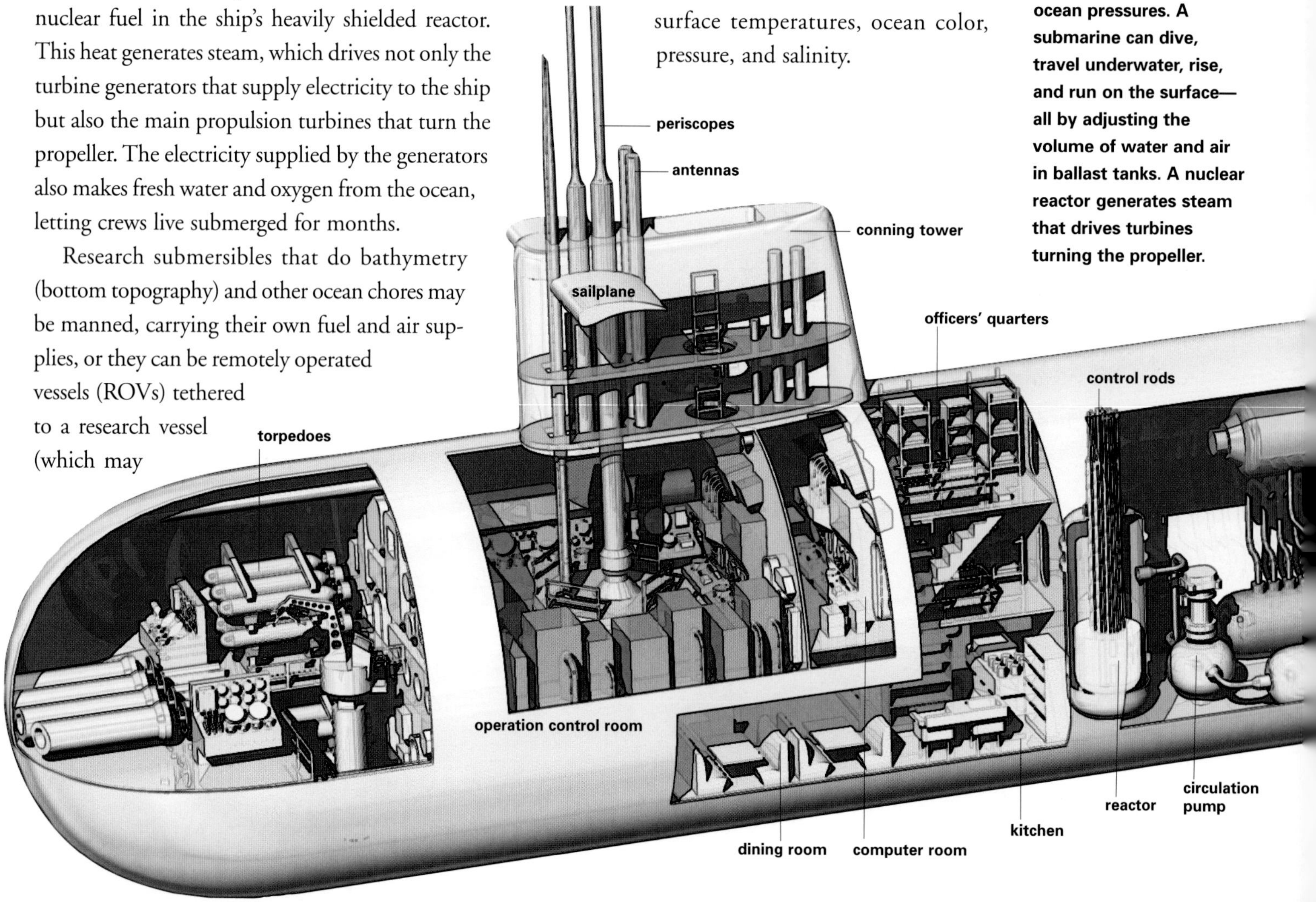

periscopes
antennas
conning tower
sailplane
officers' quarters
control rods
torpedoes
operation control room
reactor
circulation pump
kitchen
dining room
computer room

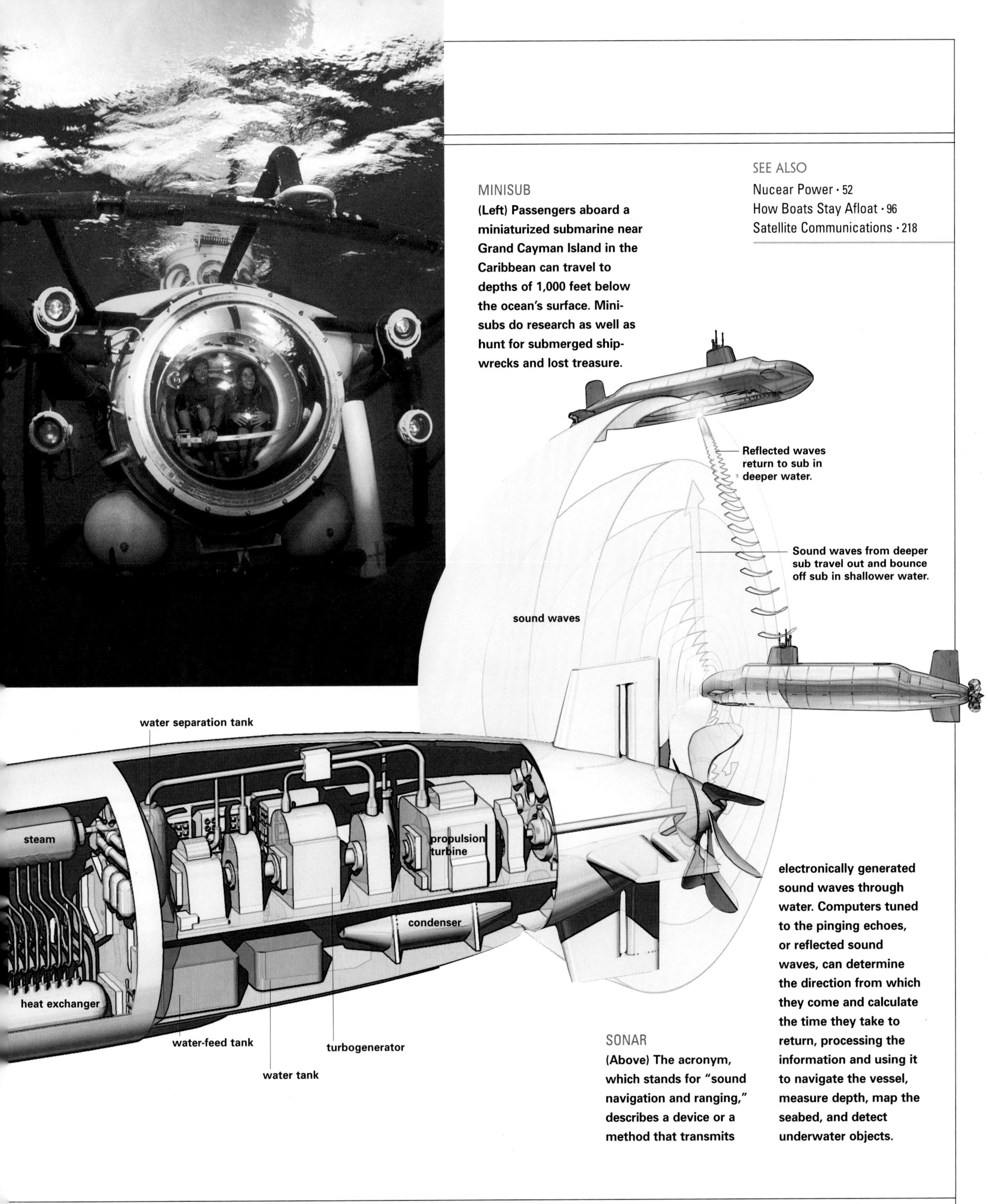

SEE ALSO

Nucear Power · 52
How Boats Stay Afloat · 96
Satellite Communications · 218

MINISUB

(Left) Passengers aboard a miniaturized submarine near Grand Cayman Island in the Caribbean can travel to depths of 1,000 feet below the ocean's surface. Mini-subs do research as well as hunt for submerged ship-wrecks and lost treasure.

SONAR

(Above) The acronym, which stands for "sound navigation and ranging," describes a device or a method that transmits electronically generated sound waves through water. Computers tuned to the pinging echoes, or reflected sound waves, can determine the direction from which they come and calculate the time they take to return, processing the information and using it to navigate the vessel, measure depth, map the seabed, and detect underwater objects.

THE PRINCIPLES OF FLIGHT

Ever since the ancient Chinese began to experiment with man-flying kites and parachutes, and perhaps even earlier, people have tried to emulate birds. Indeed, the aerodynamic principles and forces that govern aircraft are aptly demonstrated in the flight of birds. Simply put, they involve lift, drag, thrust, and weight.

Air hitting a wing's leading edge during flight and streaming over the top and bottom surfaces of the airfoils, or wings, provides the lift that raises a craft and keeps it "afloat," no easy task, considering that lift must equal the craft's weight. Enter Bernoulli's principle: Pressure is inversely related to velocity. Put another way, fast-moving air exerts less pressure than slow-moving air. During flight, the airflow over the longer upper surface of a wing travels faster than air on the underside, producing less pressure. The net force on the wing is an upward force exerted by the slower-moving stream of air beneath it. With wing design, the lifting force may be magnified beyond that caused by the impact of air.

Lift, however, must fight against drag, a force caused by friction as the plane moves through the air and by changes in airflow. Drag slows the plane, requiring the thrust of an engine to compensate and keep the plane moving. Providing thrust is expensive. According to NASA, the U.S. commercial airliner fleet consumes some 10 billion gallons of aviation fuel a year. Fuel and drag reduction are, thus, imperative, and to study ways to achieve this engineers rely on wind tunnels, which are used to study lift, drag, control, and stability. Modern wind tunnels are equipped with all the modern tools of technology: computerization, self-streamlining walls, laser and cryogenic instruments, and magnetic model suspension.

An engineless glider is a somewhat different breed of aircraft. Generally towed aloft by a plane and cut loose to glide, it requires only drag, lift, and weight. Designed to descend slowly from a higher altitude to a lower one, it searches for pockets of air (updrafts) that rise faster than the glider is descending, thus increasing its potential energy. Ground heat creates one kind of updrafts: thermals, in which a hawk can circle and gain altitude without moving its wings.

AIRLIFT

(Right) Air that flows over the top of a wing moves faster and exerts less pressure than the stream beneath; the greater pressure of the slower-moving air below the wing lifts the airplane.

LIFT

(Left) A jetliner heads for the sky using forces different from those that keep ships afloat. Planes must generate lift greater than their own weight during takeoff—and equal to it to stay aloft. Speed and wing surfaces provide the great lifting force needed to take off.

SEE ALSO
How Boats Stay Afloat · 96
Jet Engines & Propellers · 104
Helicopters · 106

LIGHTER THAN AIR

Whether they're called blimps, dirigibles, nonrigid airships, or just balloons, they all stay aloft because they're filled with something lighter than the air. Kept up by nonflammable helium or by air that expands when heated by butane or propane burners, a balloon uses the same forces that keep a ship afloat. There are differences between a balloonist's balloon and a dirigible, though. Balloonists are free-floaters, moving horizontally at the whim of the wind; their altitude can, however, be controlled by dropping ballast or venting gas. Dirigibles use engine-driven propellers and steering systems to regulate horizontal flight. None of today's airships should be confused with the now defunct zeppelin, a powered craft with a fabric-covered solid skeleton; inside the framework hung gas-filled bags called balloonets. The zeppelin was known as a rigid airship (as opposed to today's nonrigid versions), and its demise was hastened when the German-built *Hindenburg* exploded while tying up at Lakehurst, New Jersey, in 1937, killing 36 people. A combination of flammable hydrogen, doping solutions used on the hull, and static electricity apparently caused the disaster.

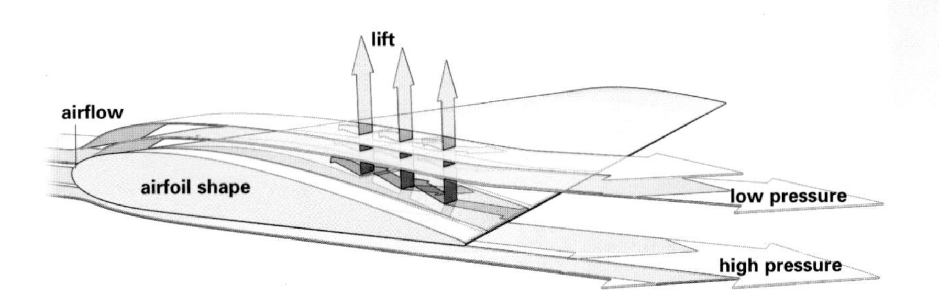

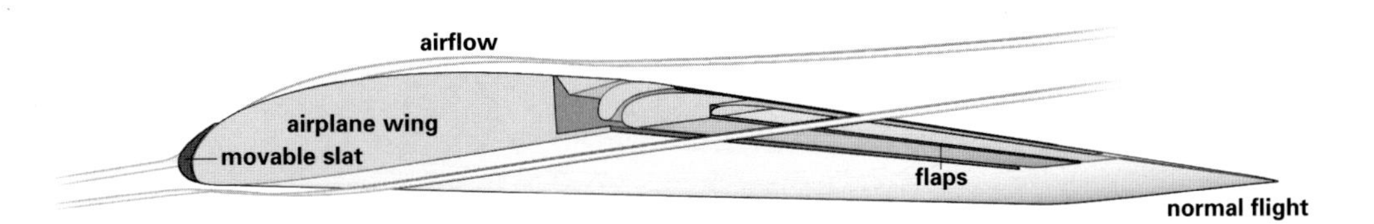

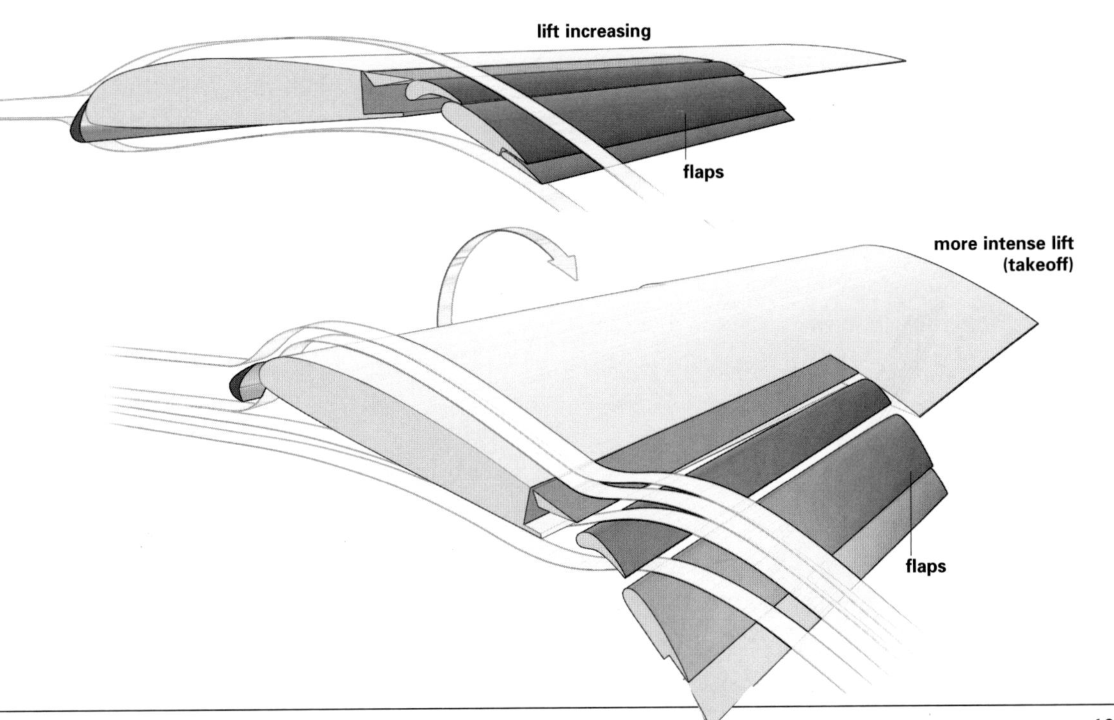

AIRFLOW

(Lower right) Movable slats at the front of a wing improve airflow, and flaps at the rear enhance wing curvature and area. Closed in normal flight, they open at takeoff and landing, increasing lift. A plane stays down after landing when other hinged surfaces rise at right angles to the wings, breaking airflow.

JET ENGINES & PROPELLERS

AERODYNAMIC RELIC

A single-engine biplane, double-winged for extra support, skims over a cloud bank. Used during World War I as fighters, biplanes with thin wings were effective, but they were soon replaced by monoplanes with thicker wings.

THE WRIGHT BROTHERS' PIONEERING FLYING machine was powered by a 4-cylinder, 12-horsepower gasoline engine; its carburetor was a tomato can. Airplanes that came along later were able to get by without tomato cans, but they also relied on simple internal combustion engines to drive propellers that converted engine shaft torque, or turning force, into thrust. Propellers have blades that are shaped like wings: The front surface of each blade is more curved than the back, so a forward aerodynamic force is produced.

A jet engine works in much the same way, although it doesn't look like a piston engine. A commonly used example explaining some of the principles behind its operation is adequate only insofar as it goes: If you inflate a balloon and release it untied, it will fizz about a room until the escaping air is depleted. A jet engine is basically an internal combustion engine; however, it uses the energy produced by combustion directly, and it does not need pistons to transmit driving power. Large quantities of air are drawn into the engine, compressed by a bladed turbine, and sprayed with kerosene fuel. When the mixture is ignited and the temperature in the combustion chamber exceeds 2500°F, the heated, expanding gases rush through an exhaust nozzle, providing the tremendous thrust that drives the plane forward.

SEE ALSO
Automobiles • 76
The Principles of Flight • 102
Helicopters • 106

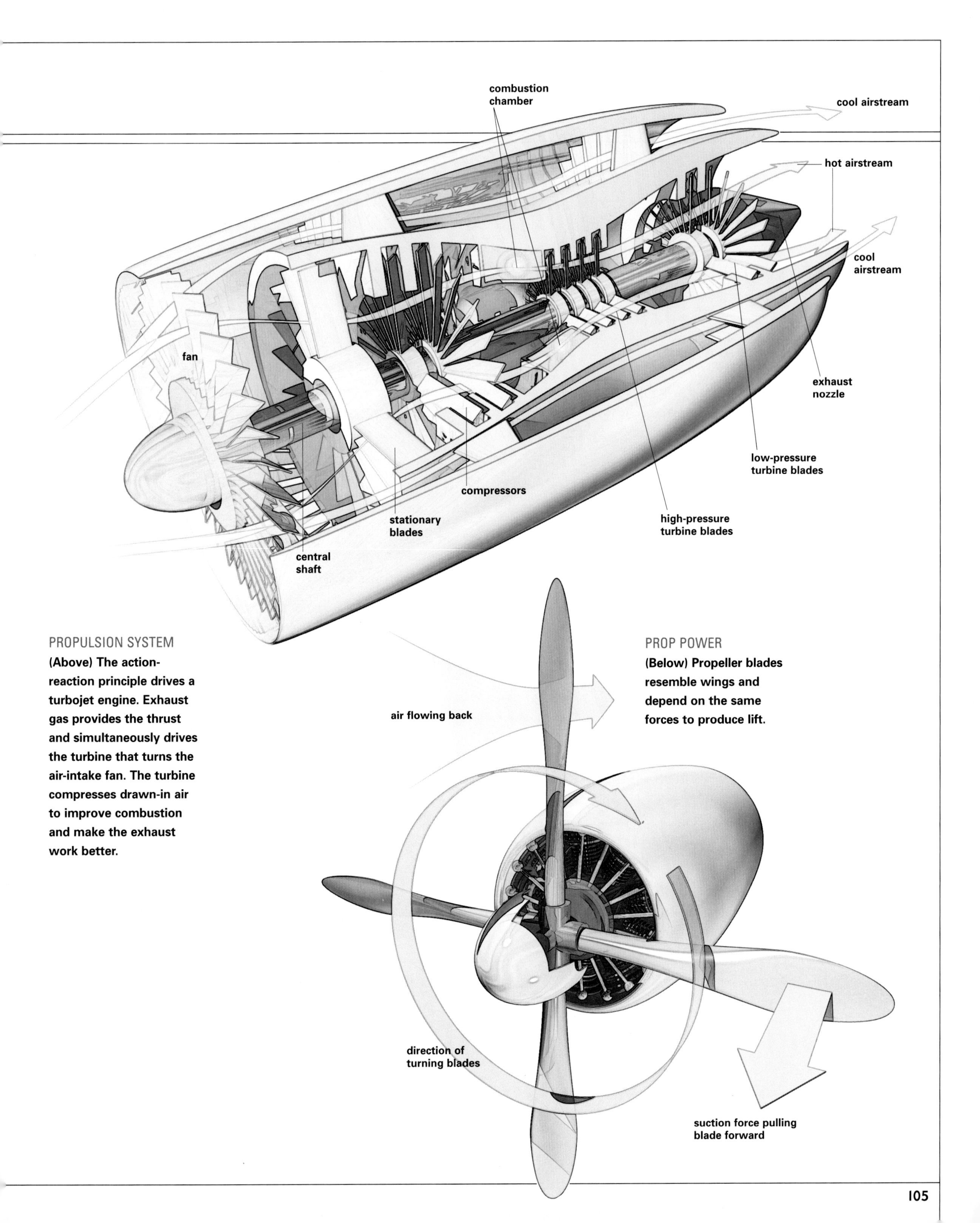

PROPULSION SYSTEM

(Above) The action-reaction principle drives a turbojet engine. Exhaust gas provides the thrust and simultaneously drives the turbine that turns the air-intake fan. The turbine compresses drawn-in air to improve combustion and make the exhaust work better.

PROP POWER

(Below) Propeller blades resemble wings and depend on the same forces to produce lift.

HELICOPTERS

LEONARDO DA VINCI IS CREDITED WITH designing a flying machine that could twirl itself skyward with the aid of a helix, or screw. He was inspired, perhaps, by "bamboo dragonflies"—the name given by fourth-century Chinese to toplike toys that could climb into the air with the pull of a string. The inventor was unable to build a working model, however, because the engine to power it had not been invented.

The first successful helicopter was invented by Igor Sikorsky, a Russian-born U.S. aircraft designer, and first flown in 1939. But the basic design was pure Leonardo, and the principles behind it were the same that govern all aircrafts. The difference, of course, is in the function of a helicopter's "wings"—actually the rotor blades that enable it to ascend and descend vertically, to hover, and to fly in any direction.

Because a helicopter's moving wings provide both lift and thrust, the angle at which each blade enters the air, called the pitch, is essential for control. As the blades turn, the pilot uses the collective pitch stick to increase their pitch equally and thus lift the helicopter vertically; when the pilot decreases the pitch, the helicopter descends. To hover, the blades are angled just enough to produce lift equal to the craft's weight.

Propelling a craft forward, backward, or sideways requires a pilot to use a cyclic control stick that tilts each blade at precise moments as it revolves. Such relative ease of operation, however, has a downside: the action-reaction law. A helicopter having only an overhead rotor would be forced in the direction opposite to the rotating blades and would spin out of control. To compensate, a tail rotor shoves air to one side and keeps the craft on the right path.

Helicopters are workhorses for search and rescue operations, battlefield engagements, load hoisting, traffic reporting, firefighting, and police work. Robot versions get guidance from global positioning satellites, or run with their own onboard computerized sensing and vision systems.

ROTARY MOTION
(Right) A tail-mounted vertical rotor produces a sideways thrust that prevents a helicopter from spinning out of control. It also helps the pilot steer: Decreasing the thrust turns the craft one way; increasing thrust turns it in the opposite direction. Tilting the main rotor blades in unison keeps a craft on a straight course, and by tilting them in sequence, the pilot increases lift on one side to move left or right.

WHIRLYBIRD WORKHORSE
(Left) A Chinook helicopter lifts a load of ammunition for the Army's First Cavalry Division. Versatile military and civilian craft, helicopters can spot traffic, serve as weapon platforms, rescue survivors, evacuate ill patients, and hoist enormous loads.

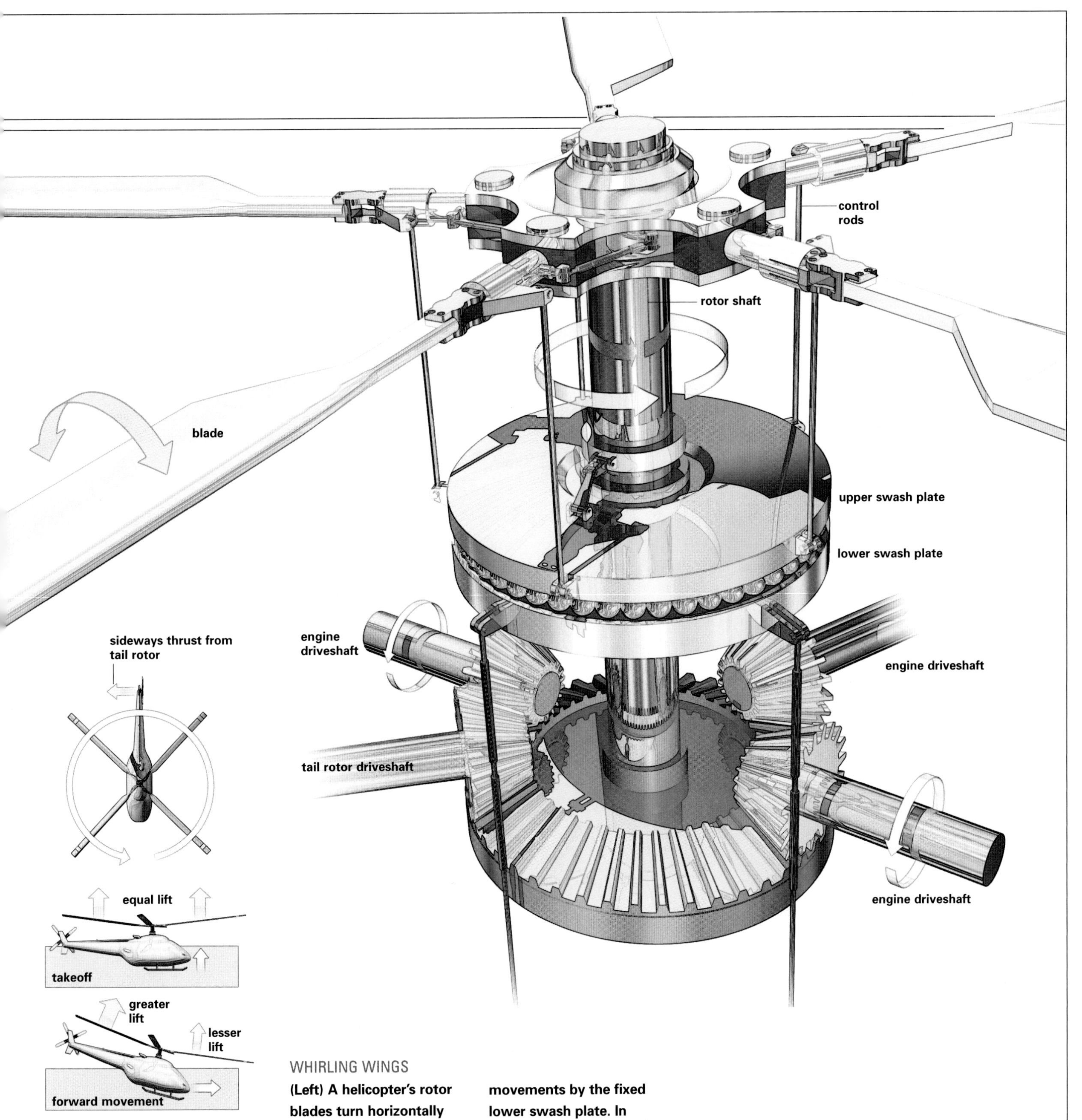

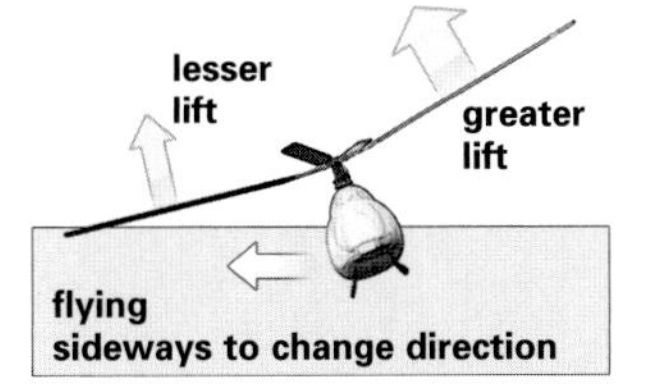

WHIRLING WINGS

(Left) A helicopter's rotor blades turn horizontally about a vertical axis, or shaft. Control rods connect the blades to a rotating swash plate that tilts, lifts, and lowers in response to similar movements by the fixed lower swash plate. In this illustration, three engines drive the main shaft via bevel gears.

SEE ALSO
The Principles of Flight · 102
Jet Engines & Propellers · 104

AIR TRAFFIC CONTROL

TODAY, FROM TAKEOFF TO LANDING, commercial pilots fly by electronics, radio beacons, and radar. From the ground, airplanes are tracked and directed by watchful air traffic controllers. Radar is still a useful tool for controllers, but it has many assists including satellite data and various alert devices that allow controllers to warn pilots of altitude problems, sequence takeoffs and landings safely, and warn of flights into restricted areas. In the air, pilots are guided and monitored by onboard computers and sensors that control aircraft movements via fiber-optic cables. Pilots use the cables to transmit coded digital signals directly to the motors moving the control surfaces.

Computers and software also establish safety parameters, check speed and direction, and correct for weather vagaries and possible pilot miscues. Weather radar in the nose sends information to the plane's data system; sensors detect wind shear and monitor fuel consumption. The cockpit gauges, which were once a bewildering array of more than a hundred dials, now consist of a relatively few glowing monitors. Under the direction of the avionics system—a blend of navigation, flight management, and digital automatic flight-control systems—only two people are generally required in the cockpit. Indeed, modern aircraft are capable of flying by themselves during and immediately after takeoff.

FLIGHT SENTINELS
(Right) Air traffic controllers at Denver's Stapleton International Airport. Although there is strong faith in the radar and computerized monitoring equipment that oversees takeoffs, in-flight travel, and landings, a controller's job comes with enormous responsibility, and is perhaps one of the most stressful of occupations.

ROTATING GROUND RADAR ANTENNAS
(Left) This array beams radar signals at an airliner and receives reflected pulses augemented with signals from the aircraft's radar beacons. Controllers use these signals to determine the approach, position, altitude, and identity of the plane.

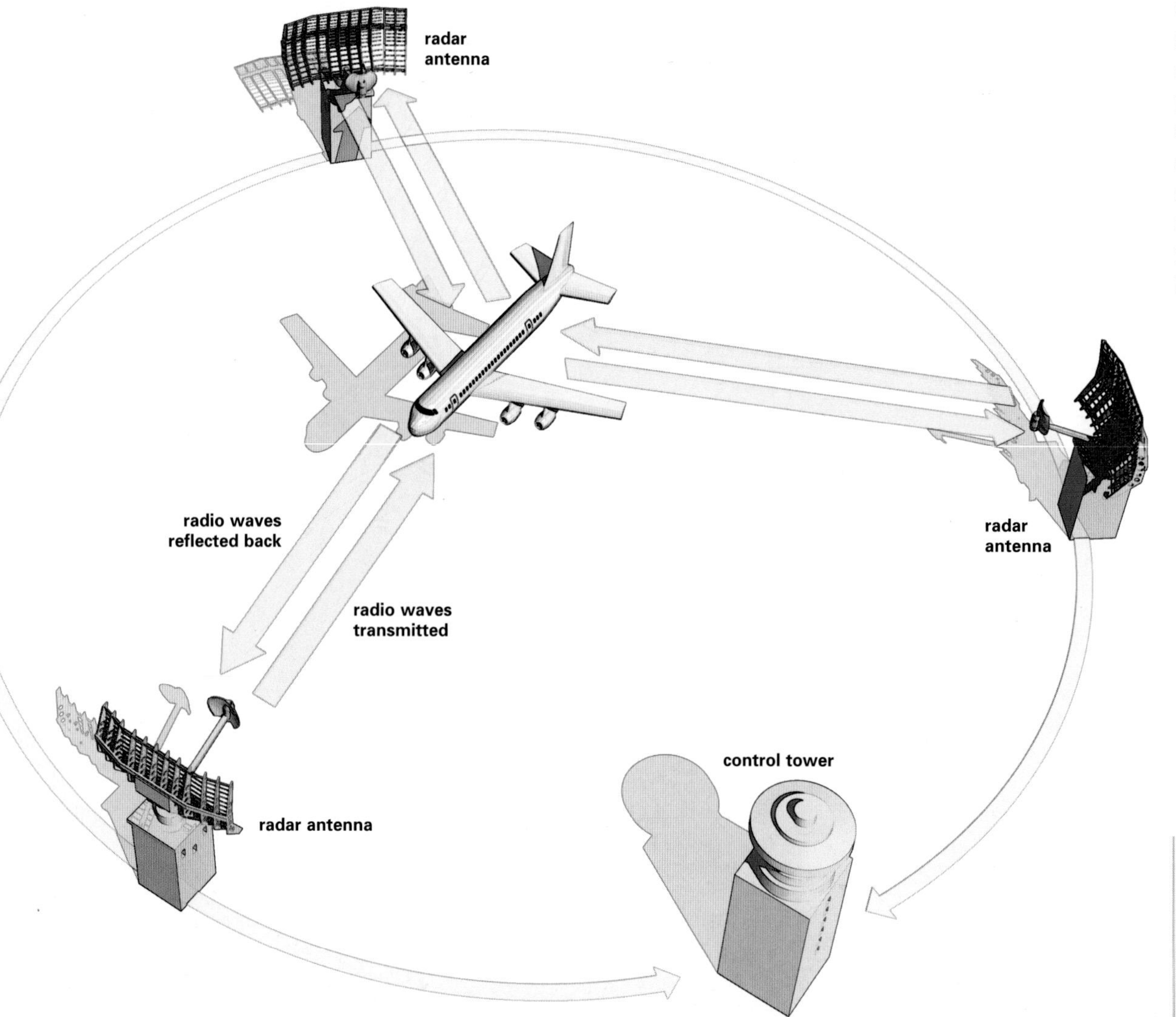

SEE ALSO
Submersibles · 100
Fiber Optic Cables & DSL · 210
Computers · 230

SUPER- & HYPERSONIC PLANES

SOUND TRAVELS AT 1,090 FEET A SECOND IN air, a figure that translates into Mach 1 in an aeronautical engineer's lexicon. The first time an aircraft flew faster than that was in 1947, when Capt. Chuck Yeager, USAF, took the controls of an experimental, rocket-powered Bell X-1. Today, while military fighters routinely break through the sound barrier, only one commercial airliner has flown at supersonic speed: the 100-seat, fuel-guzzling, ear-splitting Concorde, which streaked through the skies at Mach 2. The Concorde ended service in 2003, the victim of rising maintenance and fare costs, and is now a museum fixture.

A new generation of experimental hypersonic planes are now capable of flying at Mach 7 to Mach 10. One notable model that demonstrates hypersonic technology is the X-43, a futuristic-looking craft fueled by hydrogen and propelled by an "air-breathing" engine known as a scramjet. In a conventional rocket engine, hydrogen is burned with oxygen to supply the blast of energy needed for propulsion. This is, however somewhat disadvantageous in that heavy quantities of oxygen must be carried aboard. In the scramjet, no oxygen is carried, nor are there any fan blades like those that compress air in a normal jet engine to create forward movement. Instead, the scramjet sucks in atmospheric air as it streaks along, using it for combustion.

Another high- and fast-flying hydrogen-oxygen burner is the space shuttle that transports crew and cargo from Earth to orbit and back again. First flown in 1981, the winged shuttle is a multiengine, rocket-boosted, and reusable craft that relies on aerodynamics and the laws of physics—and a speed of 17,500 miles an hour to remain in orbit around Earth.

The shuttle is equipped with two solid rocket boosters (SRBs) that each carry a million pounds of solid propellant materials, and at launch is attached to a huge external tank loaded with more than a half-million gallons of supercold liquid oxygen and liquid hydrogen. These are drawn into the shuttle's three main engines at the aft end of the spacecraft, where they are mixed and burned in a combustion chamber; the combination of the thrust from the SRBs (which provide most of the force) and that provided by the engines lifts the entire enormous bundle of fuel, guidance and life-support systems, computers, and crew to an altitude of between 190 and 330 miles above sea level. A few minutes after launch, the SRBs detach from the orbiter and parachute into the ocean while the main engines continue to fire; minutes later, the engines shut down, the external fuel tank separates and burns up on reentry, and a pair of orbital maneuvering system engines (OMS) fire to send the shuttle into orbit.

When the shuttle returns to Earth after a week or more in space, its OMS engines shift its position, and slow it down during its high-speed reentry, a return marked by 3000°F heat produced by friction; ceramic tiles insulate the craft, protecting it and its crew. Toward the end of the voyage home, the commander takes over from the computer controls, drops the landing gear, and makes an unpowered descent and landing, like a glider would.

LIFTOFF

(Right) The space shuttle *Columbia* launches from the Kennedy Space Center, Florida. When it returns, the shuttle will make an unpowered landing, like a glider.

FUTURISTIC FLIGHT

(Above) A Horus, a German-designed spacecraft, leaves its carrier after a conventional runway takeoff and streaks for orbit on its own rockets. No longer science fiction, such rocket ships will ply space early in the new millennium.

SEE ALSO

Alternative Fuels for Cars · 80
The Principles of Flight · 102
Jet Engines & Propellers · 104

CHAPTER 5

AGRICULTURE

FORMAL PLANT CULTIVATION AND ANIMAL domestication began around 8000 to 6000 B.C., dramatically changing what humans ate, wore, and used for shelter. Once people discovered that some plants carried the germs of food in their seeds, planting may have been as simple as dropping a few pods into the ground. Or the planting may already have been done by chance, perhaps with seeds fallen from the clothes and wares of nomads, or shaken from the hair of their animals. Today, agriculture is more scientific. It is a world that relies on "pre-emerge" herbicides, "post-emerge" chemical treatments, crop rotation, and contour plowing. It produces hybrids and protein-enhanced soybeans. It uses sensors to monitor soil properties and satellites to map planting sites. It opens and cleans cultured shellfish with specialized machinery and sends fleets of computer-guided, automated farm vehicles across fields of grain.

A field-mowing combine harvester spews out the cereal grain for a nation's breakfast table.

PRECISION AGRICULTURE

FRUIT DIP
(Left) Nectarines on a conveyor belt get a thorough washing. Other modern machinery sorts fruit by color, extracts juice, and skins and divides segments for canning.

FROM SPACE TO SOIL
(Right) Mapped by a global positioning system satellite and converted to a grid, a planting field shows up on a computer screen. Such imagery helps today's farmers monitor soil types, watch for disease and insect problems, and control the application of fertilizer to exact field locations, thereby improving crop yields.

NO LONGER CAN IT BE SAID THAT THE cellular and environmental events that make for inferior crops and low yields are natural, irreversible acts. Science and technology have descended on farms.

The phrase currently in vogue is "precision agriculture." In it, satellite imaging of field conditions helps to improve crop yields. Weeds can be "mapped" with a touch of a keypad linked to a GPS receiver. The light beams of sensors analyze the ability of seed furrows to accommodate the flow of air and water. Tractor-mounted computers and satellite connections identify variations in nitrogen and pH levels, anticipating fertilizer and pesticide needs within feet of a tractor's position. "Precision agriculture helps producers apply the exact amount of chemicals they need, where it's needed, when it's needed," Deputy Secretary Richard Rominger of the U.S. Department of Agriculture has observed. "By the condition, not the calendar. By the foot, not the field. With precision agriculture, every farm, every field, every spot in the field becomes an experimental site." At this writing, dozens of satellites are orbiting the Earth in six paths, mapping out fields in a mosaic pattern.

Agriculture as a whole has taken a cue from such precision. Farmers these days are just as likely to rely on weather balloons carrying instruments that measure temperature and humidity as they are to gaze at thermometers or barometers. Other farmers stroll their fields, monitoring vegetation and soil properties through sensors that are attached to yokes strapped to their backs; microwave radiometers mounted on trucks quickly measure soil moisture; topographic maps of a field can be "drawn" and soil salinity measured with the help of satellites.

Farming methods also are more conservation oriented today. In no-till farming, for example, all or part of the current crop residue is left lying on the soil surface after harvesting has been done. This residue provides a protective blanket and a bed of organic material that conserves moisture. Later, farmers make small grooves in it and drop in seeds. To control the growth of weeds, they rely on herbicide applications rather than tillage.

SEE ALSO
Biotechnology · 118
Satellite Communications · 218
Computers · 230

MODE: STANDBY
HOLD 30.0 MPH
JOBFILE: 452340.JOB
VEHFILE: 1903.VEH
FAN RPM 0
COMPUTER
BOARD 1
BOARD 2
BOARD 3
BOARD 4
GPS OFF
EMPT
POT POTASH
BIN 1
EMPT
TSP (S PHO)
BIN 2
EMPT
POT POTASH
BIN 3
EMPT
TSP (S PHO)
BIN 4
EMPT
DESCRIPTION
BIN 5
EMPT
DESCRIPTION
BIN 6
DESCRIPTION
BIN 7
EMPT
DESCRIPTION
BIN 8
15 194 ACRES: 13.2
X-AXIS Y-AXIS
NORTH
SOILTYPE: B
ACRES DISPLAYED: 87
ZOOM: 1:1 SCALE: 1948
FEET PER PIXEL: 7.8
NAVIGATE MODE: MANUAL
90
HEADING
04/28 16:50
MAIN MENU
1 - VEHICLE FILE SETUP SCREEN
2 - PRODUCT ASSIGNMENT SCREEN
3 - DISPENSING RATES SCREEN
4 - PRIMARY SETTINGS SCREEN
5 - SECONDARY SETTINGS SCREEN
6 - SUSPEND THIS JOB
7 - CURRENT JOB SUMMARY
8 - JOB IS COMPLETE
9 - RESTART
PRESSURE
1000
RPM
FUEL

AGRICULTURE EQUIPMENT

Without doubt, Cyrus McCormick's mechanical reaper of 1831—a horse-drawn, wheeled rig that efficiently gathered and cut bunches of grain stalks—was the most important advance in the mechanization of farming at that time. What has followed is equally impressive, a technology-rich revamping of the way agriculture is done. What was once general repair work on a farm is now highly specialized maintenance since modern equipment requires more electronics and hydraulics. With the touch of a button, seed-handling machines transfer seed from hoppers to drill boxes and planters on streams of air. Combine harvesters that cut fruiting heads, thresh, and clean grain now come equipped with electronic fuel systems, ergonomic cab designs, and sensors that read changes in terrain and automatically level the machine. Corn pickers pick, husk, and send corn to shellers that remove the kernels; and cotton pickers twist cotton fiber right off the bolls. Milking a cow is robotic, as are drafting systems (drafting refers to selecting and moving animals, like cattle) that utilize a computer to select animals for a predetermined event. For example, "preprogrammed" cows may be segregated for a veterinarian's visit then sent to gated drafting units for examination.

Even the tried and true hay baler—which once meant hand-carrying the hay to the machine and then hand-tying the bales after processing—has changed. Tractor-pulled balers now have self-feeding pickups, automatic knotters that tie knots in twine or wire, and bale throwers that fling the bales into wagons. One such tosser, the Bale Flipper, works without a tractor to load bales of hay onto a cradle trailer. It automatically loads a bale into one cradle, then moves back to fill the next; when the last bale is loaded the apparatus rolls off the back of the trailer and an automatic switch turns off the motor.

There is a mechanical foliage shaker resembling a giant hairbrush, with 12-foot-long nylon "bristles" that rotate as well as shake. Pulled between rows of orange trees by a tractor, the spikes dig five feet into the tree's canopy and shake fruit onto a conveyor belt that transfers it to a self-propelled, bulk-transport vehicle. This harvester handles 300 to 400 field boxes of fruit from each acre 15 times faster than hand laborers, and it can harvest a 90-pound field box for 50 cents—a dollar or so less than the traditional cost.

HARVESTING TIME

(Right) An improvement over the sickles and scythes that reaped grain for centuries, a fleet of harvesters cuts through a Montana wheat field—an unfamiliar sight in 1910, when U.S. farms relied on 24.2 million horses and mules, and only about 1,000 tractors.

SEE ALSO
Saws & Lathes · 26
Garden Maintenance · 40

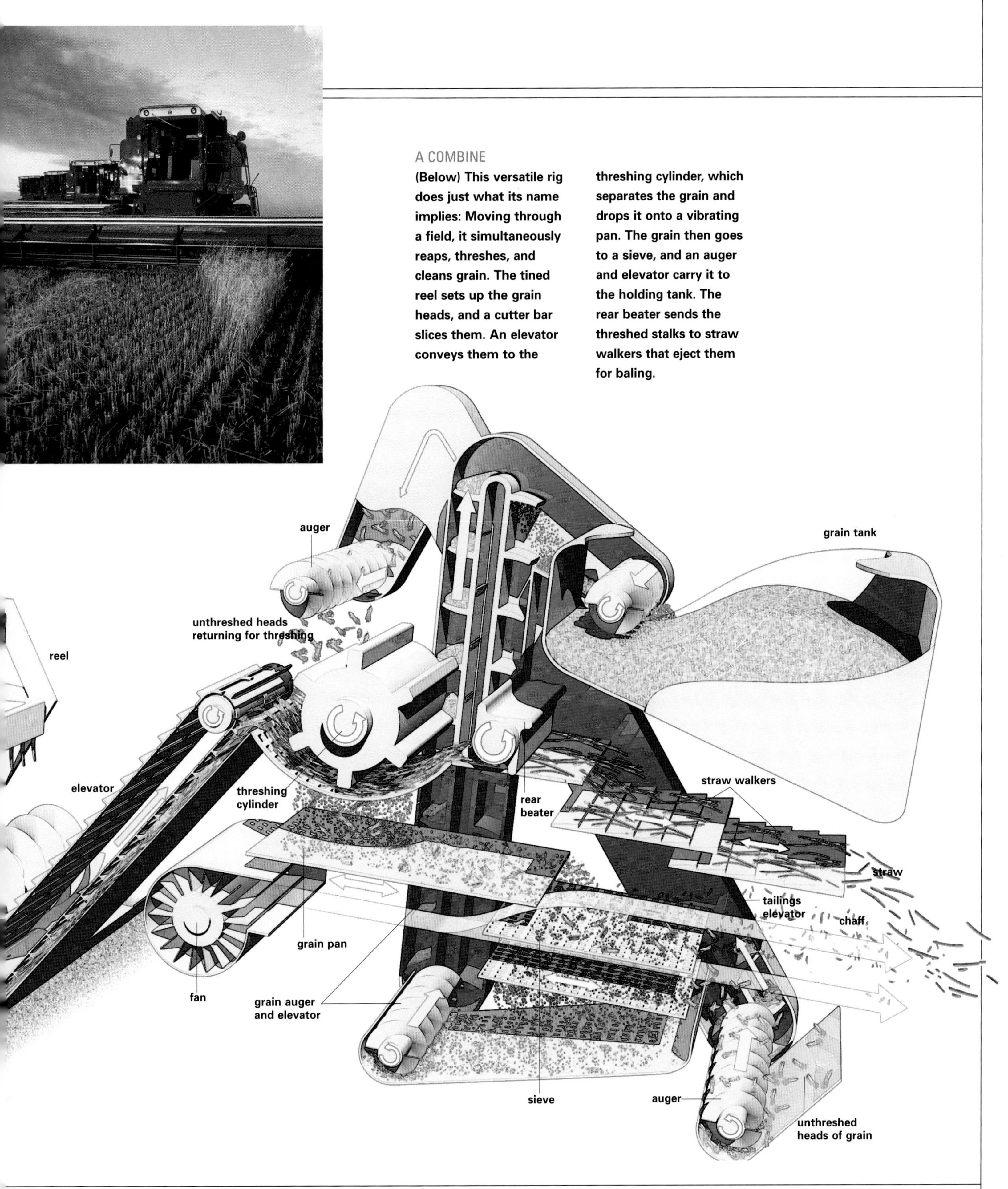

A COMBINE

(Below) This versatile rig does just what its name implies: Moving through a field, it simultaneously reaps, threshes, and cleans grain. The tined reel sets up the grain heads, and a cutter bar slices them. An elevator conveys them to the threshing cylinder, which separates the grain and drops it onto a vibrating pan. The grain then goes to a sieve, and an auger and elevator carry it to the holding tank. The rear beater sends the threshed stalks to straw walkers that eject them for baling.

BIOTECHNOLOGY

It tastes like cauliflower when eaten raw, but it is milder, sweeter, and more like broccoli when cooked. Say hello to the broccoflower, a chartreuse cross between the two members of the family Cruciferae. As every health-conscious American knows, vegetables are supposed to be good for us. First marketed ten years ago, the broccoflower—along with FLAVR SAVR tomatoes and SuperSweet onions—is a product of food biotechnology, a broad-ranging science employing a host of techniques to improve the things that we eat. Scientists can genetically alter crops by giving them new genes, thereby shielding them from disease, spoilage, and insects. The plants can then be made to grow with less dependence on chemicals. Crops also can be goaded into faster ripening and higher yields, or they can be endowed with better nutrients and more flavor.

CLONED BULL

(Above) Aptly named Second Chance, this Brahman bull was cloned from Chance, the late show bull that appeared in movies and on television. Chance's DNA got its repeat performance at Texas A&M University.

FLORAL DUPLICATE

(Right) A cloned flowering plant reposes in its test tube, the product of manipulated genes. Chinese scientists have cloned large quantities of the rare Venus flytrap, which eats insects. Others have been trying to put the scent back into flowers that have lost it due to the focus on bigger blooms.

One important example of biotechnology is selective breeding, an ages-old agricultural standby. Essentially, it involves selecting plants and animals with desirable traits and breeding them under controlled conditions, a process that has achieved much success and produced enormous yields. Selective breeding, which concentrates on entire organisms with complete sets of genes, and genetic engineering, which focuses on a few gene transfers, have given us seedless bananas, beardless mussels, leaner animals, and virus-resistant squash. Synthetic Bovine Growth Hormone (BGH), a genetically engineered drug given to cows, increases milk output by augmenting a cow's own natural production of BGH. More than half the cheese produced in the United States is made with chymosin, a biotech preparation that does away with the need for rennin extracted from calves' stomachs. Tomatoes that can be made to ripen on the vine and then can be shipped without spoiling are a definite plus when you consider that commercial tomatoes are generally shipped green and then ripened with ethylene gas.

But although biotechnology has enormous advantages, there are, as in all cases of genetic manipulation, many red flags. While proponents argue that the techniques are safe—indeed, the FDA recently concluded tentatively that milk and meat from cloned animals are safe to consume—opponents argue that gene splicing might instill lethal allergens in foods or deprive food of its nutritional value. As one analysis of the situation put it, "Whatever their interpretations of the costs and benefits of biotechnology, both proponents and opponents agree that biotechnology has the potential to fundamentally change how food is produced in the future."

SEE ALSO
Precision Agriculture • 114
Aquaculture • 120

AQUACULTURE

FISH FARM

An evening view of a fish farm's pens in Thailand. Often situated along coastlines or in estuaries, fish farms proliferate in the Gulf of Thailand, a semienclosed sea, as natural sources are overfished or destroyed by industrial development in coastal areas.

Aquaculture, the culturing of fish, shellfish, and plants in a controlled area of water, is a rapidly growing agribusiness, larger even than veal, lamb, and mutton combined. According to the U.S. Department of Agriculture, 20 percent of the fish consumed in the United States is now raised on farms, with catfish, tilapia, salmon, trout, crawfish, and shrimp among the leaders.

Fish farms may be earthen ponds, tanks, rafts, troughlike raceways, net pens, suspended cages, or bottom nets. The "seeds" are fingerlings, very young mollusks, or eggs, which are raised on commercial feed, or on natural organisms grown through water-fertilizing techniques. Because potentially polluting and disease-bearing waste accumulates in any fish farm, it must be disposed of through elaborate water-circulation systems. Some filtration systems harness bacteria that convert ammonia, which fish secrete through their gills, into nitrates that can be flushed from tanks. As an additional precaution, stocked fish may be treated with antibiotics and other chemicals, a practice that invariably raises concerns among environmentalists and consumers. One state-of-the-art technique, gene splicing, can produce catfish, salmon, and trout that not only have more resistance to disease, but also are faster growing and more immune to freezing in winter. Some feel that these fish pose a great risk to wild populations because inevitably escaped farm fish breed and produce modified offspring, which compromises the natural gene pool.

Some fish, such as salmon, which in the wild hatch in fresh water and then swim to the ocean—must be transferred from fresh to salt water. The trauma that farmed salmon suffer is called osmotic shock, and can be handled with special water and formula-feeding treatment.

Shellfish farming is also fraught with environmental hazards. Growing oysters, for example, involves fertilizing female eggs that are up to 90 microns in diameter with male sperm about 10 microns long. Large quantities of phytoplankton are fed to larvae and older oysters, which go through a process of transference from one bed to another to accommodate various growth stages.

SEE ALSO

Precision Agriculture · 114
Biotechnology · 118
Hydroponics · 122

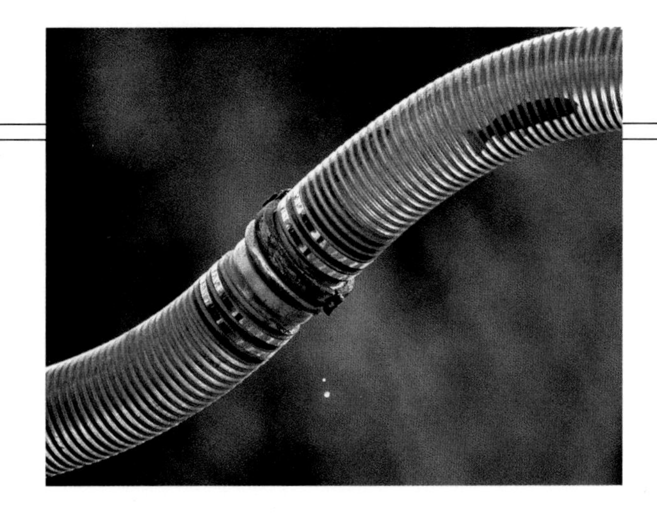

FISH HIGHWAY

A little fish flits through a hose (above) at an Idaho fish farm, while its relatives in the wild may ride a tide onto shore or travel upstream against a strong current. Unlike conventional farms, which generally consist of vast tilled fields arranged in familiar patchwork-quilt patterns, water-filled fish farms prove the ultimate in container-growing. In one model, the fish eggs fertilized in an incubator bath (right) eventually go through their growth periods in well-filtered holding tanks (below). A recirculating system cleans out fish waste, removes uneaten food, and controls the water's oxygen levels. Fish farms must also deal with the same problem that faces hog and poultry farms: How to dispose of polluting waste.

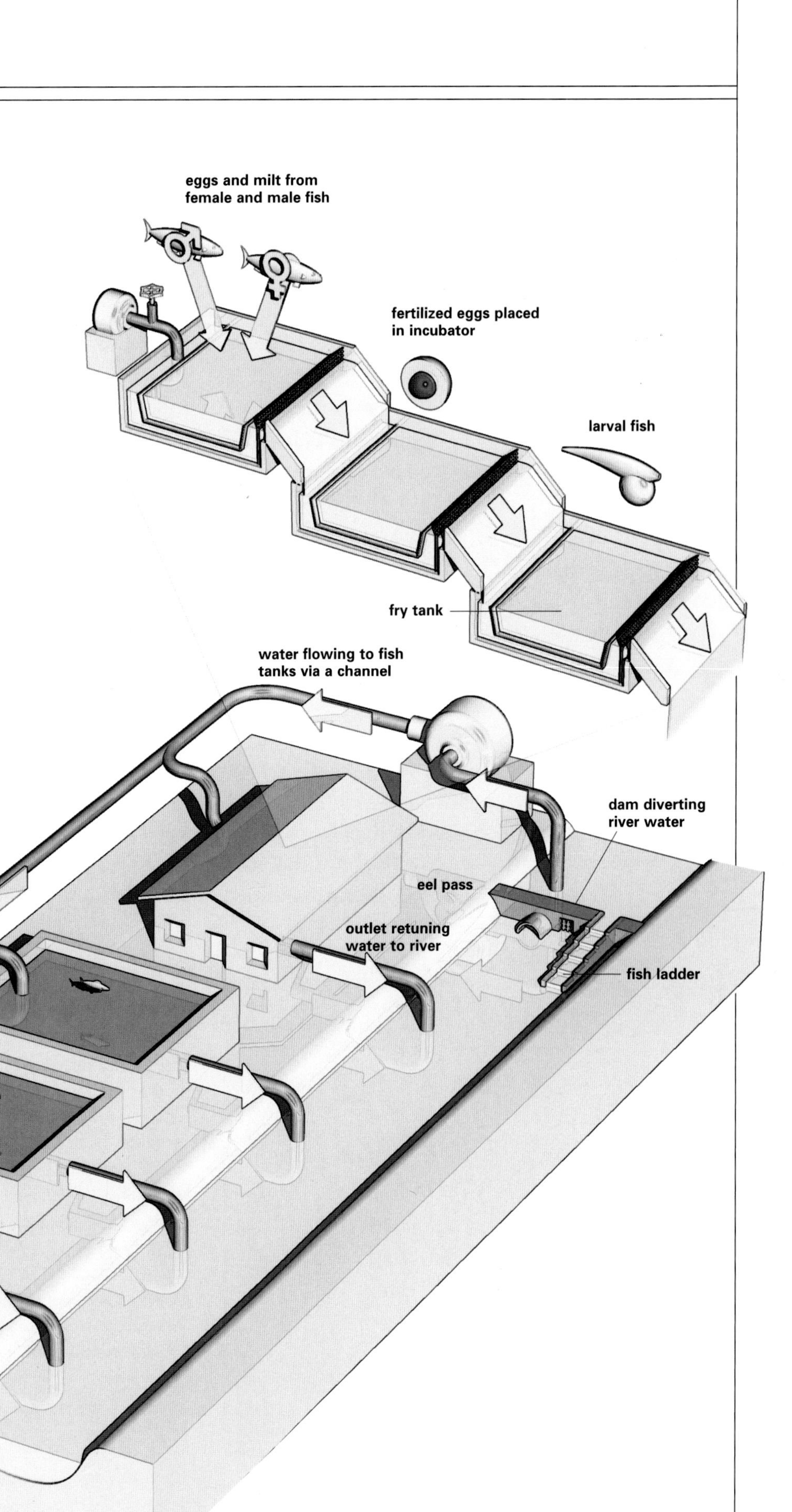

HYDROPONICS

CONSIDERING THE VULNERABILITY OF SOIL to diseases, pests, changing weather, and inadequate nutrient supply, it is no wonder that farmers long for better control of their plants and sometimes even wish they could control growing conditions. One remedy is the science of hydroponics: Plants are grown without soil in nutrient-enriched solutions, with their roots anchored in porous, nonsoil materials. Widely used in botanical research, hydroponics also grows vegetables, fruits, flowers, and herbs.

The concept is not new. Water gardens thrived along Africa's Nile River thousands of years ago. During the Second World War, the U.S. Army grew vegetables hydroponically on infertile Pacific islands. But even though crop yields can exceed the success rate of dirt farming, large-scale soilless farming remains confined to out-of-season greenhouse plants and to areas that have limited arable land.

Soilless culture begins with water enriched by the same balance of nutrient salts found in soil; when dissolved by watering, these nutrients are absorbed by plant roots. The roots themselves are supported by all manner of materials that retain air and water, including sand, gravel, glass wool, rock wool fiber, and stone. A variation of this process is aeroponics, by which plant roots are suspended in a chamber or a bag; humid air provides the proper environment while a spray mist of nutrient solution keeps the roots moist and nourished. With such midair feeding, almost no water is lost through evaporation, and roots absorb much oxygen, increasing metabolism and the rate of growth as much as 10 times over that in soil. Another kind of aeroponics, used for plants that require substantial feeding, is deep-water culture in which the plant roots are kept in water while an aerator provides the oxygenation.

NASA's interest in growing food in space has led to the development of a medium made from zeolite, a common mineral. Zeolite "soil" takes advantage of the mineral's natural properties as a "molecular sieve," allowing it to store and time-release nutrients. Mixing an additive with specially prepared zeolite creates a substance that, in laboratory tests, produces conditions almost comparable to those of the soilless "soil" in conventional hydroponics.

A HYDROPONIC GARDEN

(Right) An artificial environment provides all of the ingredients essential for plant growth: oxygen, light, heat, water, nutrients, and carbon dioxide. Inside the protective confines of a greenhouse, plant roots absorb nutrient salts from enriched water. An inert and soil-free porous medium and plastic mesh anchor the roots in the water. Sunlight or artificial "growth" lights ensure that the plants will grow rapidly and healthily. By delivering nutrients directly to the roots, a hydroponic system ensures that energy ordinarily used to produce long roots goes directly into growing a larger plant.

SLUDGE-BUSTERS

Because the nutrient solutions in the enriched, pure water that makes hydroponics possible are contained, they do not pose the same threat as runoff from fertilized soil. Tainted water pollutes aquifers and presents an immense cleanup challenge, but this is a problem that has a truly natural solution. At an experimental greenhouse in Providence, Rhode Island, sewage is piped into vats, along with a horde of plants, bacteria, snails, and fish that have an appetite for filth. In a few days, the cloudy sample of water filling the flask in the man's left hand will be as clear as the sample in the other flask. It will be nearly drinkable, virtually odorless, and ready for discharge. Bacteria have also been harnessed by scientists who used them to devour environmental pollutants, such as oil spills in the ocean. The bacteria break down the oil in much the same way that ordinary bacteria cause dead animals and plants to decompose and change into soil-enriching substances.

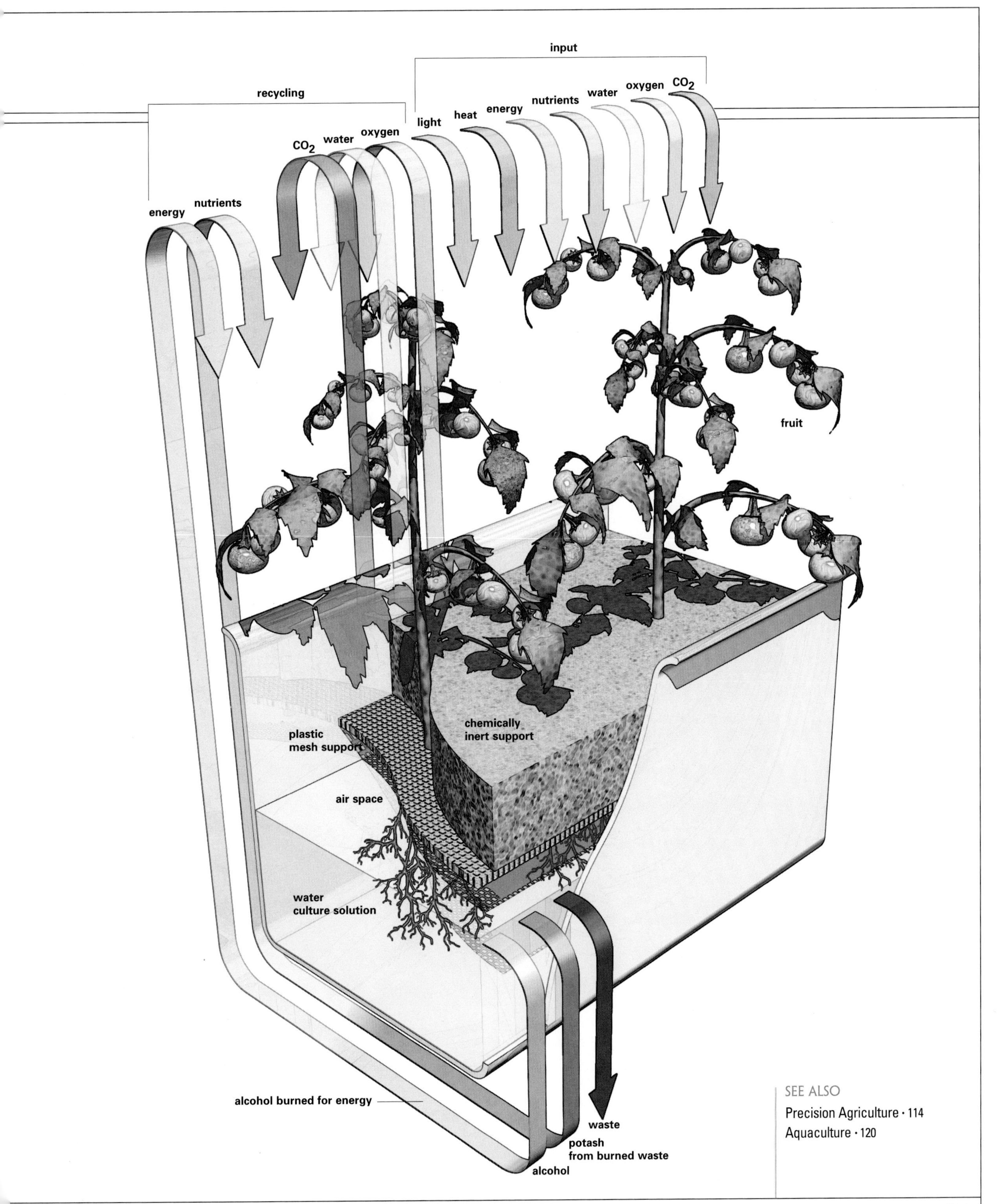

SEE ALSO
Precision Agriculture • 114
Aquaculture • 120

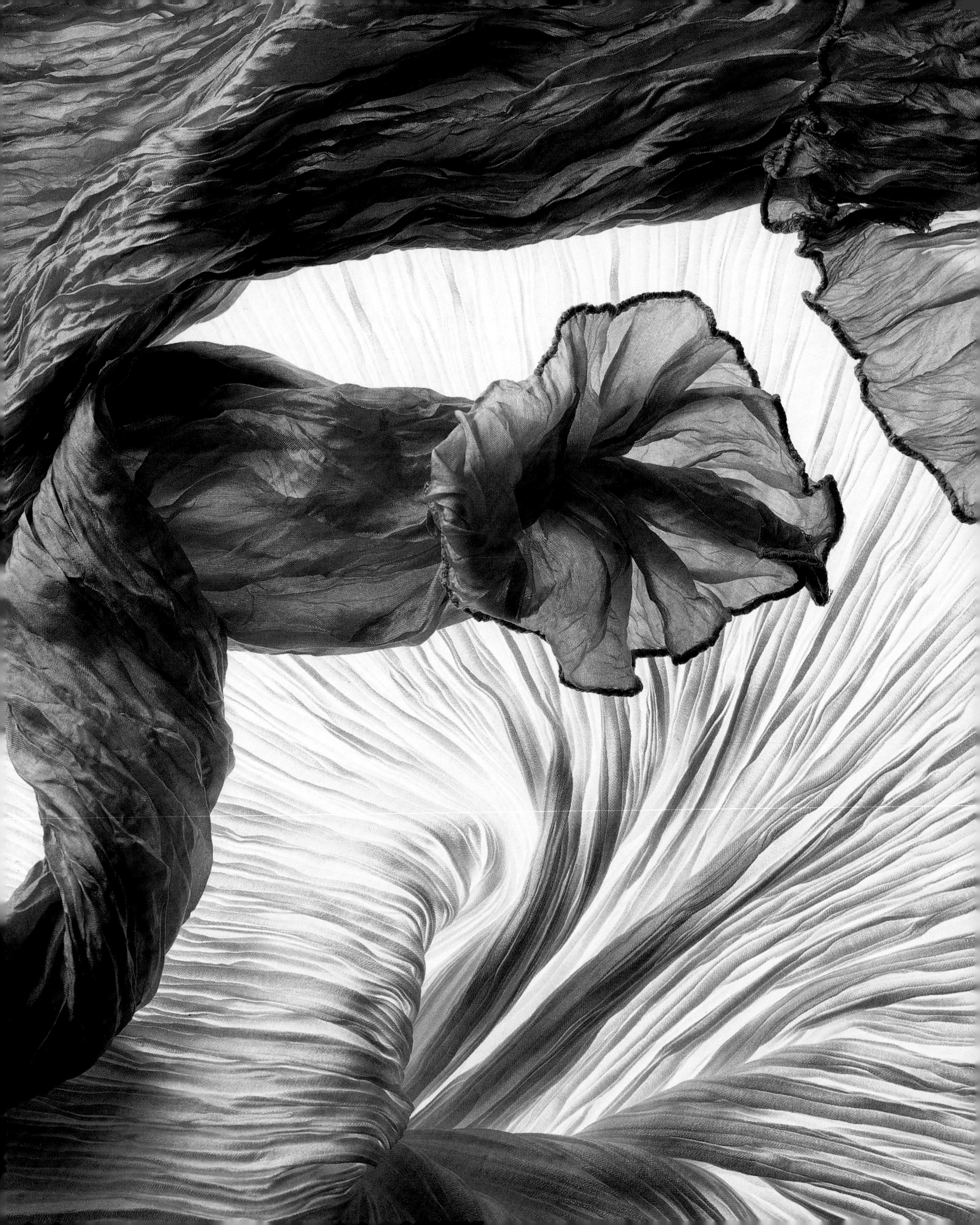

CHAPTER 6

FABRICS & FIBERS

SPINNING AND WEAVING HAVE BEEN AROUND since prehistoric times, when people began using flax, cotton, wool, hemp, and silk to make simple cloth. By the 14th century, Europeans were creating clothing and decorative fabrics from linen, silk, and wool, using virtually all of the basic weaves of today. By the end of the industrial revolution, manual processing had given way to the modernization of techniques and equipment. In today's world, the textile industry is a highly technical enterprise that integrates imaginative design with performance and function. While fibers continue to be made from plant and animal materials, they also are spun from glass and fabricated from organic polymers, which are derived from coal and petroleum. Textiles are among technology's most aesthetic and diverse achievements, and they are key elements in the economies of many nations.

Fine pleats in sheer silk characterized the early 20th-century creations of Mariano Fortuny.

THE LOOM

A HAND LOOM IS AN ANCIENT APPARATUS THAT interlaces, at right angles, two or more lengths of yarn or thread to form cloth. The longitudinal threads are the warp, and the crosswise threads are the filling weft (or woof). In its simplest form the loom is a single frame that is relatively easy to operate. Warp fiber, which is processed by spinning and then wrapped around a bobbin, is fixed to a cylinder at the rear of the frame and kept taut. A shuttle containing the filling yarn is passed alternately over and under the warp threads, and then, after each passage, a comblike device, called a reed, is used to push the filling thread against the previous line.

Larger hand looms are floor models having 4, 8, 12, and 16 harnesses that enable their users to weave a variety of patterns, colors, and fibers into the fabric. In the textile industry, weaving is still based on the principle of fiber crossing fiber, but the machinery that makes cloth is now completely automatic and often controlled by computers. Harnesses are now electromechanically activated, and controlled by computers that dictate patterns and alternate the yarn strands. Automated looms also weave noncloth materials, like fine wire mesh used to make screens; again, a computer program directs the process, which includes setting the screen's aperture sizes.

England's John Kay started the mechanization trend in 1733 by inventing a "flying shuttle" that was driven across the loom on a track by a lever blow, and then back. Massive power looms—first driven by waterwheels and steam—soon followed. Today, the shuttle is being replaced by innovations such as jet looms, which use pressurized air or water to propel filling thread through the weave.

SPOOLS OF YARN
Whether natural or synthetic, the colorful rolls underpin the textile industry. The fabrics produced from them keep the rain off our skin, the creases in our trousers, the stains out of our rugs, and winter's chill away from our bodies. "The web of our life is of a mingled yarn," Shakespeare wrote in *All's Well That Ends Well.* So, too, is the web of what we wear, sleep on, and drape over the windows of our homes.

SEE ALSO
Patterning Fabrics · 130
Synthetic Fibers · 132

A WEAVER'S TOUCH

(Left) A craftsman fashions a kimono by hand, using warp and weft in the time-honored way. Weaving involves more than just interlacing two sets of yarn: The threading of the harnesses, which determine pattern, requires experience and skill.

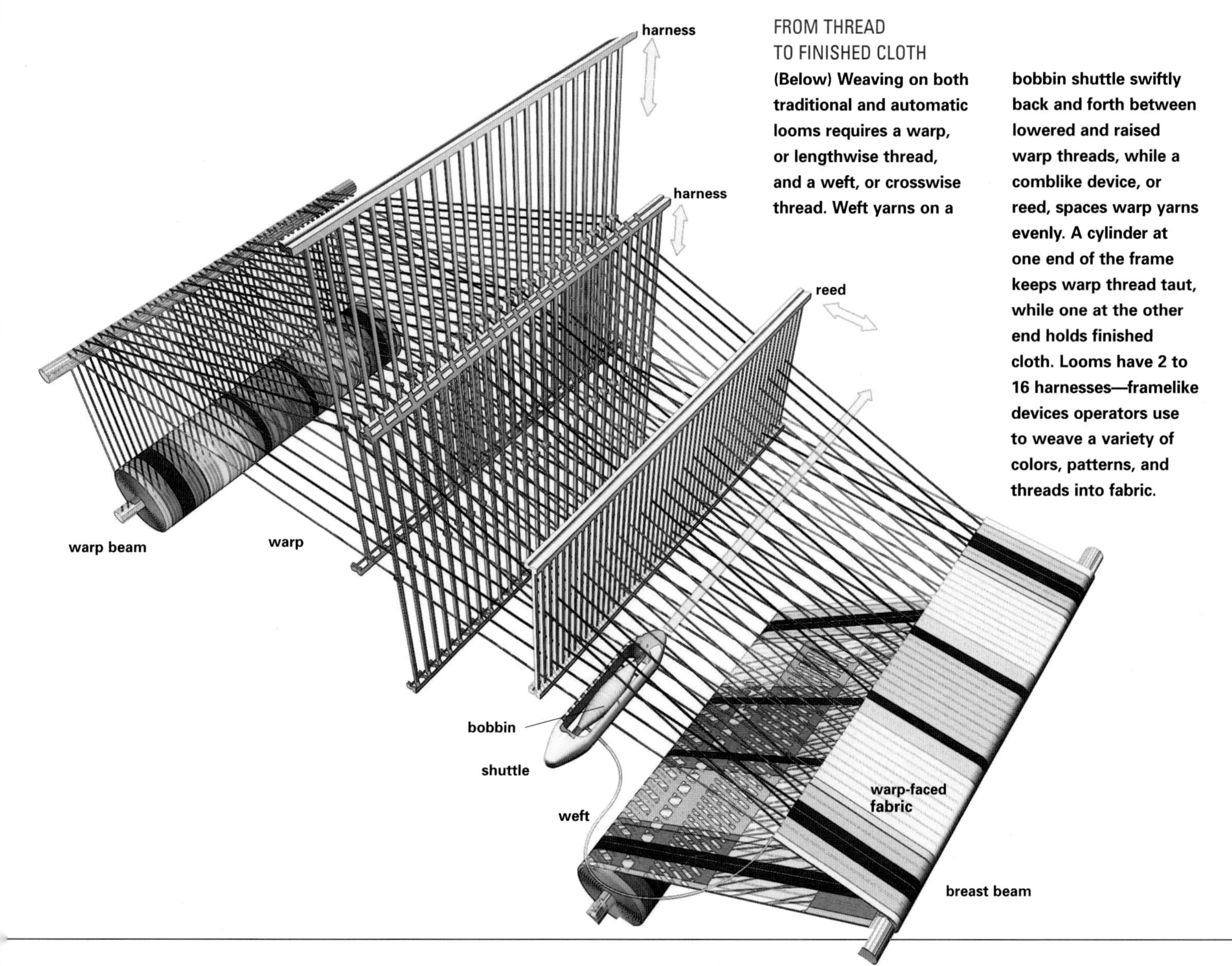

FROM THREAD TO FINISHED CLOTH

(Below) Weaving on both traditional and automatic looms requires a warp, or lengthwise thread, and a weft, or crosswise thread. Weft yarns on a bobbin shuttle swiftly back and forth between lowered and raised warp threads, while a comblike device, or reed, spaces warp yarns evenly. A cylinder at one end of the frame keeps warp thread taut, while one at the other end holds finished cloth. Looms have 2 to 16 harnesses—framelike devices operators use to weave a variety of colors, patterns, and threads into fabric.

SEWING & KNITTING

ALTHOUGH IT MAY NOT BE EVIDENT AT FIRST, sewing machines and knitting needles rely on the same simple principle: the loop. A machine-sewn seam holds fabric together by loops of thread, while knitting needles make rows of interconnected loops to create fabric. Sewing and knitting machines, both remarkable examples of inventiveness, do the job faster, but the idea is still the same.

The sewing machine dates from 1790, and in the mid-19th century Elias Howe, an American inventor, patented one that contained many of the features of the modern machine. Whether it is powered by an electric motor or an operator's foot on an antique treadle—the 1851 model named after inventor Isaac M. Singer had a treadle—a sewing machine needs a grooved needle threaded through an eye near the point. It also needs a thread-filled, spool-like bobbin that rotates in the machine beneath the fabric. As drive belts push and pull the needle through fabric, the thread in the needle forms a loop around the bobbin thread to make a tightened lockstitch. Another driveshaft operates a so-called feed dog that moves the fabric along. Computerized sewing and embroidery machines that can be linked to a home computer are also available; these devices automatically set stitch length and width, shift the needle into position, make buttonholes, and store stitching patterns and commands.

The knitting machine was invented in 1589 by an English cleric, William Lee, to whom Queen Elizabeth refused a patent because the device was a threat to hand knitters. The machine later progressed from a simple stocking knitter to huge warp knitters, which make fabric for undergarments and outerwear out of fibers threaded from spools through guides. To select needles, the hand-operated machines use punched cards that work like the key-activation mechanism in old-fashioned player pianos. High-speed electronic machines, driven by pattern software, display knit designs on an LCD and have a computerized system that directs needle patterns and machine speed and automatically shuts down the machine. Knitting machines now produce an incredible range of wearables, including artificial fur.

STITCHING TIME

(Right) The needle, thread, and bobbin work together to make simple stitches. From top to bottom, a threaded needle heads down through cloth and forms a loop of thread (red) caught by a rotating hook (blue). As the needle rises, the hook passes the loop over thread (yellow) from a bobbin (blue), and the threads join in a lockstitch. An arm at the top of the machine tightens the loop and draws the stitch up into the cloth.

ZIPPER

The word "zip" sounds like a fast-moving object and connotes speed. And indeed, zipping a zipper is much faster than fumbling with a row of buttons. A zipper contains two rows of interlocking teeth—metal or plastic—which are sewn on strips of tape attached to each side of an opening in a bag or an article of clothing. A zipper also has a sliding tab with wedges on each side. When a person pulls on the tab, the wedges force the rows of teeth together, closing the opening in the bag or garment. Pulling the zipper's tab in the opposite direction allows an upper wedge to force the interlocked teeth apart.

SEE ALSO
Patterning Fabrics · 130
Synthetic Fibers · 132

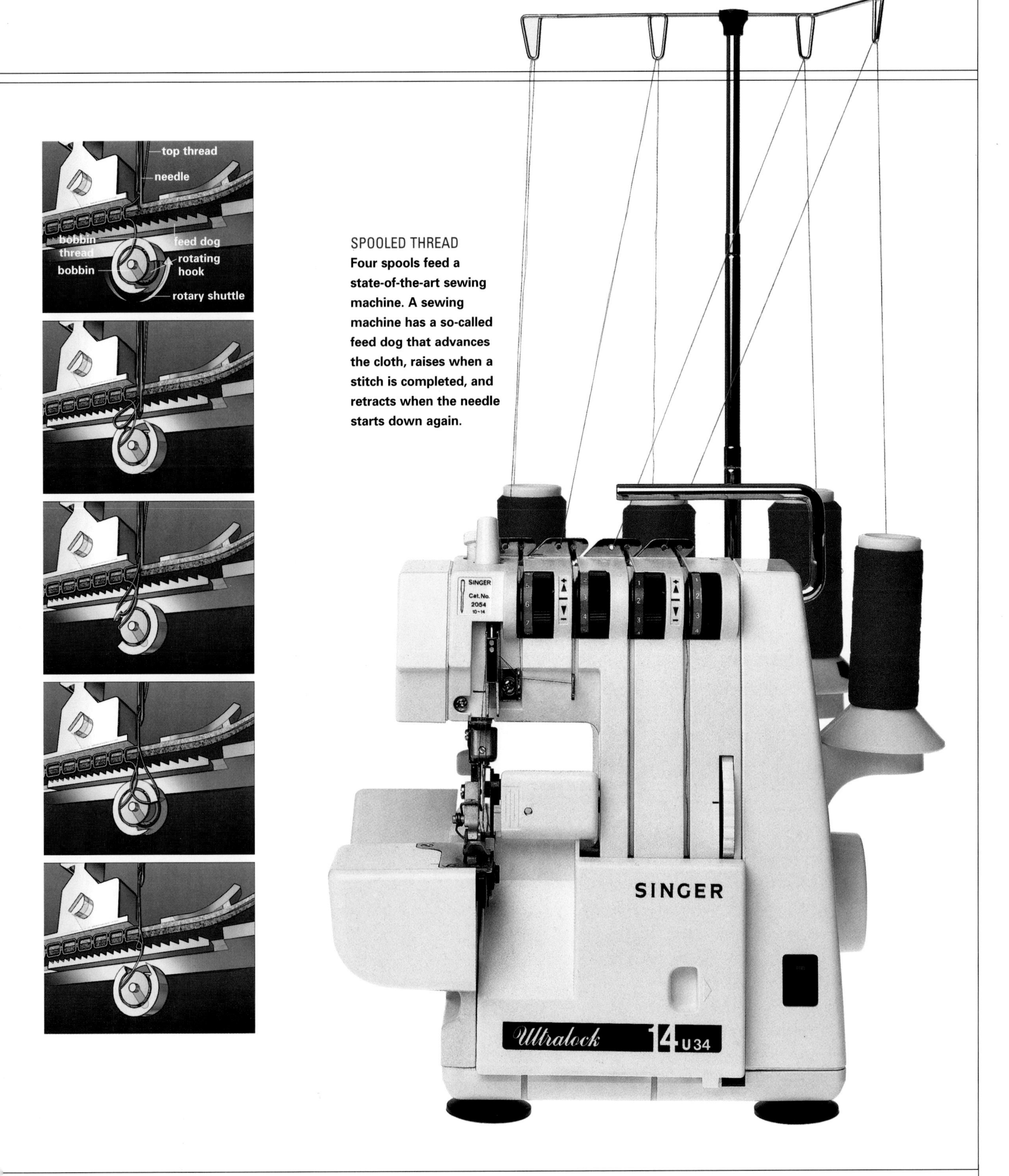

SPOOLED THREAD

Four spools feed a state-of-the-art sewing machine. A sewing machine has a so-called feed dog that advances the cloth, raises when a stitch is completed, and retracts when the needle starts down again.

PATTERNING FABRICS

Weaving a bolt of plain cloth has become a simple mechanical operation, but designing a fabric of distinction requires much creativity. More than 4,000 years ago, weavers in Peru knew how to manipulate patterns, textures, and colors to make extraordinary fabrics. Using plant and mineral dyes, they tinted yarns before weaving, alternating each strand on looms to create handsome designs, or they colored the whole fabric afterward. Both methods are in use today, although mass production is also employed.

In digital printing, a computer helps design and generate fabric prints; analog printing requires screens and printing plates. Electrostatic transfer technology may be used to create images with charged particles that attract toner—a process that is similar to one used in copy machines. Another method, called ink-jet transfer, sends streams of ink droplets to fabric mounted on a rotating drum. The thermal wax method, used for printing T-shirts, heat-fuses colored wax film onto fabric. In the method known as intaglio roller printing, designs are etched on separate rolls of copper—one for each color—and smeared with printing paste; images are transferred when fabric moves through the rolls. Designs can also be transferred to fabric by squeezing color through a stencil on a flat screen.

Distinctive patterns and effects may be woven into fabric. To make twill, filling threads pass over one and under two or more warp threads for a diagonal appearance. Satin's sheen comes from light reflected off warp yarns passed over exposed filling yarns. Pile comes from ordinary weave that has had its filling cloth pulled out into loops that can be cut, as in velvet, or left alone, as in a towel.

Technology has also enabled manufacturers to perform the difficult task of fabric inspection, a tiring process when done by a human inspector. Defects can now be discovered quickly with automated systems that scan fabric and catalog the defects.

A RAINBOW OF COLOR
(Above) Mainstays of the textile industry, aniline dyes come from coal tar, as do all aniline products, including sulfa drugs, explosives, and polyurethane.

SEE ALSO
The Loom • 126
Synthetic Fibers • 132
Copiers • 224
Computer Printers • 226

ROTARY COLOR
(Right) Technology and color combine to produce a textile pattern. Rotary printer rollers, each carrying a single color, leave their individual imprints on the fabric to produce a variegated design. More than ever, the designs of modern textiles require their creators to have a comprehensive understanding of the chemistry and technology of materials and dyes, as well as a keen appreciation of art.

SYNTHETIC FIBERS

Synthetic fibers were first made in the late 19th century, when chemists learned that cellulose could be extracted from wood pulp and formed into thread through nozzles. Originally called artificial silk, the synthetic eventually was produced commercially as rayon.

While rayon is actually not a synthetic fiber but a reconstituted one, it is one of a large family of nonwool, noncotton, and nonsilk substitutes that seems to be in everything we wear, from socks to watch straps to fake furs. True synthetics are made from organic polymers, many of which soften if they are heated. Nylon, the first commercially successful synthetic, is made into clothing, rope, and parachutes. It also is cast and molded to make zippers, gears, and bearings. The range of such fibers is enormous. Dacron, a polyester, has a nonstretch quality; polyurethane elastomers stretch; Quiana, a silklike nylon, holds a crease and remains wrinkle free; carbon fibers reinforce; and high-strength aramid and polyethylene can resist bullets.

Most synthetic fibers are made into thread by melting the polymer and forcing it through tiny holes in a spinneret. A core construction technique spins a wrap of cotton or polyester around a continuous filament of polyester fibers, and then two or more single yarns are twisted together to form thread; another method textures the filament and heat-sets it. Air-entangled construction, used in heavy denim jeans, entangles fibers by passing them through a high-pressure air jet before twisting, dyeing, and winding. Monocord construction, used in the threads for making shoes, bonds nylon filaments together.

Faux furs, favored by those who object to the animal fur industry, or who would rather not pay a higher price for the real thing, also rely on what is essentially plastic for their backing and "furry" pile. In a synthetic fur, the backing is generally woven from acrylic or polyester, and the pile from long fibers of acrylic, or Modacrylic, a related plastic polymer. Dyes are applied to the fibers during manufacture to produce fur look-alikes for seal, fox, leopard, and just about every member of the animal kingdom whose fur keeps them, and humans, warm.

SHEER BEAUTY

(Right) Exemplar of the many uses of nylon, women's hosiery is also a fashion statement and, encasing a pair of shapely legs, the frequent subject of photographers. Early stockings were seamed up the back by hand, but seamless hose became available in the 1940s when nylon replaced silk.

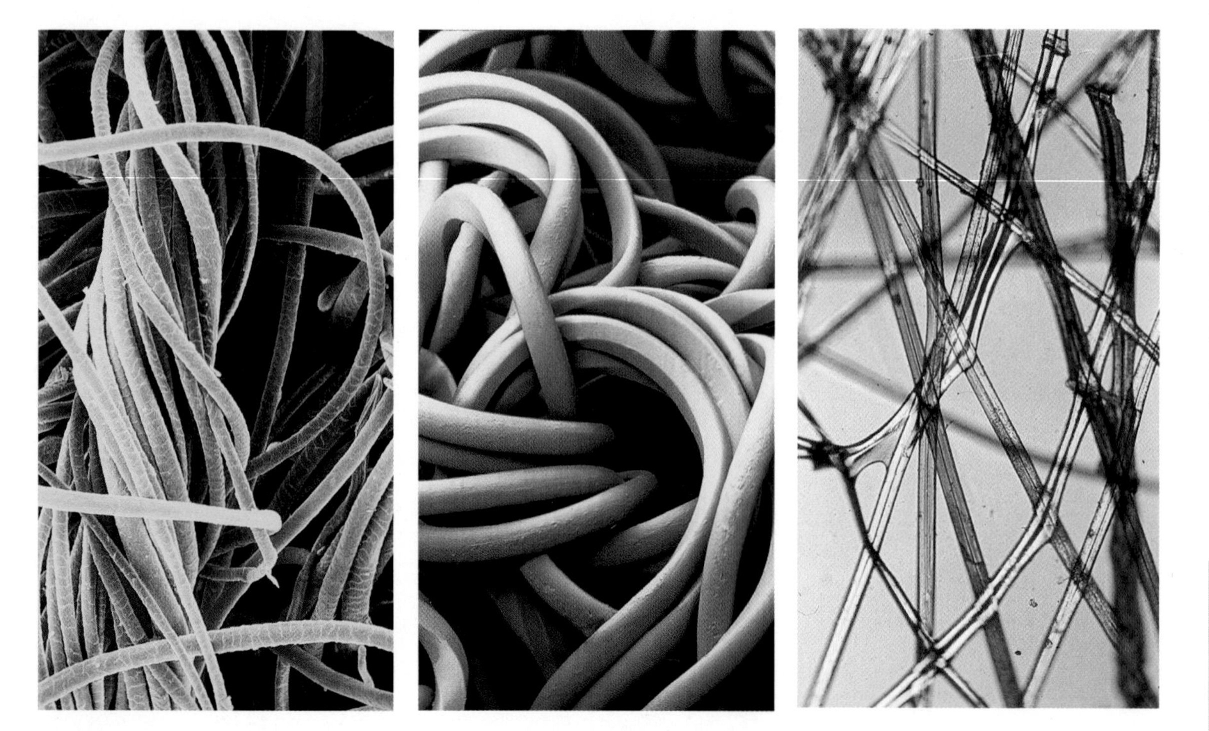

FIBERS

(Left) Shetland wool fibers (far left), taken from a knitted pullover, reveal scaly surfaces in an electron micrograph. A color-enhanced electron microscope image shows nylon fibers (center) in a pair of low-quality, unstretched stockings. Spun by polyphemus moth larvae, greatly magnified silk fibers (near left) look smooth and flat in polarized light.

SEE ALSO
The Loom · 126
Incredible Fabrics · 134
Lumber · 172
Plastics · 184

INCREDIBLE FABRICS

For protection in battle, Persian and Roman warriors clothed themselves in overlapping metal scales attached to linen or leather. In the Middle Ages, the Crusaders wore chain mail made up of interlaced metal rings over their clothing. This flexible armor was effective, but it was still unwieldy, heavy metal.

Today's flak jackets and other types of bullet-resistant vests worn by soldiers and policemen are not the product of a blacksmith's shop. They were created by a new "thread" science that fabricates tough, lightweight clothing able to protect wearers from fire, cold, and water. Fabrics can be reinforced and strengthened with carbon fibers produced from simple coal tar, or by using heat to chemically change rayon or acrylic fibers. Polyester or nylon can be treated with resins that form tough, smooth protective coatings on thread. Warm, fleecelike fabric can be made from flaked and melted plastic bottles. Cloth can also be woven with glass fiber or fine metal wire.

Among the noteworthy and widely used innovations is Kevlar, a DuPont trademark for a highly crystalline polymer that is dissolved in a solvent and ultimately drawn into incredibly strong fibers used in bullet-resistant vests, military helmets, ropes, and gloves. Another well-known, special fabriclike material is Gore-Tex, a waterproof, windproof, and "breathable" product that is a trademark of W. L. Gore and Associates, Inc. Gore-Tex fabric is actually layers of polymers fashioned into membranes bonded to a special fabric. The membranes are full of microscopic pores that are thousands of times smaller than a raindrop. They prevent water from entering but let sweat vapors leave, which allows the wearer to stay drier and cooler in warmer, wet conditions.

Even more forward-looking are the "smart" fabrics used in thermal clothing. In one design—a spin-off from materials used to keep astronauts' gloved hands warm during space walks—microencapsulated "phase-changing" materials, called microPCMs, do what traditional insulation methods that rely on trapping air within a fabric cannot do: They absorb and store heat, then release it in response to temperatures next to the skin. Another new fabric, called Holofiber and manufactured by Hologenix, LLC, is billed as a body-responsive textile. It is designed to increase oxygenated blood flow, which in turn can increase circulation and build strength. In diabetics, for instance, an improvement in skin oxygenation can accelerate wound healing. All this textile wizardry will eventually offer suits that change shape when mood or temperature changes; clothing may be wired with locating devices; battle-wear may warn of and deactivate toxic chemicals and germs; and camouflage fabrics may one day change colors and patterns to match the background.

TAKING THE HEAT
(Left) Enveloped in protective clothing, a steelworker pours a liquid fire of molten metal. His wardrobe might include flame-resistant underwear, aluminized outerwear, heat-shedding gloves, a face shield that reflects heat waves away from the face and eyes, and steel- and Kevlar-reinforced shoes.

VESTED INTEREST
(Right) Protected by hat and slicker, a cowboy in the rain manages chores. Ever since it became known that creatures like sea otters had waterproof fur, and that water rolls off some butterfly wings, science has searched for ways to waterproof what we wear. What have emerged are non-breathable materials (which unfortunately retain perspiration) made of vinyl plastic compounds or other materials coated with them; and more comfortable breathable products that transfer moisture vapor from inside a garment to the outside.

SEE ALSO
Synthetic Fibers • 132
Plastics • 184

CHAPTER 7

ENTERTAINMENT

HOWEVER WE DEFINE IT AND WHEREVER WE seek it out, entertainment serves as humankind's leavening agent and safety valve. Over the years, some of its tools have changed little in appearance and function—Chopin would have little difficulty at the keyboard of a modern piano—but the vast majority have been transformed and improved, and many more have been created. Although most of the principles have remained essentially the same, the apparatuses and the way they process information have changed a great deal. Cameras and records are now digitized. Radios come microsize. Music can be made electronically. Video games let us sink and raise the *Titanic*. From the new, metallurgically engineered golf clubs to digitized movies to the ever more exciting roller coaster, entertainment's accoutrements have come under the powerful influence of technology.

Pleasing to eyes as well as ears, the flashy compact disc, or CD, symbolizes entertainment.

MAKING MUSIC

MUSIC, AS THOMAS CARLYLE SAID, MAY WELL be the "speech of angels," but it is also the product of human hands, throat, mouth, and lips. It is the result of someone manipulating the vibrations and the increases and decreases in air pressure that are at the root of all sound. Everything that makes sound sets up vibrations that disturb the air and, in turn, change the air pressure against our eardrums. The eardrums vibrate, and a sound message is carried to the brain, which differentiates between cacophony and an angelic choir. Playing a musical instrument transmits the vibrations that move through the air as audible waves.

Sound's loudness or softness depends on the amplitude of sound waves, while its pitch (highs and lows) depends on their speed of vibration. Humans cannot hear sounds having less than 30 or more than 40,000 vibrations per second. Loud, sharp sounds, such as rifle shots or police whistle shrieks, are at high frequency; they create considerable compression in the air, while a brass horn emits waves of lower frequency. Music is composed of notes, which differ from one another in their pitch, and what we know as a musical tone is actually a sound of definite maintained frequency.

The instruments that play the notes create vibrations differently. String instruments create sounds when tightened strings are vibrated. Air blown into a woodwind vibrates a thin, flat reed that makes the air inside the instrument also vibrate, producing notes that can be changed by opening and closing holes. Percussion instruments send out vibrations when a tightly stretched skin is struck with sticks, hands, or a mallet.

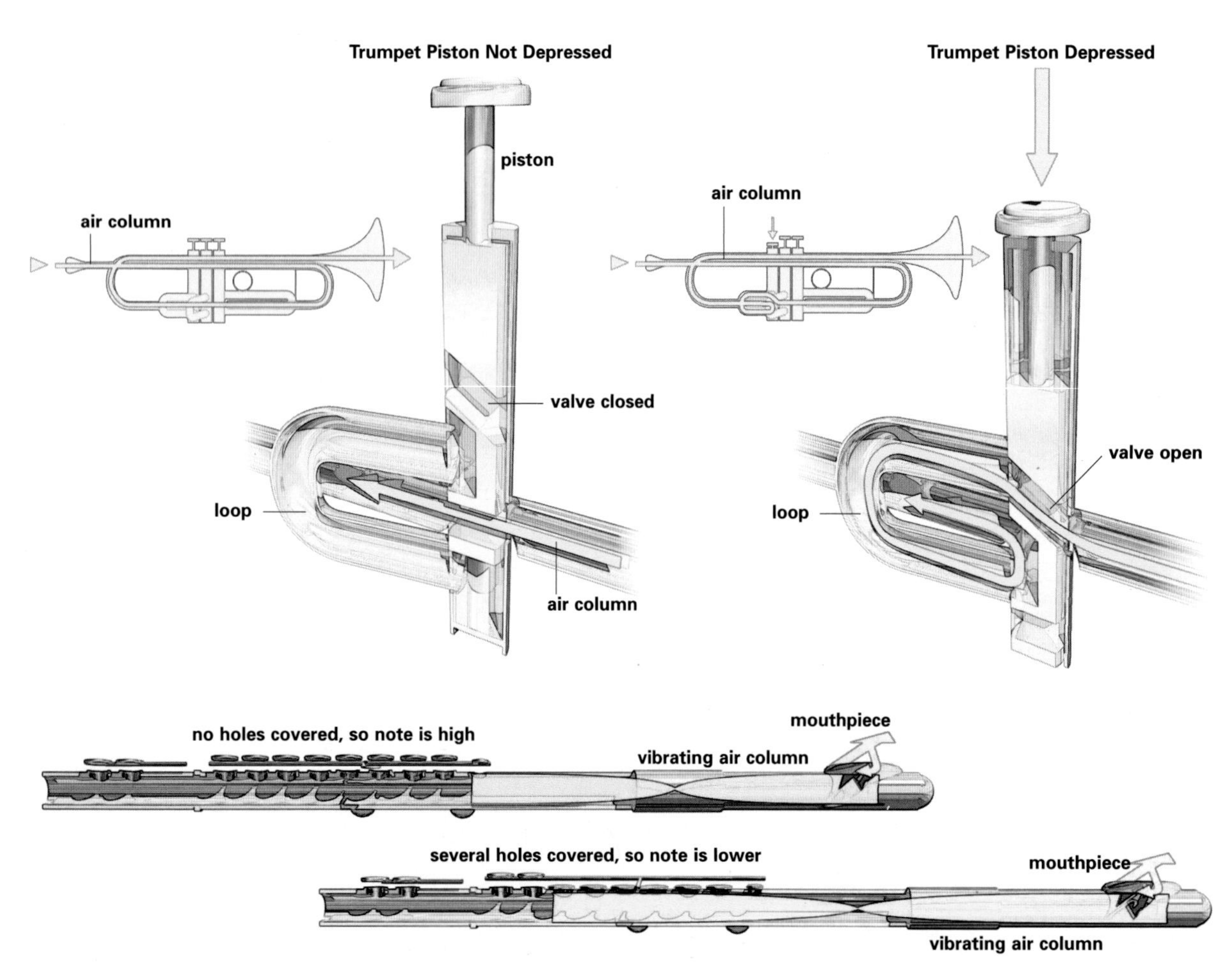

STRINGED INSTRUMENT

(Right) The late, great trumpeter John Birks (Dizzy) Gillespie with his trademark upturned instrument, beret, and well-developed cheek muscles.

SOUNDING THE TRUMPET

(Left, top) Pressing and releasing a piston on a trumpet causes a valve to open and close an extra section of tubing. In the "up" position, the valve shuts off the loop attached to it, and the air goes straight through. In the "down" position, the valve opens the loop and diverts the air through the extra section. Using various combinations of the three pistons, a musician blowing with tensed lips can create as many as eight different notes.

WIND INSTRUMENT

(Left, bottom) When a musician blows air over the mouthpiece of a flute, the air column in the instrument vibrates, producing a note. To create new notes, the flutist covers and uncovers tone holes along the length of the tube.

SEE ALSO
The Piano · 140
Radio · 142
Fiber Optic Cables & DSL · 210

THE PIANO

Music's most versatile instrument may be the piano. Not only can it produce a wide range of notes, but it can also vary the loudness and duration of each one, permitting a variety of interpretations and moods. These expressive effects are possible because the strings, or wires, are struck by felt-covered hammers connected to keys that respond to the touch of a pianist.

When the earlier harpsichord was played, its strings were plucked by small picks, giving the player little control over the quality of the tone. The piano, on the other hand, could be played softly or loudly, a feature that was recognized in an 18th-century name for an early form of the instrument: pianoforte, which is Italian for "soft-loud."

The piano's action mechanism does not attach a hammer rigidly to its corresponding key. Instead, it drives it through a series of levers that achieve different tonal qualities. When the key is pressed, force is transmitted through intermediate levers called the wippen and the jack, and against a roller mounted on the underside of the hammer. This forces the hammer up to strike the string and then lets the hammer fall back immediately so that the string can vibrate. At the same time, a damper lifts from the wire. When the pianist releases the key, the damper drops back and cuts off the note.

The electronic piano, which only remotely resembles the stringed version, has a keyboard with touch-sensitive keys that select the notes, but inside its casing is an array of transistors and microprocessors that play digitally produced sound that is very much like conventional piano sound.

THE GRAND PIANO
(Right) Derived from dulcimers, and preceded by clavichords and harpsichords, a concert grand piano uses the same hammer action invented by Cristofori (1655-1731), an Italian harpsichord maker. With its metal frame cast and braced to resist the enormous tension of the strings, the piano produces gradations of loud and soft tones beyond the capability of the harpsichord.

BEHIND THE KEYBOARD
(Left) A piano relies on lever action that begins when a musician presses the keys. The operating mechanism—a complex system of levers known as the action—and one of its key components, a metal wippen, controls the sounding of the strings. The levers also raise and lower the dampers, devices that stop the strings from vibrating and control the duration of the tone.

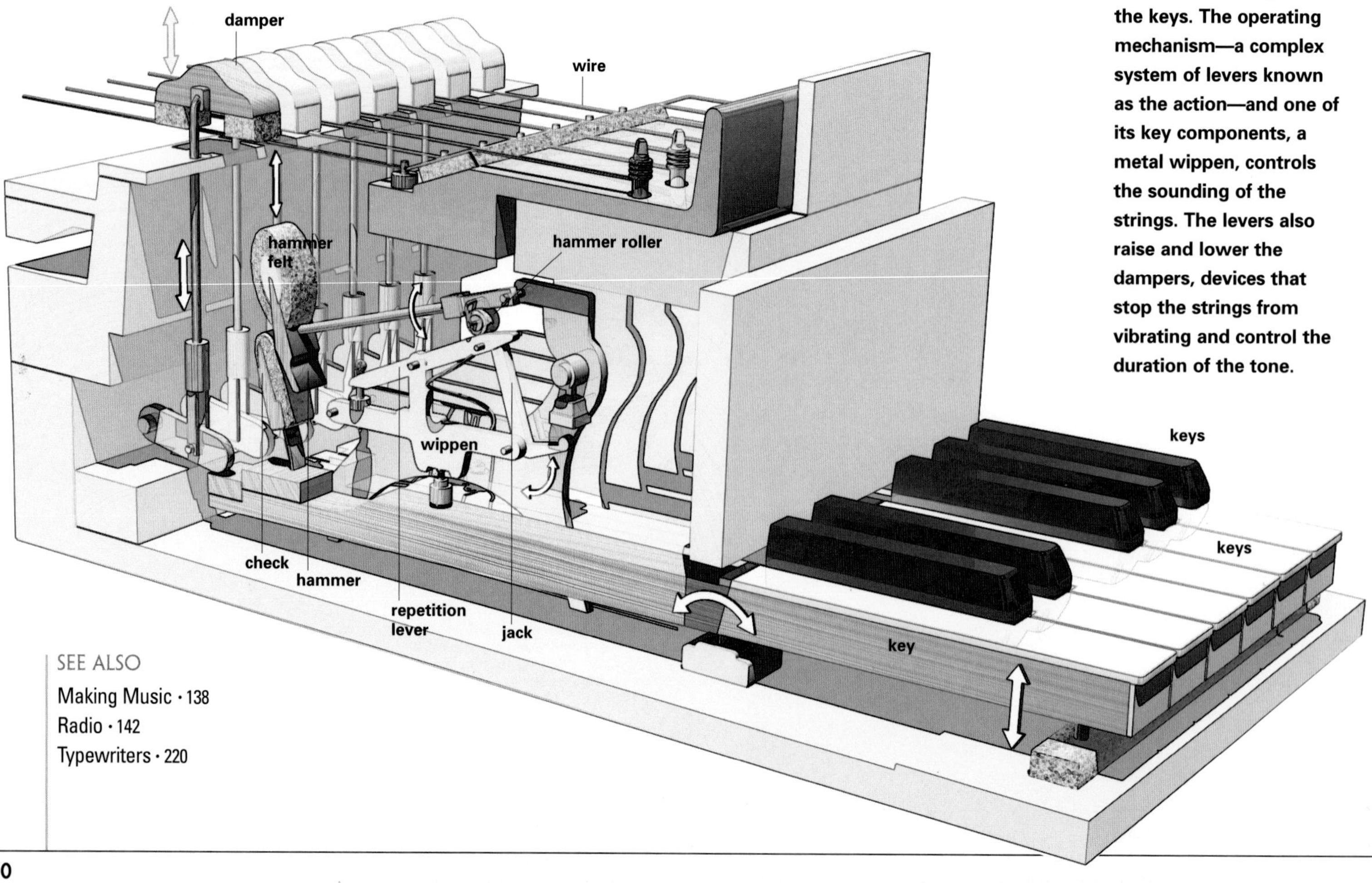

SEE ALSO
Making Music · 138
Radio · 142
Typewriters · 220

STEINWAY&SONS
NEW YORK HAMBURG
STEINWAY

RADIO

Guglielmo Marconi may have invented the wireless radio, but his feat drew on the research of Heinrich Rudolph Hertz, the German physicist who first demonstrated the existence of radio waves. Hertz proved that the waves could be reflected and refracted, as light is, and that they could be sent through space. He gave us hertz and megahertz, units that are used to measure the frequency of electromagnetic radiation, which includes radio waves. The units (MHz) correspond with a station's number on the radio dial and also measure the speed of computers.

Radio waves are made to carry signals by changing, or "modulating," the waves. Some radio stations send their signals by changing the size, or amplitude, of the radio waves. These AM (amplitude modulation) stations broadcast on frequencies that are measured in thousands of cycles per second, or kilohertz. Other stations broadcast by making small changes in the frequencies of their radio signals. These FM (frequency modulation) stations are assigned frequencies in millions of cycles per second, or the megahertz range. Television makes use of both kinds of waves, with pictures carried by an AM signal and the sound by an FM signal.

In AM broadcasts, sound vibrations in the form of amplified electrical signals are impressed onto electrically generated carrier radio waves by adjusting the amplitude of the carrier waves to keep them in tune with the audio signals. Frequency modulation gives clearer transmission and reception, but it does not affect the amplitude of the carrier wave. Instead, it varies the wave's frequency in accordance with the sound to be transmitted.

When a station is selected on a radio, a tuning circuit picks one and tunes out all the others by permitting current to oscillate at a single frequency. The two conducting plates of a capacitor, or condenser, store energy as electricity, while a coil to which they are linked stores energy as a magnetic field. The magnetic field collapses and sends an electric current to recharge the capacitor, which discharges again through the coil, instigating an oscillating current of one frequency. Essentially, the capacitor blocks the flow of direct current while allowing alternating and pulsating currents to pass.

MUSIC MASTER

(Right) Equipped with headset and settled behind his microphones, a disc jockey in Lomé, Togo, spins music into the radio waves that will carry it to distant locales via a central transmitter.

MAKING WAVES

(Below) Radio waves demonstrate characteristics associated with their frequencies. For example, longer wavelengths have lower frequencies. Impressing sound waves onto radio waves involves amplitude modulation (AM) or frequency modulation (FM). In AM, the signal is carried by varying the amplitude of the radio wave; in FM, the signal is carried by varying the frequency along with the sound signal.

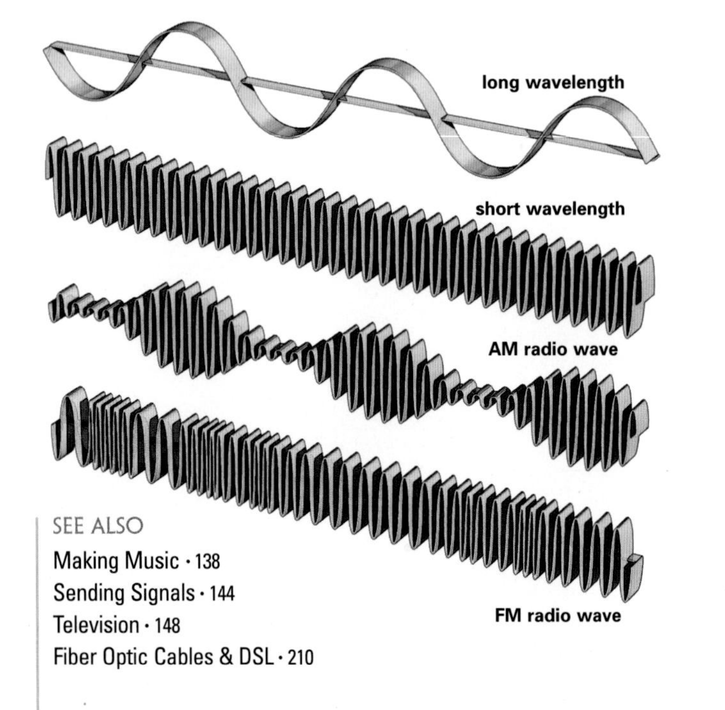

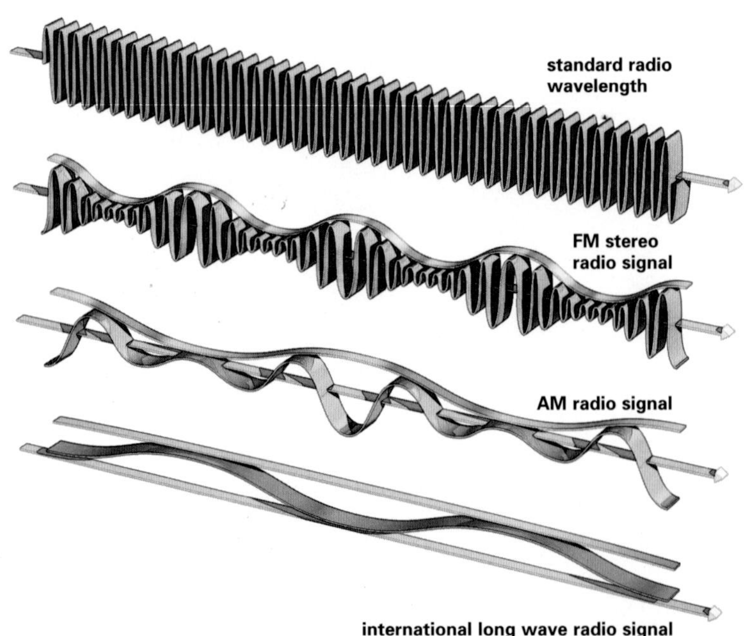

SEE ALSO
Making Music · 138
Sending Signals · 144
Television · 148
Fiber Optic Cables & DSL · 210

SENDING SIGNALS

THE MUSIC AND THE VOICES WE HEAR COMING from our conventional radios have traveled great distances and in many guises. They begin, as they do in a telephone transmission, at a microphone; only this time the microphone is in a broadcasting station, where vibrating sound waves are converted into relatively weak electrical pulses. Sound waves vibrate the diaphragm of a microphone, which turns the acoustical energy into a weak electrical signal. The weak signal is amplified and then added to a carrier wave so it can be broadcast. Each radio station is assigned a carrier wave with a different frequency. An antenna at the top beams the audio-carrying radio waves—faster and more powerful than audio signals alone—at the station's assigned frequency. The distance the radio waves travel is determined by their frequency and by electrical atmospheric conditions.

After the waves leave the transmitter, they are picked up by the antenna in your radio. A tuner then selects the program by matching the receiver to the station's transmitting frequency. Weakened by the distance traveled, the radio wave signals are turned into electrical signals that are amplified and then turned into audio signals. These are sent to the radio's speaker, where the electrical waves are converted back into sound and amplified again.

SKY EARS

(Left) A dishlike array of radio telescope receivers in New Mexico gathers and amplifies distant signals for astronomers.

RADIO WAVES

(Right) When waves leave a transmitter, the path and distance they travel depend on their frequency and on electrical atmospheric conditions. At the top: Essential to worldwide communication, indirect—or short—waves reflect between the sky and Earth's surface. In the middle: FM radio and TV signals, called surface or direct waves, travel almost parallel to the Earth. They cannot pass the horizon, so relay towers extend their range. At left in the bottom illustration: Very high frequency (VHF) waves in FM, police, and citizens-band radios reflect off the ground and stay in the line of sight. At bottom right: Medium waves bounce off the ionosphere, and distant points can pick them up.

SEE ALSO
Radio · 142
Television · 148
Fiber Optic Cables & DSL · 210
Telephones · 212

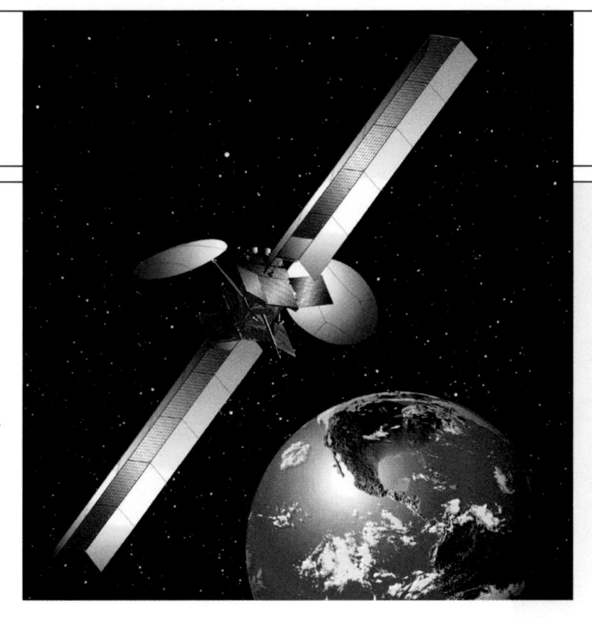

XM-SATELLITE RADIO

This radio innovation is a 100-channel broadcasting system advertised as faster than cable TV and CD and MP3 players. Unknown until only a few years ago, satellite radio has attracted more than a million listeners, who pay a fee for coast-to-coast, static-free news and entertainment. From a ground station, programs are transmitted to satellites like this powerful Boeing-702 that bounce their digitized signals back to portable radio receivers or to receivers installed in a car. The receivers decode the signals for the listener, and along with the audio come song titles, names of artists, and other information, all displayed on the receiver's screen. Aside from multichannels generally free of commercials, satellite transmission allows a driver to cross the country without worrying about a station fading.

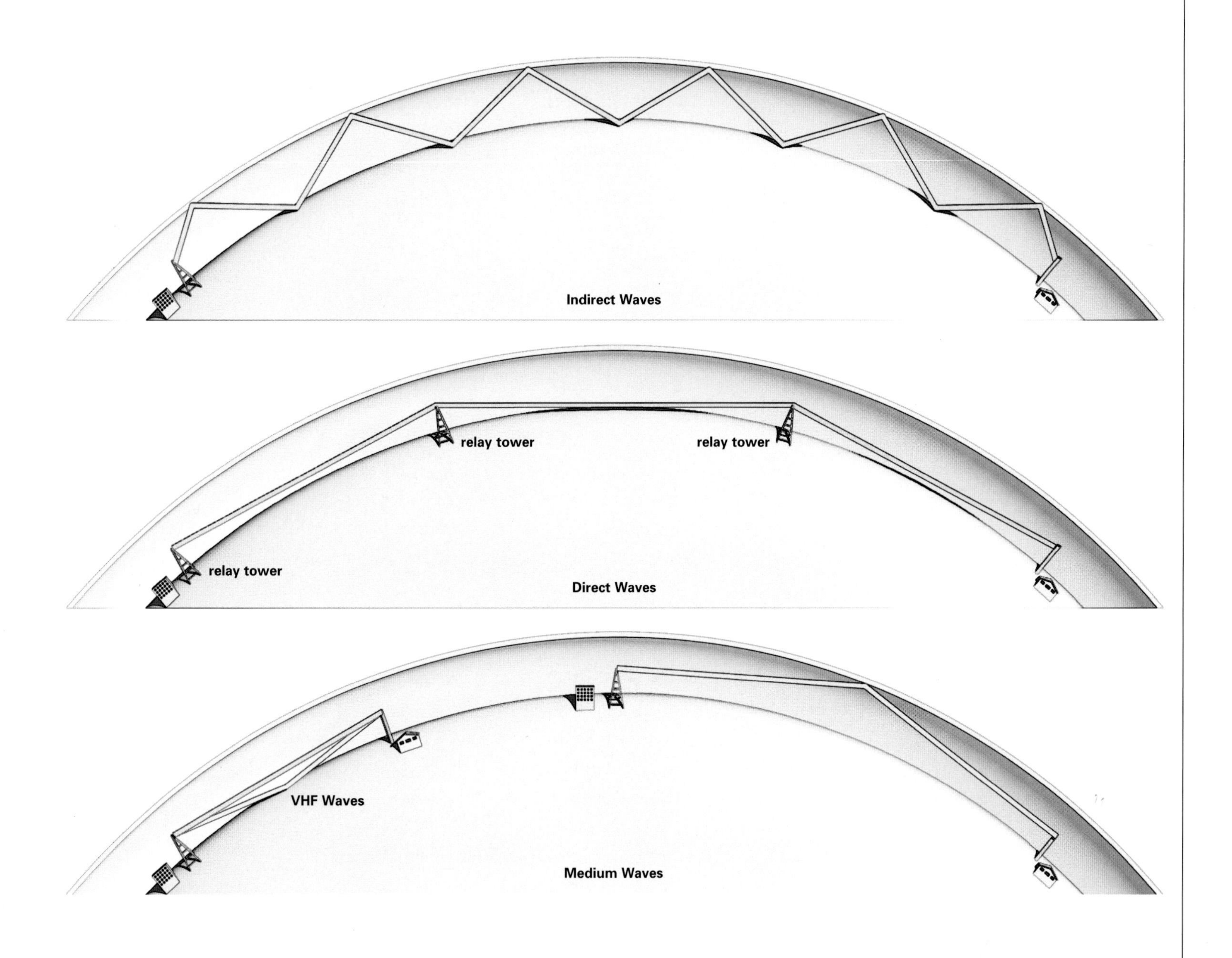

DIGITIZED MUSIC

SOME BIRDS THAT HAVE THE RIGHT VOCAL CORDS and throat formation can imitate spoken words with a fair degree of accuracy, but it takes electricity and electronics to produce a nearly limitless array of sounds with great authenticity.

In the now almost extinct phonograph, a diamond stylus, or needle, played back sound vibrations that had been scribed through a microphone into grooves on a vinyl record. As the record turned, the needle picked up mechanical vibrations, converted them into electrical signals corresponding to peaks and troughs in the sound waves, and sent them to loudspeakers that converted the signals to exact replicas of the original sound. Sound's electrical counterparts may also be embedded magnetically on a specially coded plastic tape; to reconstitute the recorded sound, the tape is run past a recorder's electromagnetic playback head, which sorts out the coded sound and allows it to be amplified and sent to the speakers.

What has revolutionized music storage and music-making is the conversion of sound from its vibratory, analog-wave state into a digital format of numbers, and then back again to an analog wave that can be amplified and heard through speakers. The compact disc (CD), along with its computer counterpart, the CD-ROM, are read by a thin laser beam that probes sequences of microscopic pits and flat regions for binary codes that will be reconverted into sound.

If the CD was a breakthrough in sound recording, the blending of computers, the Internet, and MP3 players has totally transformed the way people acquire, listen to, and share music. The MP3 is an audio compression format that shrinks digital files by reducing the number of bytes; and it does so without affecting the original sound quality. MP3 audio files can be downloaded from the Net and played with software installed on the computer's hard drive, converted into CD files, stored on data CDs, or loaded into portable players.

Music can also be produced entirely by electronics. A synthesizer, for example, is a keyboard system of waveform generators and computer connections; it simulates the sounds and overtones of a variety of instruments by creating the appropriate electrical sound signals. Another system, known as MIDI, for Musical Instrument Digital Interface, has virtually erased the boundary between the real and the artificial. MIDI allows synthesizers to link up to other synthesizers, computers, and to sounds from different instruments. The electronic keyboard—along with hardware, software, controllers, sequencers, power amps, and speakers—lets a musician design and layer sounds, dub and overdub, fill, edit, and play. In portable form, a keyboard lets a musician use it like a high-tech scratch pad that understands MIDI's arcane language.

Under MIDI's direction, an entire music studio's output can be controlled, and a single musician—a keyboardist, for instance—can become a one-person band. The future may even hold "hyperinstruments" that understand the performer's intentions and enhance musical expression.

A CD BARED

Scanned by a electron microscope, the cracked surface of a compact disc displays the musical layer below (at center). Made of plastic, the disc has been pressed with a series of fine depressions, or notches, which represent a digitized musical signal capable of being read by a laser.

SEE ALSO
VCR & DVD · 150
Laser Surgery · 204
Calculators · 228

compact disc
focusing lens
semisilvered mirror
cylindrical lens
beam
light sensors

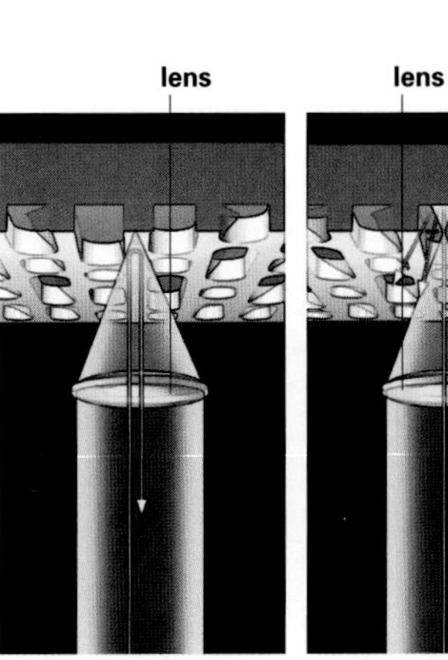

INFRARED LASER BEAM

(Above) An infrared laser beam replaces the styluses used on old LPs. It scans the track of a CD's pits and flats with pinpoint accuracy. When the beam hits a flat surface, it reflects as an "on" signal, or a 1; when it strikes a pit, the light disperses, for an "off," or 0. The ons and offs then reconvert into the sound originally digitized into the CD from a magnetic tape.

INSIDE A CD PLAYER

A 4.75-inch compact disc system outshines its predecessors—the 12-inch, black vinyl LP and belt-driven player. Inside the CD player, laser light bounces off a mirror and flashes through a focusing lens to the disc's underside, where it reads digitized sound. Reflected back to sensors, the light converts to electrical signals and then into sound. Microscopic pits and flat areas (see inset), encoded with 1s and 0s representing sound, dot the reflective aluminum underside of a compact disc.

TELEVISION

TELEVISION WAS NOT—ALTHOUGH THE POINT may be argued—developed by one individual. Most of the credit, though, goes to two inventors of the 1920s. Britain's John Logie Baird developed the picture tube and was the first to televise moving objects. Vladimir Kosma Zworykin, a Russian-born U.S. engineer, developed an electronic scanning device and an electronic "television" camera called the iconoscope. As Zworykin conceived it, a picture could be transmitted between distant points by shining it into a mica disk covered with a mosaic made of photoelectric material. The disk would be scanned by a thin electronic beam to search for weak and intense emissions. Modern television transmission borrows from Zworykin's idea and from radio transmission, which uses radio-frequency carrier waves to transport information.

Essentially, a TV camera changes light from an image into a video signal by breaking the subject into an arrangement of 525 to 625 "lines of resolution," which are then scanned electronically, line by line. Arriving by cable at the TV set, the sequence of pulses is turned back into scan form. It is then reconstituted as an optical image on a screen that is coated with chemicals sensitive to the three primary colors of light: red, green, and blue.

Technology has transformed the traditional boxy television set and personal computers with their bulky picture tubes into sleek and slim panels that screen extraordinarily vivid images. For one thing there is the shift from the familiar analog transmission with its 525 scan lines to digital transmission with more than 1,000 lines, as in high-definition digital television, or HDTV, which projects movie-quality images. In an HDTV set, digital signals, transmitted by cable or satellite to a home, are decoded for viewing.

A drawback of the cathode ray tube in a traditional set is that it needs to be fairly long to produce a wide picture. Flat panel sets, either plasma or LCD (liquid crystal display) provide a solution. Plasma displays, which are about six inches thick and come in large sizes, work by lighting up minuscule fluorescent lights—red, green, and blue—to produce an image. Plasma, a key component in fluorescent light, is an electrically charged and conducting gas, which has been referred to as a fourth state of matter. In the form of xenon and neon in a plasma TV set—where the gas is sandwiched between two glass plates in tiny cells—its atoms release light particles, photons, when excited.

BIG SCREEN

A wide-screen HDTV plasma monitor showcases to advantage a blockbuster film at an electronics store in Niles, Illinois.

LCD sets rely on liquid crystals (substances that flow like a liquid but retain some of a crystal's characteristics) to reflect an image. Familiar components of calculators, they can be readily affected by electromagnetic fields, changes in temperature, and various chemicals. They also manage to react to electrical current and are able to affect how light behaves when it passes through them. Inside a screen, liquid crystal-filled cells—again red, green, and blue—get an electrical charge that shifts the crystal to a certain angle which, when illuminated by a light behind the screen, controls the images.

SEE ALSO

Radio · 142
Sending Signals · 144
VCR & DVD · 150
Fiber Optic Cables & DSL · 210

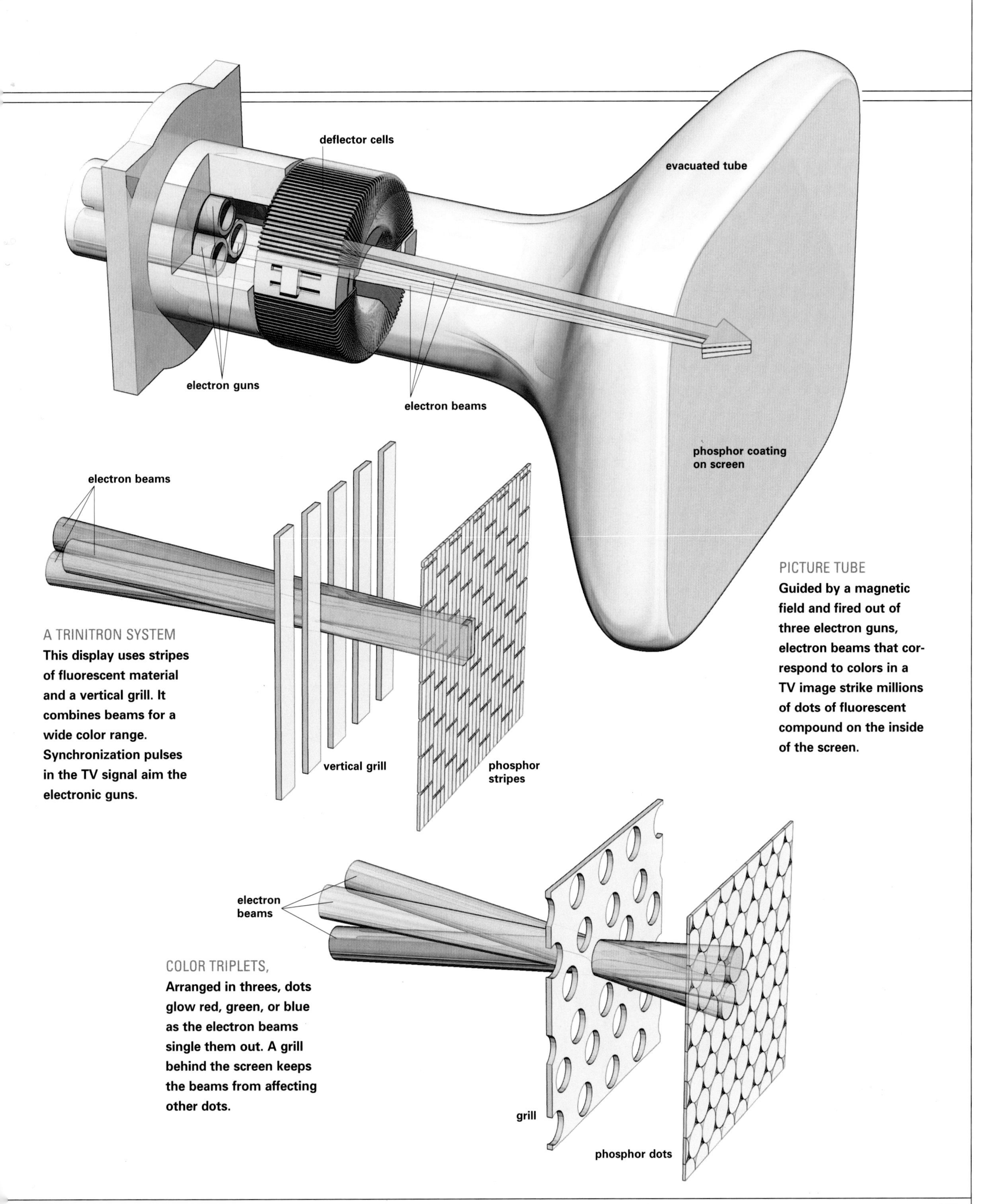

PICTURE TUBE
Guided by a magnetic field and fired out of three electron guns, electron beams that correspond to colors in a TV image strike millions of dots of fluorescent compound on the inside of the screen.

A TRINITRON SYSTEM
This display uses stripes of fluorescent material and a vertical grill. It combines beams for a wide color range. Synchronization pulses in the TV signal aim the electronic guns.

COLOR TRIPLETS,
Arranged in threes, dots glow red, green, or blue as the electron beams single them out. A grill behind the screen keeps the beams from affecting other dots.

VCR & DVD

VIDEOCASSETTE RECORDERS (VCRs) AND digital video discs (DVDs) both record and play, but though each stores pictures and sound, they work in vastly different ways. The VCR, like a television set, picks up video signals directly from a cable or antenna, but instead of projecting them onto a screen, it stores them on a reel of plastic. The reel is magnetic tape that does not need to be processed as conventional camera film.

Moving images can also be captured by a camcorder, a downsized video camera combined with a tape recorder. It focuses images onto a light-sensitive chip called a charge-coupled device that, in turn, converts the colors, shadows, brightness, and sounds of the scene into electrical signals. The vast amount of encoded information, which requires videotape to be wider than audiotape, is passed on to tape heads and stored as patterns on a tape that is coated with magnetized iron oxide.

When a videocassette is inserted into a VCR connected to a TV monitor, the tape plays back the images, regenerating the stored electrical signals for synchronized pictures and sound on the screen. When the "record" button is pressed, TV images that flow in via the cable or antenna connection are fed into the VCR, which passes them on to tape. Looped between two reels in the cassette (as in an audiotape recorder), the tape is drawn between two sets of electromagnetic heads: One set records sound; the other set records pictures.

The DVD looks and operates like a music CD in that a laser is used to read encoded digitized data stored in tiny pits in the disc's track. But a DVD, which is classified as an optical storage medium, has a much larger storage capacity, and because of its digital format it can store not only music (eight hours of it per side), but also text, full-length movies, and other images. DVDs can also "read" CDs, but not vice versa. With peripherals, videotapes can be transferred to DVDs, as can live TV programs. DVD drives can also be hooked up to PCs to run software or play music and movies.

Another recording device built for television is the TiVo, a computer-like box that plugs into a TV like a VCR and stores hundreds of programs in a hard drive. It can record selected programs, pause, skip commercials, and record just about everything arriving at a TV set, whether or not the viewer wants it.

TALE OF THE TAPE
A videographer tapes a service in a church. Recorder of people and events, video cameras have become as common as still cameras.

SEE ALSO
Digitized Music · 146
Television · 148
Video Games · 152
Film Camera · 156

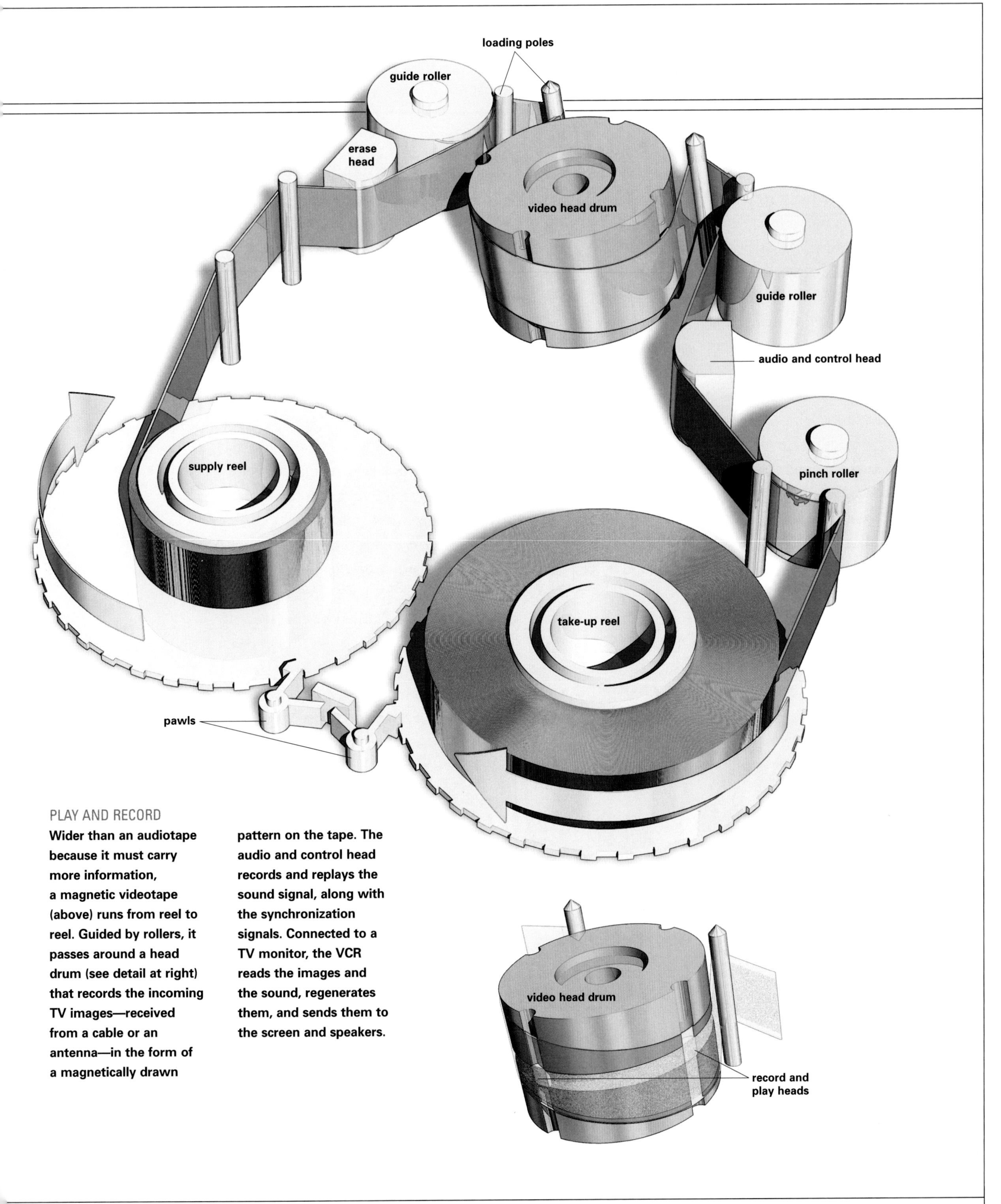

PLAY AND RECORD

Wider than an audiotape because it must carry more information, a magnetic videotape (above) runs from reel to reel. Guided by rollers, it passes around a head drum (see detail at right) that records the incoming TV images—received from a cable or an antenna—in the form of a magnetically drawn pattern on the tape. The audio and control head records and replays the sound signal, along with the synchronization signals. Connected to a TV monitor, the VCR reads the images and the sound, regenerates them, and sends them to the screen and speakers.

VIDEO GAMES

When people operate video games, chances are they don't know what creates the special effects. The machine may be a stand-alone console in an arcade, a compact hand-held model complete with a microprocessing minicomputer inside, or a personal computer turned into a game center by binary imagery stored on a CD-ROM. Whatever its form, a video game is a collaboration of animators, 3-D graphic artists, audio programmers, interactive designers, software specialists, and electronics technicians.

Game concepts often begin with a design outline on a sheet of ordinary paper. An array of penciled connect-the-dots represents electric signals that stand for commands, screen action, background, sound, color, and movement of the images. In a process called rendering, video game programmers "draw" images on a screen with the help of 3-D graphics microprocessors or special software programs. Animation is achieved frame by frame by using computer programs that create and play back the artwork for editing. This process is an extension of the non-computer technique of creating, say, animated cartoons by filming a series of images hand-painted or drawn on plastic "cels." Color and brightness are controlled by manipulating pixels (picture elements), the tiny bits of colored light that make up a video display. After images have been created, they are installed in a computer-generated background and stored in binary code, which can be resurrected and directed with a flick of the wrist.

The devices that enable video games are fairly complex. Game consoles are fairly bursting with such high-tech stuff as microprocessors, memory chips, wireless controllers, and analog and digital audio-video outputs. Some devices, like those that let kids do digital microscopy, for example, are plugged into a PC that displays magnified images from a special digital microscope and camera. Games that are plugged into a television set use game cartridges or have controllers containing games and necessary hardware built in. Others are interactive, like one that has an electronic home plate, ball, and a plastic bat that allows the player to simulate hitting a baseball. The batter faces the TV screen standing above the electronic home plate, which emits an infrared beam, and swings at a simulated ball; an electronic system in the bat sends the swing action to the home plate, and the result is displayed on the screen.

TARGET PRACTICE
Virtual shooting at a video game arcade. Marksmen and markswomen can hone their target skills with a realistic display and a make-believe gun that fires electronic "bullets."

PORTABLE FUN
Tibetan children enjoy interacting with a video game. Hand-held video games complete with sound effects and liquid crystal color graphics can be played on batteries just about anywhere, and what's under their hoods often mimic some of what's inside a home PC. Palm-size, button-operated portables let one play games anywhere. Others are larger and more sophisticated. One device, Game Boy Advance, has a speedy central processing unit, more than 65,000 colors on its palette, a stereo sound generator, and playing time of up to 20 hours. Peripherals can connect the game to another console, allow four players to participate, and even connect the game to the Internet via a mobile phone to download other games and trade data. One unique accessory, called an e-reader, reads data about the game's various characters that has been stored on special cards. By scanning bar codes, a player learns more about the characters' methods of operation, and how to deal with them.

SEE ALSO
Digitized Music · 146
Television · 148
Motion Pictures · 160
Computers · 230
Backup Storage · 236

MULTIMEDIA

INCREASINGLY, AS THE COMMUNICATIONS AND entertainment industries seek new means of disseminating their wares and services, they have become inextricably intertwined. Once separate entities with their own clearly defined, one-task missions, communication and entertainment systems have blended into multimedia arrangements. The term the industry uses for these new applications that place the features and functions of various electronic devices into a single apparatus, or a network, is "convergence technology," a cutting-edge wrinkle that can include uniting the computer, television set, and telephone for one-stop shopping for data, video, voice service, and everything the Internet offers.

One example of the new technology is the cell phone, which at first was just a mobile way to make an ordinary call from anywhere outside the home without reliance on the wired desk-bound telephone. Now, a cell phone, with its color screen and weighing less than a deck of cards, does much more: It transmits e-mail and photographs, downloads games, connects to the Internet, and provides digital camera add-ons, voice dialing, and a variety of information-management features.

But what is arguably the strongest example of integrated technology is Interactive TV (ITV), a system that has already dramatically changed the role of a television set. ITV allows a viewer to combine standard television usage with actual interaction through TV programming and a network, like the Internet—all with a remote-control keypad. With an ITV hookup, which generally involves plugging a receiver into a TV set and phone line, installing software, and linking up to a server—a "couch potato" can communicate via e-mail, send and receive pictures, and access the Web, just as he or she would do with a computer. But more than that, the interaction allows viewers a startling number of other options, all of which enable one to not only see what's on the screen but also to respond to it and even change, store, and replay what is viewed. A viewer might switch camera angles for sports events, obtain additional information about a program, reduce an event to a smaller window so other programming data can be displayed, get instant access to product information, customize news to suit one's interests, and cast a vote during or after a program poll. An ITV user may also compete in games, participate in videoconferences, and interact with televised advertisements to order, for example, a pizza or a set of knives, or participate in live, interactive "distance education."

The computerized brain of the ITV system is the "set-top box" (STB), also known as a decoder, a converter, a receiver, or a smart encoder—which receives digital signals, containing all the interactive content, from a satellite, an aerial, or through a cable. It then converts the digital information to analog signals and displays them on the TV. The STB also accepts commands from the user's keypad—commands that enable interaction—and sends them back to the network through a "back channel," which can be a phone line.

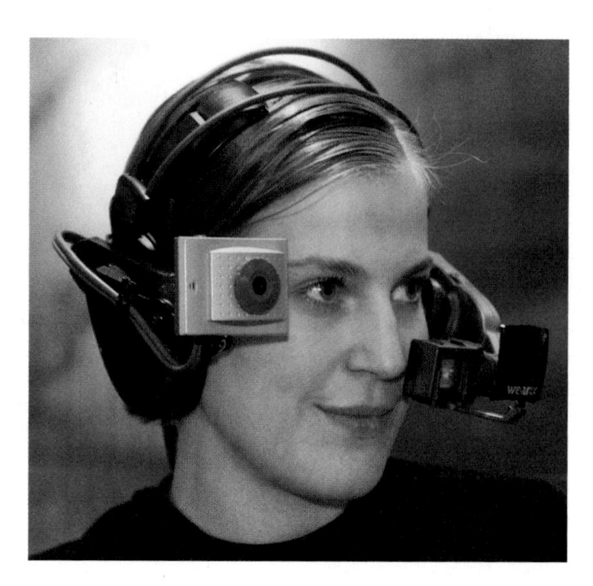

PLUGGED IN

(Left) Wearing a "Web Reporter Headset," an observer covers an event from all the angles with a combination camcorder and microphone that can transmit photos, video, sound, and text over a built-in wireless telephone.

VIDEO LEARNING

(Right) Interactive schooling, without a traditional classroom and a "live" teacher, is a valuable multimedia product of our information age. Sitting at individualized workstations linked to a distant center via a learning network, students in effect work with "digital textbooks" and audio and visual materials, interacting with their instructor to exchange and access information, and take tests. Online schools already offer high school diplomas and a variety of college degrees, including Ph.D.'s.

SEE ALSO
Television · 148
Fiber Optic Cables & DSL · 210
Telephones · 212
The Internet · 238

Preset
Program
Henrico Count
Public School
The Virginia
Department of
Education
presents

FILM CAMERA

Aristotle and Leonardo da Vinci were well acquainted with the concept behind the camera obscura: Light rays from an external object enter a darkened chamber through a tiny hole in one wall, converge and cross, and project an inverted image of the outside scene on the opposite wall. Make the room smaller and add a lens, mirrors, prism, shutter, and film, and you have a no-frills camera.

A standard camera (that is, not a digital one) is basically a lightproof container that focuses, through a lens, light from a scene onto light-sensitive film inside. Photographs are produced by light and the control of light. Too much of it means overexposed, washed-out images; too little results in underexposed, dark ones. To regulate the amount of light that strikes the film, the shutter speed is either manually or automatically adjusted; in modern cameras this normally ranges from timed exposures of several seconds to 1/1,000 of a second. Cameras also have automatically adjustable apertures that affect a picture's sharpness and contrast. Peering through the viewfinder of a standard camera, a photographer sees what was an upside-down camera obscura image that is now righted by a mirror reflecting through a prism. When the shutter release button is pressed, the mirror slides out of the way, the shutter opens, and the right amount of light bathes the film. Light exposure can also be controlled by adjusting the period of illumination. An electronic strobe, for example, flashes from a relatively slow speed of 1/100 of a second to less than 1/500,000 of a second, fast enough to catch a bullet in flight.

GETTING THE PICTURE:

Light gives life to the art and processes of photography. In a camera, a mirror and a prism correct an inverted, reflected-light scene, and a viewfinder lets a photographer see it through the camera lens, which actually consists of a system of optical lenses. Adjusting the distance between the lens and the film brings an object into clear focus. When the photographer presses the shutter button, a spring-activated device opens and closes; it keeps light out except during exposure. With the aid of a diaphragm—a fixed or adjustable component forming a large or small opening—the shutter allows the correct amount of light to come in through the lens. As the light strikes the light-sensitive film at the rear of the camera, it leaves an imprint of the scene.

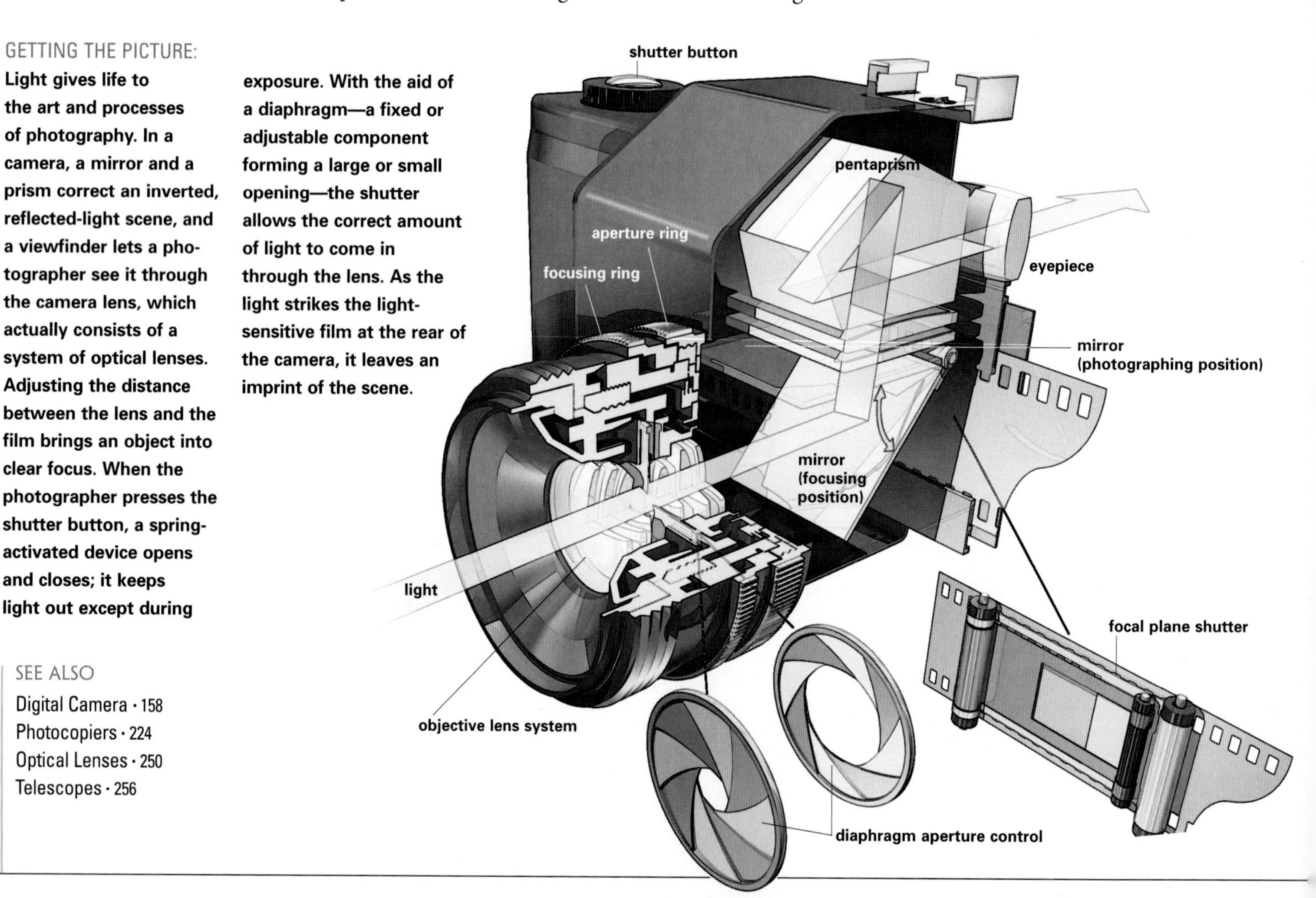

SEE ALSO
Digital Camera • 158
Photocopiers • 224
Optical Lenses • 250
Telescopes • 256

PROCESSING

Photographic film is basically a strip of plastic coated with an emulsion of gelatin and crystals of light-sensitive salts called silver halides. When black-and-white film (near right) is exposed to light, the silver compound changes, and a latent image forms where the light strikes. Later immersed in a reducing agent, or developer solution, the silver turns black only in the places struck by light. A fixing solution removes the unchanged silver, making a permanent negative image. To produce a photograph, light is shone through the negative onto a light-sensitive paper. Color film (far right) contains sensitizing dyes in three layers of emulsion, making each layer sensitive to a specific light color when exposed. Latent images form on the layers, and combining them brings out the true colors. In making or enlarging a photograph in a darkroom, a "safe light" of a color and intensity that does not harm sensitive film is used for illumination. Contrast in the photo is controlled by the amount of light that is allowed to strike the paper and by the exposure time; the grade of paper also regulates contrast.

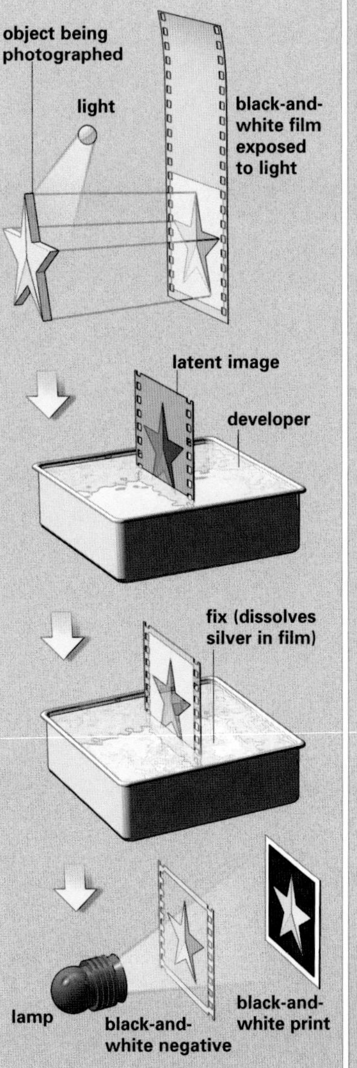

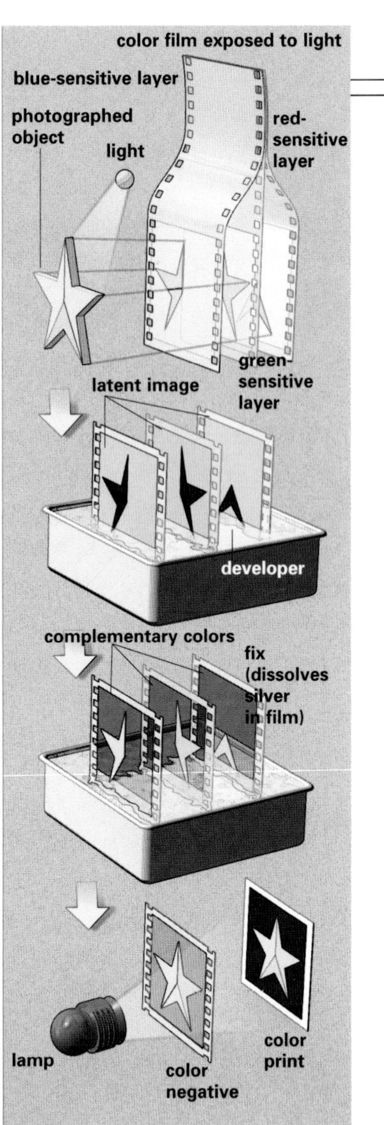

CLOSE UP

A photographer trains his camera and flashes on a treetop frog in the Borneo rain forest. Lenses have varied distance-focusing and magnification capabilities, and so-called macro lenses, along with special filters, allow close-up work.

DIGITAL CAMERA

When Louis Daguerre, the French painter and theatrical designer devised the first practical way of producing a permanent photograph in the 1830s, his method was pure chemistry: images produced on polished silver through the action of vapors from iodine, bromine, and mercury.

Chemistry is fast giving way to computerized photo-imagery in the filmless pictures taken by a digital camera, a device that is smart both in its level of electronic intelligence and in its appearance. Pictures are snapped in the usual way on a camera that somewhat resembles a film camera. But instead of using film to capture the image, an arrangement of charge-coupled devices—light-sensitive semiconductors that can store electrical charges—does the job. These digital sensors—composed of pixels, or picture elements—transform the light from the scene snapped into an analog signal that is then converted into a digital version. An electronic filtering system regulates color and other picture components, while another system reduces the picture. The image is finally sent to a temporary storage area, and then onto a memory card.

The images are transferred to a computer, where they can be enhanced or cropped, printed, attached to e-mail, inserted into Web pages, or used to illustrate computer-generated reports. One obvious advantage of the digital camera is its storage capacity. A standard 35-millimeter film camera shoots around 36 pictures, while a digital can handle hundreds before the card is full. Another advantage is the camera's ability to screen the image just taken on a tiny video monitor on the back.

DIGITAL CAMERAS

Armed with digital camera-loaded mobile phones, Japanese fans, some with conventional cameras, shoot singers arriving for a music awards event. Digital cameras have not yet replaced conventional cameras, but they are becoming more than photographic "toys." These cameras are smart in their level of electronic intelligence, as well as in their appearance. They save snapped images on a card or diskette that a computer can read. After people transfer the images to a computer, they can attach them to e-mail, insert them into Web pages, or use them to illustrate reports.

INSTANT PHOTOS

An Indonesian aborigine displays a Polaroid photo of himself. In 1947, American physicist Edwin Land found a way to produce a positive film image without having to handle a negative. His invention was the Polaroid camera, an optical system that created a finished black-and-white photograph in one minute. Early Polaroid models work like other cameras, but their film packs contain a special positive white paper that is not light sensitive. The film packs also hold jelly-like developer-fixer reagents (stored in pods) and negative film. After exposure, the film is pressed against the paper by steel rollers; the reagents squeeze out of the pods and are spread between the film and the paper; a positive image is formed by diffusion on the receiving sheet. To produce a color photograph, the film is coated with color-responsive layers containing silver halide crystals—salts in film's emulsion that are activated by light—and layers of dyes and developers.

SEE ALSO
Digitized Music • 146
Multimedia • 154
Film Camera • 156
Photocopiers • 224

MOTION PICTURES

IN 1872, LELAND STANFORD BET A FRIEND $25,000 that a running horse had all four feet off the ground at some time in its stride. When this railroad builder and former California governor asked photographer Eadweard Muybridge to help him prove it, Muybridge hit on a novel solution. He lined a racecourse with cameras, attached a string to each shutter, and stretched the strings across the track. As a horse galloped along, it broke the strings, releasing the shutters. The photograph won Stanford his bet by portraying a series of still shots of the stride of a galloping horse—something that artists had never done accurately.

Modern motion pictures are technology driven and awash in special effects, but they are nonetheless generally dependent on film, lenses, and single pictures. A conventional movie camera captures movement on a strip of film that is drawn past the lens aperture and stopped for a fraction of a second at timed intervals. When the film stops, the shutter opens quickly and exposes one frame, or picture; when the shutter closes, the film advances to the next frame. This sequence repeats at the rate of 24 exposures a second. Projected on a screen, a movie is an illusion of motion made possible by the fact that the human eye briefly holds onto each one of the single images until the next one replaces it.

Movie film still has the best quality, at least for the moment, but in the view of many filmmakers digital-camera moviemaking will eventually win out. Already film editing and special effects are digital, filmmaking schools teach digital techniques, and independent filmmakers, who don't have budgets for more expensive standard filmmaking equipment, are producing digital movies for viewing over the Internet and via cable TV.

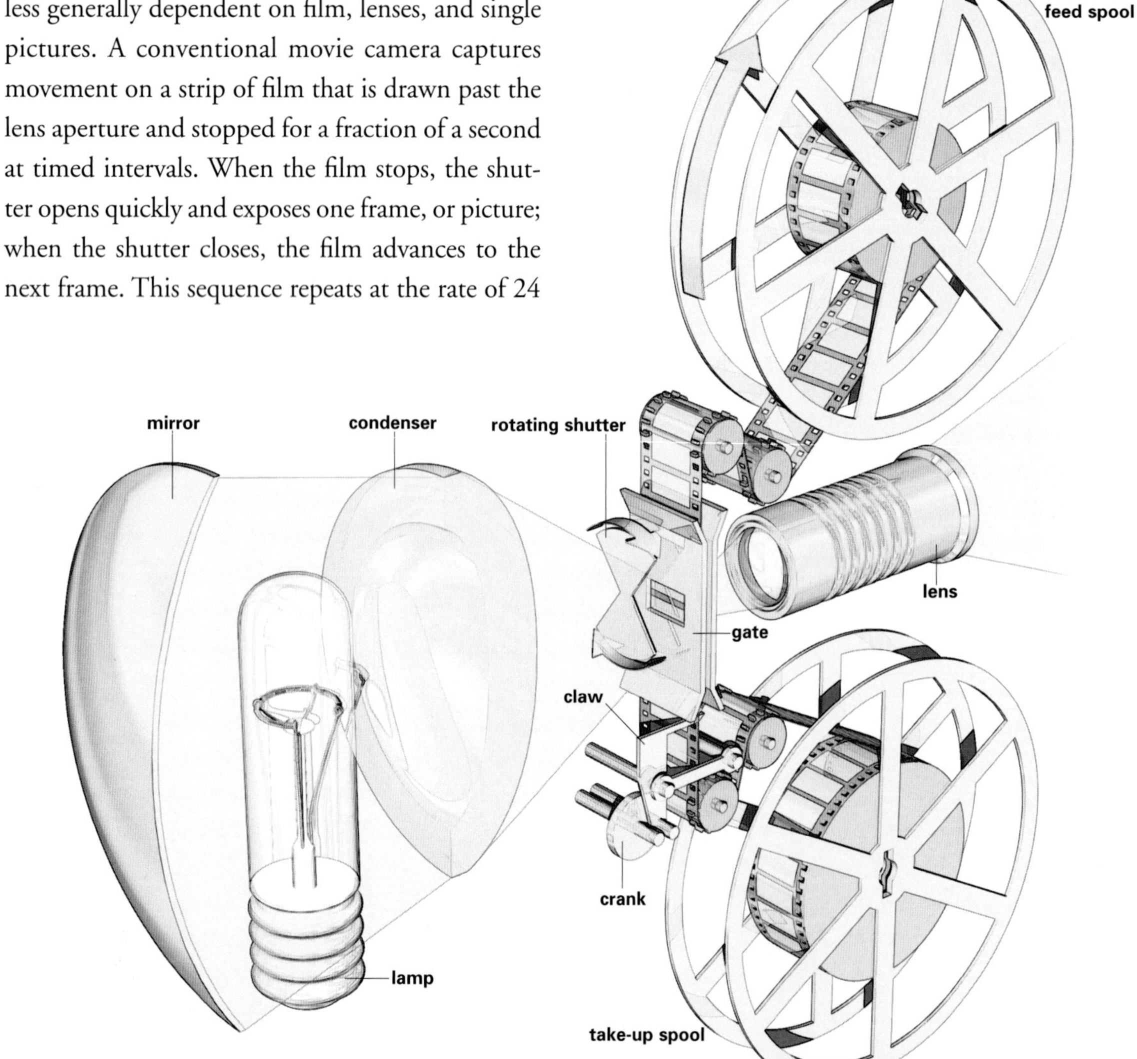

MOVING PICTURES
In a motor-driven movie projector, a claw moves a perforated reel of film past a powerful lamp and a mirror-reflector behind the film. A shutter blocks out light while the film moves along frame by frame, from reel to reel, at exactly the same speed as the camera that took the pictures. When the film stops briefly, the shutter opens and the lens focuses the image on a screen.

SEE ALSO
Television • 148
VCR & DVD • 150
Film Camera • 156
Digital Camera • 158
Computers • 230

DIGITAL FILMMAKING

Special digital effects created this image of the *Titanic* for moviegoers, including the water, people, birds, and smoke. Entire movies can also be shot digitally. Indeed, some digital films, like documentaries, can be shot by one person for a few thousand dollars. Some of those opposed argue that the widespread use of digital camcorders might trivialize moviemaking, and that digital imagery lacks the warmth that film can convey. Those in favor argue that going digital affords directors and cinematographers far more opportunities to manipulate scenes that are usually frozen on regular film, or change an actor's appearances and location. While digital films are sure to arrive in full force one day, only one notable big-screen movie was shot entirely on digital video, *Star Wars: Episode II, The Attack of the Clones.* The movie was, however, generally transformed into a 35-mm format for screening in theaters. In the near future, digital movie projectors will undoubtedly replace conventional projectors.

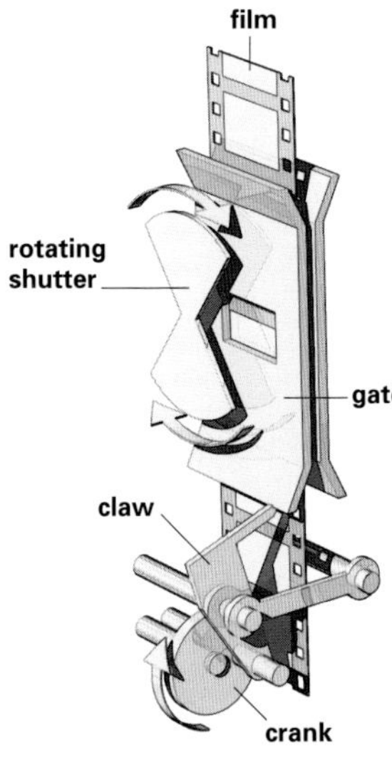

Light shines through film and image is projected on screen.

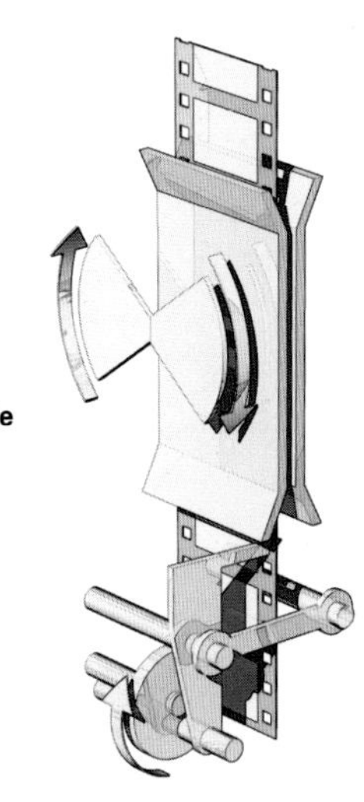

Shutter moves between light source and film.

Process repeats.

FRAME BY FRAME

As film passes through the projector, the rotating shutter intermittently projects an image. It opens for a fraction of a second and then cuts off light by closing in front of the film.

OLD GAMES, NEW MATERIALS

IN THE PAST, GOLF CLUBS WERE MADE OF persimmon, brass, and steel. Baseball bats were made of hickory and ash, and tennis rackets were made from maple and strung with silken gut. Football helmets were leather, and running shoes contained plain rubber and canvas. Vaulting poles were bamboo.

Today, high-tech metals, polymers, ceramics, and a host of composites are the norm. Pole-vaulters used to manage 10 feet or so with bamboo; now they soar more than 19 feet with fiberglass poles. Some bike seats have inflatable air chambers. Lightweight tennis rackets are made of graphite, a composite material reinforced with carbon fibers. Walking shoes are shock-absorbing beauties of encapsulated EVA-foam with thermoplastic urethane heels. Even the classic birchbark canoe has metamorphosed into a hot item having a lightweight polyethylene hull and vinyl gunwales.

Professional baseball bats are still made of wood. But college bats and other models are made of a lighter and stronger aluminum that, according to enthusiasts, will turn a .250 hitter into a .310 one. A newer version of the aluminum bat also promises to put more spring into a swing: It has a pressurized chamber loaded with nitrogen. Golf clubs, too, have improved. Wood club heads are virtually gone, replaced with aerospace titanium and sometimes steel or aluminum; shafts are made of graphite fibers bound by epoxy resins.

MODERN TWIST

(Right) The kayak illustrates how new materials have changed the appearance and construction of sports equipment. Once a frame covered with skins, kayaks are now made of molded plastic, fiberglass fabrics, aluminum, wood strips, resins, and gelcoats. They have air chambers made of welded urethane. Some are inflatable.

BASEBALLS/GOLFBALLS

Game balls are made of sturdy stuff that resists the punishment they suffer. Baseballs, for instance, have a small core of cork or rubber in a rubber shell that is wrapped tightly within layers of wool, cotton, and polyester yarn. A latex adhesive saturates the wrapped ball, and over that is a hand-sewn, tight-fitting leather cover. The ball is then shrunk even tighter in a dehumidifier room and finally rubbed down.

While baseballs have remained relatively unchanged over the years (if one doesn't count the research-type radar ball with its embedded liquid crystal and accelerometer that calculate how fast it's pitched), golf balls have evolved from a 15th-century leather sphere stuffed with boiled goose feathers, to balls with innards made of twisted elastic bands, to today's mixed bag of aerodynamic specials. They can be constructed of two, three, or four pieces—made up of layers of natural or synthetic rubber, resins, and sometimes thread windings—and have relatively soft, dimpled covers. The dimples on a golf ball are important because they control the distance a ball will travel as well as its flight pattern: A smooth golf ball will not travel as far as one with dimples, and its trajectory probably will not be as true.

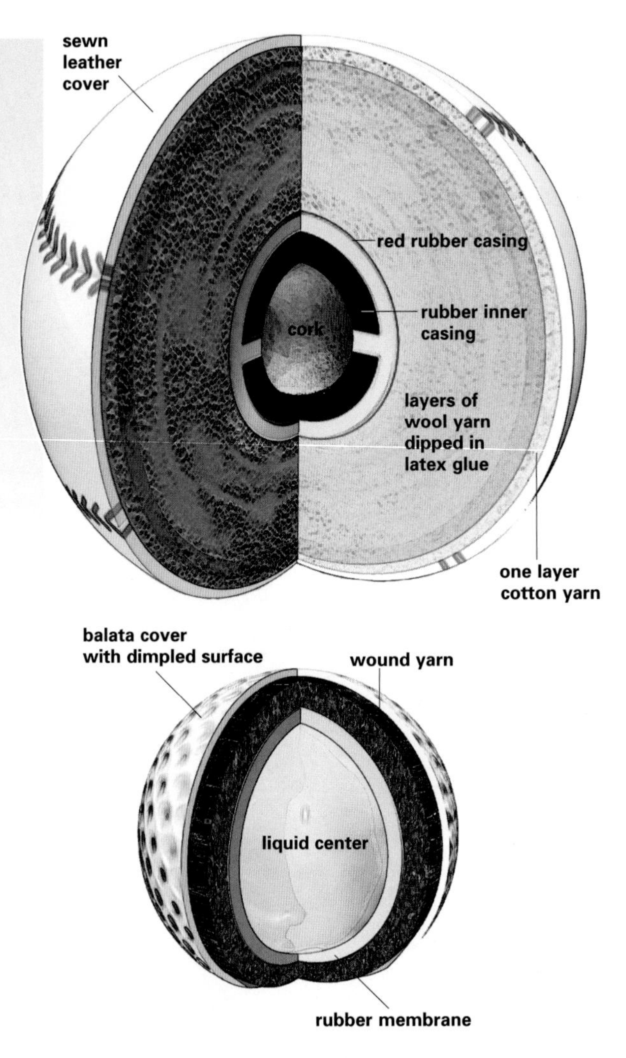

SEE ALSO
Synthetic Fibers · 132
Incredible Fabrics · 134
Plastics · 184

THE ZAMBONI

WHILE ITS NAME MAY SEEM TO DESCRIBE an Italian beverage—and it does go well on ice—a Zamboni is something quite different. It can be seen at ice rinks, and it is often as much fun to watch as the sport it serves. This innovative machine resurfaces the ice in skating rinks, and its presence at a hockey game or a figure-skating performance is certain to raise a question about how it works.

Driven around a rink, a Zamboni scrapes the rough ice surface with a 77-inch-long blade fixed at a 10-degree angle beneath the machine. The shavings are gathered up by a screwlike attachment and collected in a "snow tank" as water is fed from another tank and sent to a squeegeelike conditioner that smooths the ice. Dirty water is vacuumed, filtered, and returned to the tank, and clean hot water is spread over the ice by a flat dispenser, or "towel," which is located behind the conditioner. Zamboni drivers may not be in a class with Indy-500 racers, but on slippery ice they need a fair amount of skill. While steering with just one hand, they must operate controls that not only break up ice and snow jamming the conveyor unit but also adjust water flow, raise and lower the blade, and convey shaved ice to the dump tank.

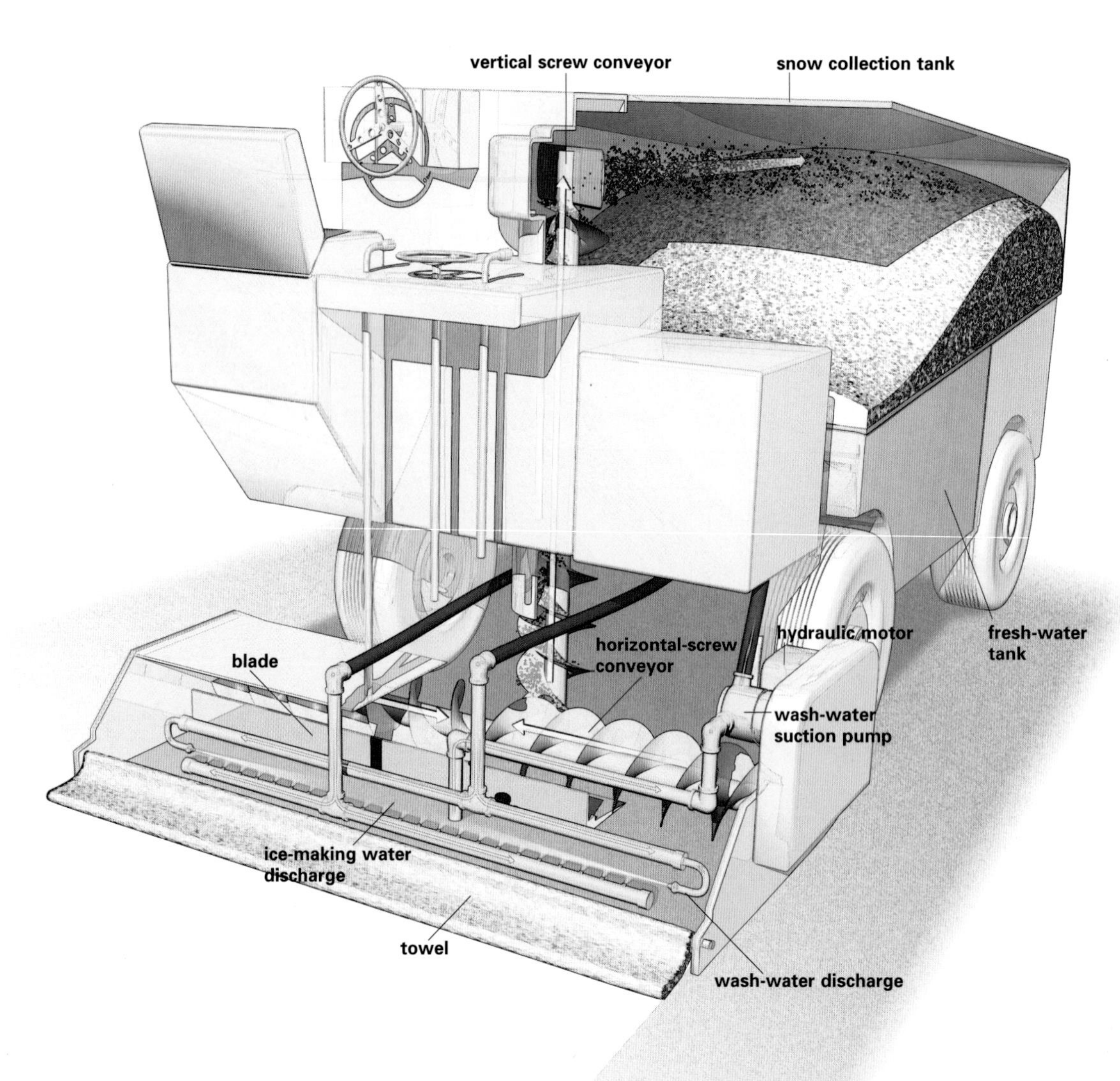

ICE PHYSICS

(Right) What is it about ice that enables Michelle Kwan to glide on it so effortlessly, as she does during this 2004 figure-skating championship event in Atlanta? The answer is that ice has a very low coefficient of friction, which makes it slippery enough for a person to skate or iceboat on. So why can't we skate on glass? Because the weight of the skater's body melts the ice slightly under the skates, providing a thin film of water to glide on, a film that refreezes when the skater leaves the spot.

THE ZAMBONI

(Left) Invented by Frank Zamboni, this machine shaves time off old methods of restoring ice surfaces. While an operator drives it around a rink, its razor-sharp blade scrapes the ice surface. A horizontal screw gathers up ice shavings, and a vertical one propels them into the snow tank. The water tank feeds a squeegee-like conditioner that smooths the ice, while suction removes dirty water. A large "towel" spreads clean, hot water, renewing the ice.

SEE ALSO
Washers & Dryers · 22
Home Heating & Cooling · 32
Garden Maintenance · 40

ROLLER COASTERS

DEFYING GRAVITY

(Left) A roller coaster hurtles through its paces in Gurnee, Illinois. Created to satisfy the human craving for thrill rides, roller coasters operate with relative safety: According to the International Association of Amusement Parks, a rider has only a 1-in-4 million chance of receiving a serious injury and a 1-in-300 million chance of a fatal injury.

WHIRLWIND RIDE

(Right) The faces of these thrill-seekers say it all—which is that rides that twist, jerk, shake, and otherwise stress the human body may be scary to some, but for others they offer a nerve-tingling escape from immobile boredom.

THEY HAVE INTIMIDATING NAMES: MEAN Streak, Millennium Force, Dragon Kahn, Cyclone, and Beast. They can move you hundreds of feet up a ramp and propel you to the bottom at a 65-degree angle in a stomach-churning, bone-shaking plunge that takes only seconds. Roller coasters are the ultimate thrill ride, a mode of g-force transportation in open cars reaching one hundred miles an hour, and they are designed for one purpose only—to give riders an adrenaline rush.

The car-trains that slowly drag you up the tracks of a wooden or steel scaffold in anticipation and apprehension, and then cause you to free-fall around pretzel-like twists and turns in pure fright, have their origin in sledlike, hollowed-out blocks of ice that raced down ice slides in 15th-century Russia. Then, as now, they fulfilled an age-old lust for heart-stopping sensation.

The first roller coaster built in the United States was conceived in 1884 by LaMarcus Thompson, who called it the Gravity Pleasure Switchback Railway. Operated at Coney Island, New York, its cars were hauled manually to the top of a 45-foot incline before being released on a six-mile-an-hour ride; passengers climbed a flight of stairs to get to the cars.

Today's roller coasters are basically still gravity machines that follow Isaac Newton's first law of motion, which dictates that a body in motion keeps moving until someone, or a force such as friction, applies the brakes. Drawn to the top of an incline by a motor and an attached chain, a roller coaster rolls along on the sheer momentum—the kinetic energy—from the first descent. The ride ends when friction and wind resistance slow the cars. In an effort to create the illusion of higher speed as the roller coaster slows down, ride engineers have added sharper curves at the end to take advantage of residual energy. Linear induction motors are being used to achieve faster speeds and help the roller coaster maintain speed as long as possible. One of the world's fastest is a Japanese machine that hit 107 miles an hour on its maiden run, which included a close to vertical drop of 170 feet before hurtling around a bend.

SEE ALSO
Steam Engine Trains · 84
High-Speed Trains · 90

FIREWORKS

Fireworks have elicited oohs and aahs from delighted watchers ever since the ancient Chinese gave the world an early form of rockets, Roman candles, and fiery pinwheels. The Chinese may have invented pyrotechnics for use as weapons or to scare away demons, but not much time went by before they realized that fireworks would play very well as outdoor entertainment.

Even in the tenth century, fireworks technology was not so different from that of today; it relied on gunpowder mixed with various chemicals to provide color, and it included metal shavings that would create the sparkle effect. Fireworks now are generally tube-shaped or spherical paper containers packed with explosive powder and a time-delay fuse. The containers also hold strategically placed packets of metallic salts: Lithium or strontium produces red; barium nitrates make green; copper compounds result in blue; sodium creates yellow; charcoal and steel produce sparkling gold; and titanium makes white. Powdered iron, aluminum, or carbon produce the sparks and other special effects.

A pyrotechnist places a fireworks shell in a mortar or a special gun and lights the shell's main fuse, using either a road flare or a computer-controlled electric switch. The lift charge ignites and propels the shell high into the sky. Next, the delay fuse ignites the various compounds, creating patterns and shapes based on the placement of the chemicals within the shell. Nowadays, a computer in a firing station might transmit electrical ignition orders every millisecond to fuses that light the rocket and control the bursts of sparks; the speed of the firing signals that light one rocket after another in such rapid succession is what creates the seemingly endless luminescent showers. Not all fireworks are launched, however. Some of the more complicated ones stay on the ground, or are attached to poles or trees, and are spun about on wheels, their motion produced by the recoil as the fire escapes from cases; so-called floral fountains send a shower of sparks upward, and the familiar hand-held sparkler sprays its fire as pyrotechnic materials coating a stick burn from the top down.

No matter how the displays are generated, however, one thing is certain: The imprint in our mind's eye lasts much longer than the life of the shell and its beauty.

BRILLIANT BLOOMS

(Right) Floral fire paints the sky as a time-delay fuse detonates fireworks in a carefully timed, multicolored light show that never fails to evoke oohs and aahs.

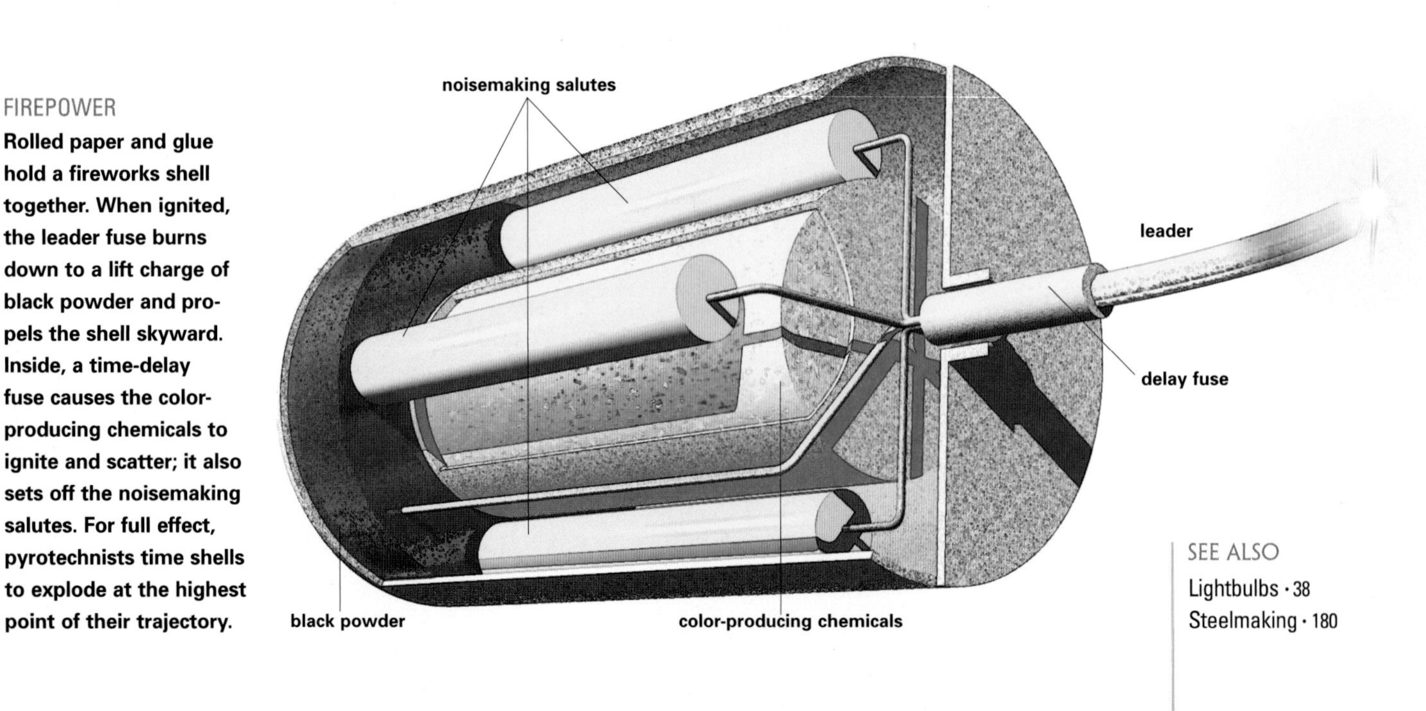

FIREPOWER

Rolled paper and glue hold a fireworks shell together. When ignited, the leader fuse burns down to a lift charge of black powder and propels the shell skyward. Inside, a time-delay fuse causes the color-producing chemicals to ignite and scatter; it also sets off the noisemaking salutes. For full effect, pyrotechnists time shells to explode at the highest point of their trajectory.

SEE ALSO
Lightbulbs • 38
Steelmaking • 180

CHAPTER 8

MINING & MANUFACTURING

"IF YOU HAVE GREAT TALENTS, INDUSTRY WILL improve them," English portrait painter Sir Joshua Reynolds once said. "If you have but moderate abilities, industry will supply their deficiency." His remarks came during a 1769 address to students of the Royal Academy, and his use of "industry" referred to diligence in any pursuit. But it also could have applied to the industrial revolution, which introduced power-driven machinery to late 18th-century England and forever changed the workings of the factory system. Today, the industrial world employs the modern tools of technology and science to manufacture products, extract value from natural resources, and create new versions of the tools themselves. Increasingly, manufacturers are relying on information technology to help them improve efficiency and become more competitive in the global market.

Coal-mining equipment glows at dusk during a worker's watch.

LUMBER

ACCORDING TO AN OLD SAYING, HE WHO SPLITS his own wood warms himself twice. This adage seems a major understatement if you consider the things lumber does for us. In the United States, for example, we take more than 30 billion board feet of lumber from our forests to build 95 percent of our homes. The average citizen uses about 750 pounds of wood-derived paper each year, while many thousands of tons of cellulose-based textiles and chemicals are used throughout the world. Without wood, our planet might not even be environmentally sound: For every ton of wood a forest grows, it removes 1.47 tons of carbon dioxide—which in excess contributes to global warming—and replaces it with 1.07 tons of oxygen.

Industries and individuals own some 70 percent of the nearly 750 million acres in U.S. forests, and the net growth in those forests exceeds harvesting by around 35 percent if you combine the number of naturally regenerated trees with that of seedlings planted each day. The sawmill industry depends on harvesting, which is the starting point for the many conversions wood undergoes. After timber has been logged and debarked, it is sawed into boards that are trimmed and edged, graded and dried, and planed and treated with decay-preventing chemicals. Steam-heated logs are spun on lathes and skinned to make veneers that, in turn, are glued together to form plywood. Particleboard and fiberboard are made from wood residues mixed with resins and wax. Wood chips cooked into pulp make paper, and digested cellulose eventually becomes rayon, proving that "wooden" doesn't always mean stiff and lacking flexibility.

CLEAN CUT
(Left) A circular saw is but one of the many cutting tools that loggers and carpenters use to transform forest wood into a vast industry.

LUMBER
(Right) Felled trees become mountains of logs at a lumber mill in Oregon. Transformed by industry—pulped for paper, dressed into boards and timbers, and processed into rayon—wood and its products help clothe, house, warm, and inform us.

SEE ALSO
Saws & Lathes • 26
Elements of Construction • 60
Synthetic Fibers • 132

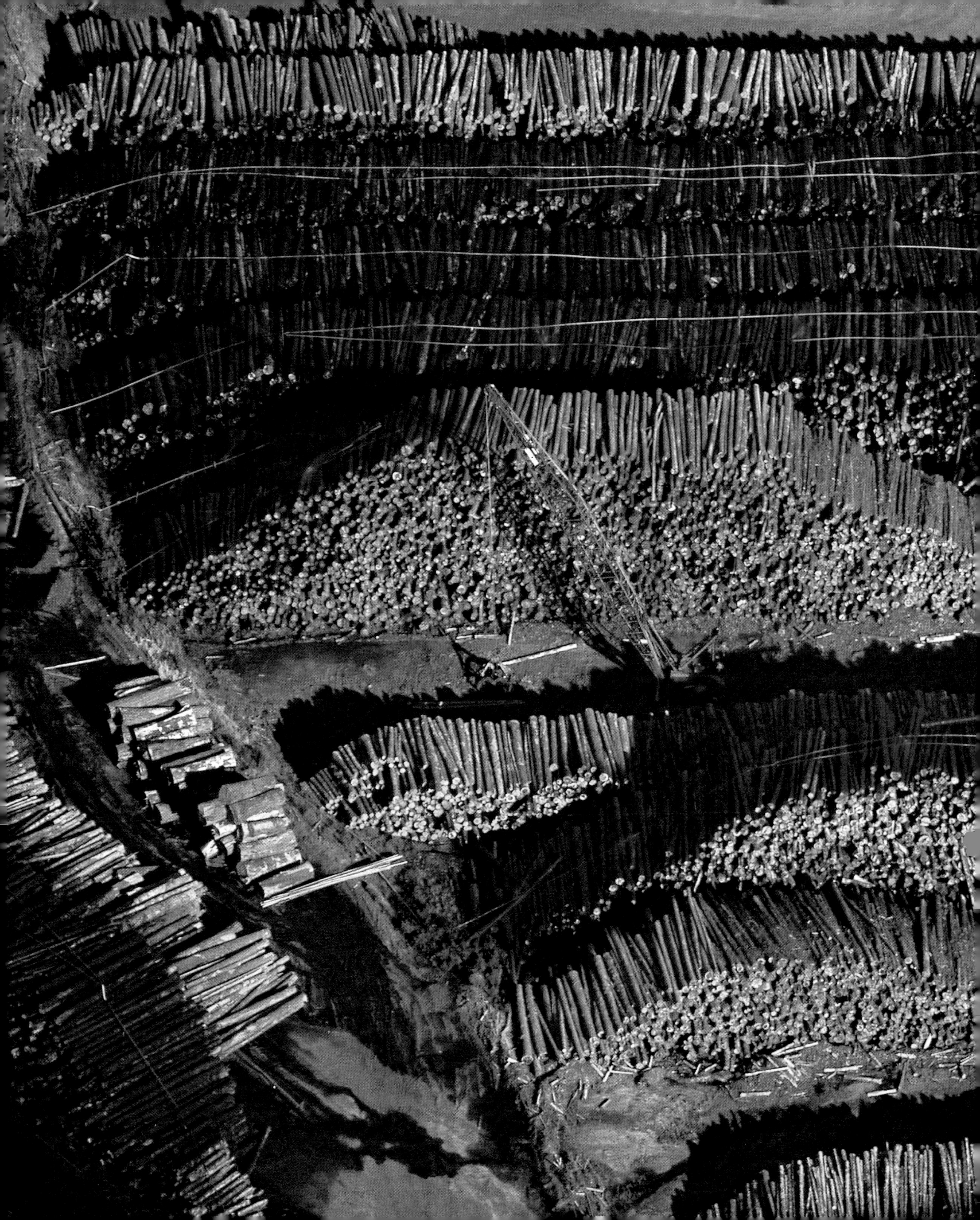

PAPERMAKING

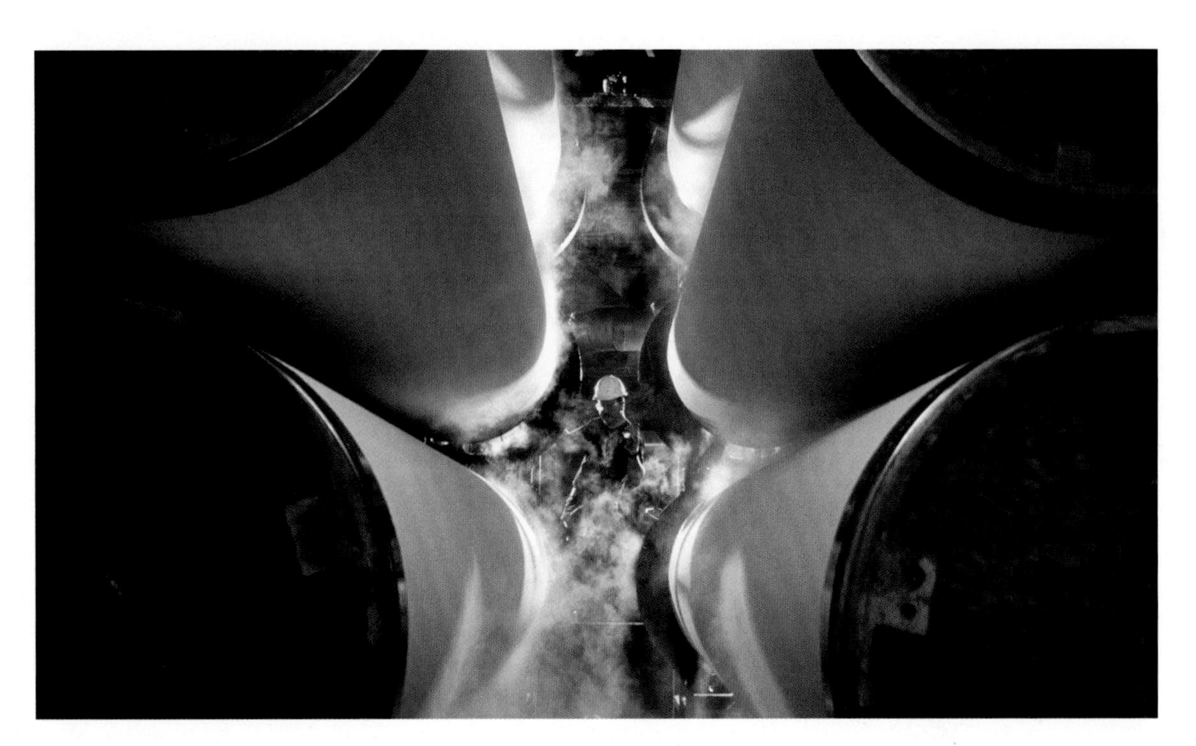

ROLLING STOCK

(Left) Dwarfed by the huge rollers in a paper mill, a worker checks some of the elements needed to make the perfect paper: temperature, pressure, proper drying, coating, and smoothing.

QUALITY CONTROL

(Right) A paper-mill specialist meticulously examines what may be an understatement given its enormous size—a "sheet" of paper. Before newly made paper is removed from its take-up rolls and cut and trimmed into the down-sized sheets we all know, it must pass stringent specific requirements for color, texture, weight, and strength. Some rolls may be sold directly to a customer after inspection, while others of a different grade go through another smoothing and polishing process, which might include custom embossing.

THE OLD ADAGE "NOT WORTH THE PAPER IT'S printed on" pays homage of sorts to one of the most valuable, capital-intensive, and multipurpose products of manufacturing. Paper still rules the high-tech world despite repeated dreams and promises of a computerized, paperless society. Even "legal tender for all debts, public and private" is made from it. Derived from the word papyrus, a tall water reed that the ancient Egyptians layered, pressed, and dried to make sheets for writing, paper as we know it today was actually invented by the Chinese around the second century B.C. Made from a boiled mush of plant fibers, discarded fishing nets, and rags, primitive paper was often used as clothing, and in lacquerware, kites, and currency.

Modern papermaking relies on the same general principles favored by the ancients, except that today's sprawling mills employ a variety of mechanical and chemical methods to prepare an oatmeal-like slurry from "cooked" hardwood and softwood chips (rags may be used for high-quality paper, and waste paper for recycled pulp). During the process, the pulp is "digested" to remove unwanted materials that can discolor paper, then bleached, more often now with environmentally safe, chlorine-free agents. The pulp is then beaten in machines called refiners to improve its quality and strength, rinsed, and spread onto a moving belt made of wire or plastic mesh that shakes off the water and mats the fibers.

As the now drying web of paper moves from the belt's "wet" end to the "dry" end, it passes through rollers and then through heated pressure drums that remove the remaining moisture. Various coatings—dyes, sizing agents, and fillers—may be applied at this stage to impart shade and color, gloss, or other characteristics. The mass of intertwined fibers, now paper, is then sent through another set of metal rollers for "calendering," a process that presses and smooths it. At the end of the line, the paper is wound into rolls, cut into sheets, and trimmed by automated equipment. Packed as reels or reams, the finished product is stored on pallets in light-protected areas, where it is eventually sorted for delivery by computerized machinery.

SEE ALSO
Saws & Lathes · 26
Elements of Construction · 60
Synthetic Fibers · 132

GLASSMAKING

We peer into glass to see ourselves and through it to see others, for this material is both reflective and transparent, depending on how it is treated. Its basic composition can be reduced to three essential ingredients melted together at high temperatures: sand, soda, and limestone. In more elaborate chemical terms, this translates into silicates and an alkali flux, with metallic oxides added for color.

Glass is similar to most other solid materials except on a microscopic level. There, it lacks the orderly molecular arrangement of true solids, a disorder that makes it resemble a liquid. Indeed, it has been termed "the rigid liquid," a reference to its high viscosity—the property of a fluid by which it resists shape change or relative motion within itself. Glass is transparent because its atomic arrangement does not interfere with the passage of light. (A glass mirror reflects because a thin layer of molten aluminum or silver is applied as backing.)

The actual manufacture of glass dates back to 3000 B.C., when Egyptians glazed ceramic vessels with it. Much later, the Romans made glass for utilitarian and decorative purposes. The art of stained glass, made with metallic oxides fused into it, flourished in the Middle Ages. Today, glass is made in large crucibles in furnaces where the melting temperature reaches 2900°F. Skimmed of impurities and cooled, molten glass may be poured into molds and pressed, blown, cast for mirrors and lenses, or "floated" or drawn to produce window glass or tubing. Shaped glass goes through an annealing process to fix colors and to remove internal stresses and make it less brittle. (Some products that require high strength, such as glass doors or eyeglasses, are specially tempered. This rapid cooling process is the reverse of annealing in that it induces high, permanent stress.) The cooled glass is then ground, polished, bent, laminated, or decorated. Computer-operated cutters slice the glass into different shapes, while other automated equipment grinds edges, drills holes, bevels, and inscribes. Production speed of many items is striking: Machinery can now produce over a million glass bottles a day.

BOTTLES

(Right) Glowing from their heat treatment, new glass bottles take shape on a production line. Except for automation and new materials, the manufacture of glass bottles has changed little since people in Egypt and Mesopotamia made glass vessels many centuries ago.

GLASS

(Left) Under the curious eyes of tourists, a glassblower in her shop in Cannon Beach, Oregon, plies a time-honored trade. The ancient Romans and Syrians, among others, learned how to make glassware by firing and melting one end of a glass tube and then blowing through it to form a bubble that could be shaped. Later, metal blowpipes were used to inflate a "gather" of molten glass at the end to produce various shapes while the glass is still soft, or to blow the liquid glass into a mold.

SEE ALSO
Plastics · 184
Recycling · 186

MINING COAL

OPEN PIT MINE
(Left) Resembling an ancient amphitheater, an open pit, or surface, mine like this one is advantageous when the coal or metal ore is relatively close to the surface. Open mines provide more than 60 percent of U.S. coal.

COAL-MINING FACILITY
An aboveground installation contains life-support systems, processing and loading equipment, and a head gear for lifting machinery and coal. The mining operation begins when a road-header—a vehicle having large cutting heads—opens a path to coal seams. Leaving columns to support the roof, continuous-mining machines dig out the coal, and conveyor belts carry it to coal skips, open cars drawn to the surface through shafts. Companies estimate how much coal a seam will produce by using electronic coal-sensing probes to measure the thickness of the deposit.

THE WORLD'S VAST STORES OF COAL WERE probably laid down some 300 million years ago during the Carboniferous period, when the remains of plants decomposed and compressed to form the hard, black substance we burn as fuel. The U.S. alone has trillions of tons of this organic rock, and it is enough to take us, at the current rate of consumption, well into the 24th century.

One way coal is extracted is by excavating through strip mining, a hotly debated process that removes surface material from mountaintops and other areas to expose coal seams or beds. The equipment is mammoth: A typical truck's bumper might be as high off the ground as a worker's helmet, and the load carried may be about 250 tons a trip. Critics of this type of mining argue that it not only mars the landscape but produces acidic runoff and loads streams and other nearby water with debris.

Underground mining is the other major method of extraction. Entered by tunnels or shafts dug into the ground, miners, paid by the tons of coal they loaded, had to muscle out their coal carts. Today, new roof-bolting technologies are used, and mines are drained, ventilated, and fitted with mechanical and electronic equipment to make them safer.

A typical underground mine is excavated with the room-and-pillar method. Rooms generally 20 to 30 feet wide are cut into the coal bed, leaving huge pillars, or columns, to support the roof and control the airflow. The cutting is done by a machine called a continuous miner, which obviates the need for blasting and drilling. When the cutting reaches the end of the line, workers begin "retreat mining" in which they pull as much coal as possible from the remaining pillars until the roof collapses.

Coal is not only pulverized for power generation, but coal gasification partially oxidizes coal to produce a gas that can be used to fire a gas turbine to produce electricity. Coal liquefaction converts the coal into liquid fuel by removing carbon or adding hydrogen; refined, the liquid can be used in the manufacture of plastics and solvents.

SEE ALSO
Power Stations · 44
Tunnels · 70
Steam Engine Trains · 84
Drilling for Oil · 182

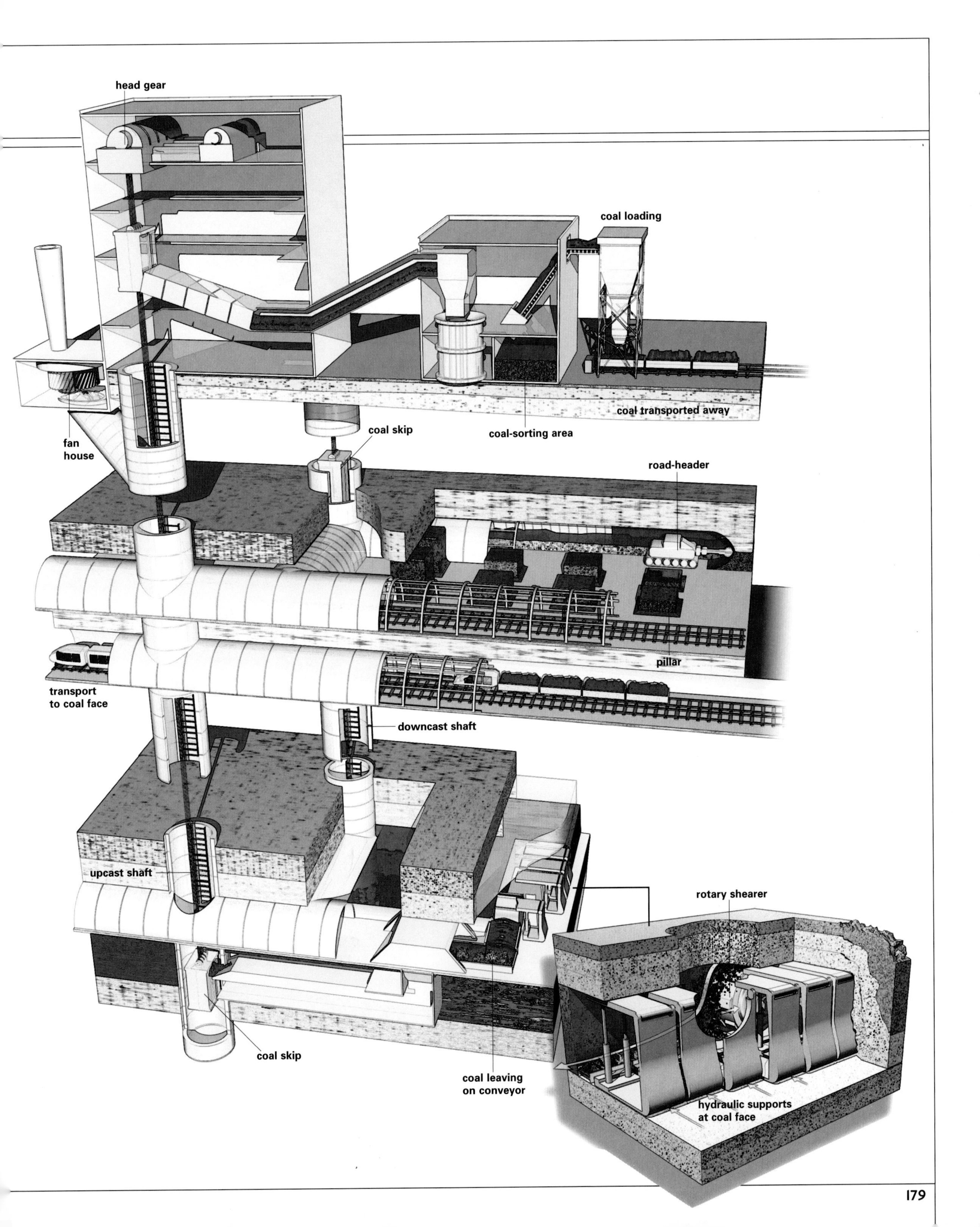
head gear
coal loading
fan
house
coal skip
coal-sorting area
coal transported away
road-header
pillar
transport
to coal face
downcast shaft
upcast shaft
rotary shearer
coal skip
coal leaving
on conveyor
hydraulic supports
at coal face

STEELMAKING

Few industries can match the massive, complex equipment and the awesome spectacle of steelmaking. From raw and dirty lumps of iron ore to sheets of shiny, corrosion-resistant stainless steel, the process is a meld of intense heat, blasts of high-pressure air, molten metal, violent boiling and bubbling, arcs of electricity, and the din of forges and rollers. First mass-produced in the 19th century, steel is a mainstay of ships, automobiles, and skyscraper frames.

If the process of steelmaking could be reduced to a few terms, those terms would include the blast furnace, Bessemer converter, and open-hearth furnace.

Steel is an alloy of iron and a tiny amount of carbon and other elements, a mix that causes the iron atoms to bind together tightly and produce a material that is even harder and tougher than iron. To create steel or to make cast or wrought iron, the iron is first extracted by smelting oxide ores mixed with coke (a form of carbon) and limestone in a blast furnace—a towering, cylindrical stack where a blast of air fuels combustion. In molten form and full of impurities (pig iron), it can be made into cast or wrought iron.

To convert the pig iron to steel, it can be refined in a Bessemer converter, which uses hot compressed air to remove impurities. An open-hearth furnace may be used instead. Stainless steel, a long-lasting alloy containing at least 10 percent chromium, is generally produced in an electric-arc furnace, an apparatus that produces batches of molten steel known as "heats." Carbon electrodes supply energy to the furnace's interior, where iron is melted so that it can be mixed with stainless steel scrap, chromium, and other elements, such as nickel and molybdenum. The mix is cast into ingots or a slab and then hot-rolled, cold-rolled, or forged into final form.

INDUSTRY SPARKS

(Right) Sparks and smoke define the process of turning molten metal into steel. Melted iron seethes at a temperature of 2800°F, when it will be poured from a cauldron into a steelmaking furnace. Manufacturers can use Bessemer converters, or electric-arc or open-hearth furnaces.

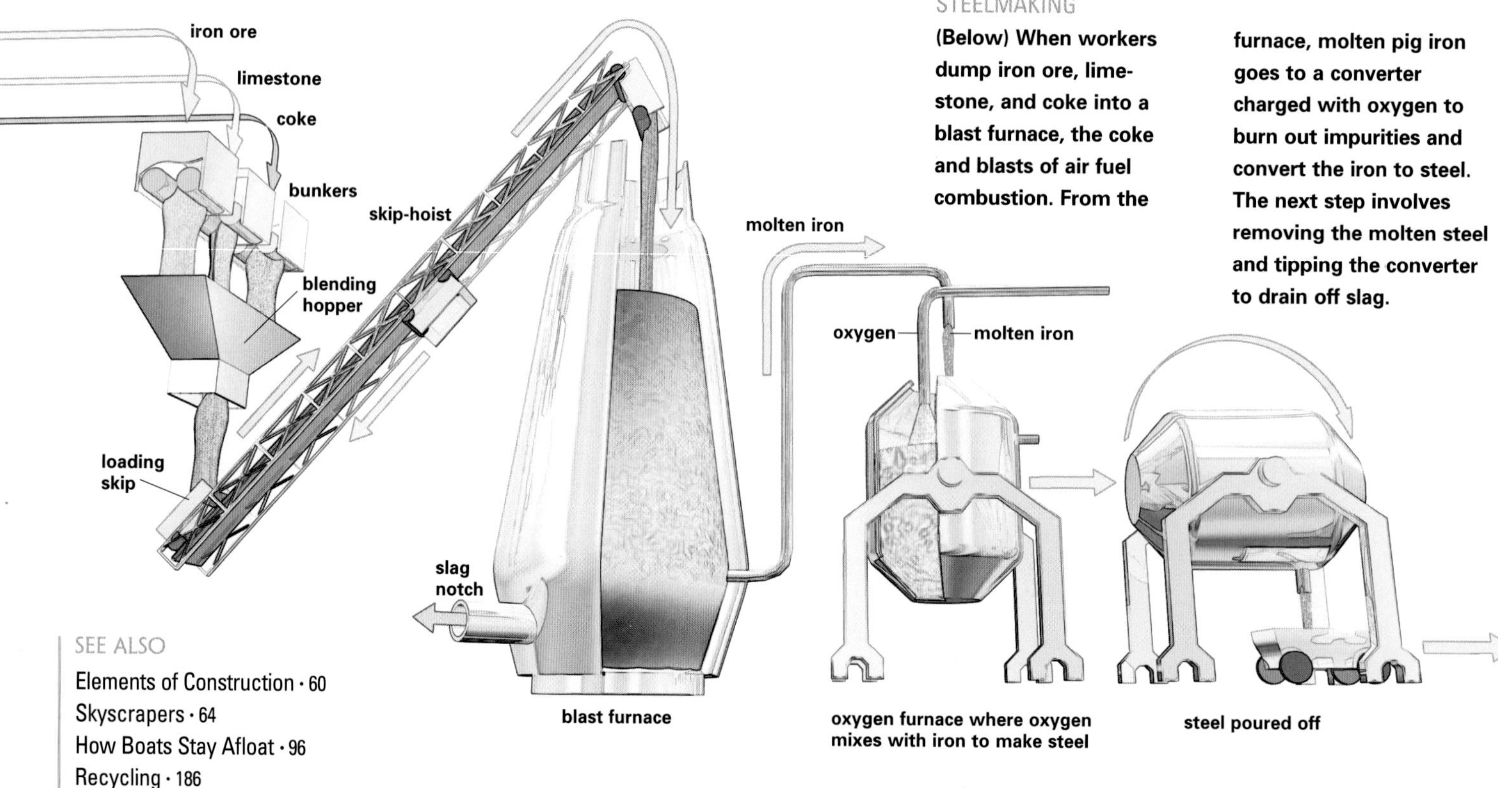

STEELMAKING

(Below) When workers dump iron ore, limestone, and coke into a blast furnace, the coke and blasts of air fuel combustion. From the furnace, molten pig iron goes to a converter charged with oxygen to burn out impurities and convert the iron to steel. The next step involves removing the molten steel and tipping the converter to drain off slag.

SEE ALSO
Elements of Construction · 60
Skyscrapers · 64
How Boats Stay Afloat · 96
Recycling · 186

DRILLING FOR OIL

BRIGHT LIGHTS

A gleaming petrochemical plant and its dark towers dominate a landscape. The stuff of an endless range of products from plastics to pesticides, detergents, nitrogen fertilizers, and explosives, petrochemicals make up a large group of organic and inorganic compounds spawned by oil and natural gas.

A COMPLEX MIXTURE OF HYDROCARBONS THAT formed from the organic debris of long-dead plants and animals, petroleum is the oily, flammable liquid of power and petrochemicals. Pooled deep in the ground, within layers of rock under intense pressure, it is extracted by a drilling rig, a series of rotating pipes supported by a derrick. But first, oil must be found. After sensing devices guide drills to where oil is located, the underground reservoir is tapped; oil tends to burst out explosively, a tendency that used to provide scenes of black gushers spewing into the air. But today's technology and environmental sensitivity have made such sights a memory. Drilling rigs—which probe to depths as great as 25,000 feet—pump a flushing mud into the ground to carry debris to the surface and prevent oil from erupting.

After a well is drilled, the oil usually flows up under its own pressure and into a separator, where gravity helps remove the oil from the mix of briny water, natural gas, and sand that came along with it. The cleansed crude oil then is sent through an array of pumps, compressors, and dehydration towers to make it suitable for refining; at the refinery, petroleum undergoes fractional distillation, which produces gasoline, kerosene, diesel and fuel oils, lubricants, and asphalt.

In the early days of prospecting for oil, a reservoir often was discovered by chance when, say, a farmer spotted oil mixed with the water in his well. Nowadays, geologists examine soil and rock with the help of satellite images, gravity meters, and magnetometers, which analyze changes in the Earth's gravitational and magnetic fields that suggest the presence of oil flow. Electronic "sniffers" also pick up the smell of hydrocarbons, and seismometers assess shock waves created to detect hidden deposits. Ultrasound waves provide color-coded seismic data that show images of rock formations in the ground or seabed. In addition, when oil is found, it is transported far more easily than in the past: Each day, for example, the Alaska pipeline brings a million barrels of crude oil over 800 miles of wilderness to tankers that can carry several thousand metric tons apiece.

SEE ALSO
Home Heating & Cooling • 32
Power Stations • 44
Alternative Fuels for Cars • 80
Plastics • 184

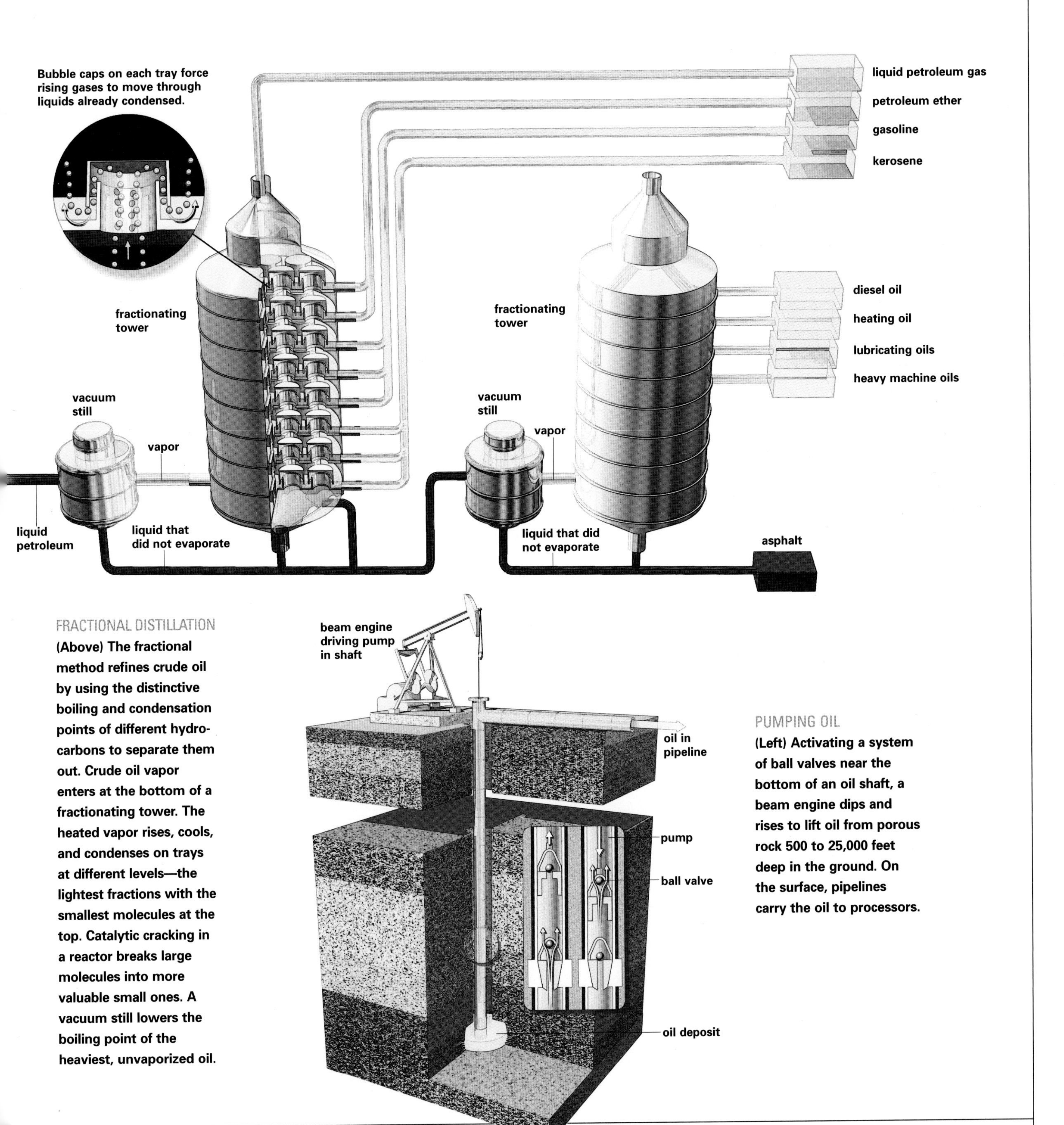

FRACTIONAL DISTILLATION

(Above) The fractional method refines crude oil by using the distinctive boiling and condensation points of different hydrocarbons to separate them out. Crude oil vapor enters at the bottom of a fractionating tower. The heated vapor rises, cools, and condenses on trays at different levels—the lightest fractions with the smallest molecules at the top. Catalytic cracking in a reactor breaks large molecules into more valuable small ones. A vacuum still lowers the boiling point of the heaviest, unvaporized oil.

PUMPING OIL

(Left) Activating a system of ball valves near the bottom of an oil shaft, a beam engine dips and rises to lift oil from porous rock 500 to 25,000 feet deep in the ground. On the surface, pipelines carry the oil to processors.

PLASTICS

If wood is one of nature's great genuine articles, plastic is the welcome, artificial usurper, a take-charge material with a seemingly endless capacity for changing its shape and function. Most plastics are by-products of petroleum or coal, which are organic materials, but they are considered artificial because they are made rather than grown.

Plastic came into being more than a century ago when a $10,000 prize was offered to the person who could create a substitute for ivory in billiard balls. John Wesley Hyatt of New York, an inventor-entrepreneur, entered the competition, and although he did not win, he came up with semisynthetic Celluloid, a mixture of nitrocellulose and camphor put under heat and pressure. Hyatt's invention was a commercial success that found immediate use in dental plates, eyeglass frames, combs, and men's collars.

Today, more than 50 varieties of plastic stand in for a host of materials, and it is hard to imagine anything spun or molded without a plastic presence. As Lucite or Plexiglas, plastic substitutes for glass in aircraft windows, car taillights, boat windshields, clock faces, and camera lenses. As Bakelite, it replaces rubber as an electrical insulator. As Corian it passes for marble, and as Styrofoam, for insulation and disposable food containers. Polyvinyl chloride (PVC) substitutes for metal in drainpipes; Teflon keeps food from sticking to skillets; nylon and other synthetic fibers replace wool and cotton.

Plastics generally start out as powders of polymers, the binder in a mix that includes plasticizers, fillers (talc or glass), and pigments. A recent advance developed at Penn State vastly improves various properties of plastics: natural clay. According to Evangelos Manias, a materials scientist at the university, clay particles are incredibly tiny, no more than three to five atoms thick; when these are dispersed in a polymer, they increase its strength and resiliency and improve its ability to keep gases and liquids in or out, a boon to food-packagers. The raw materials that go into plastics are compressed in molds under heat and pressure or poured into cold molds to harden. They can be squeezed between rollers or extruded through a die to be cut in lengths or coiled. Because they do not decay, plastics present environmental problems. Often maligned, they are nonetheless remarkably useful creations that are here to stay, for better or for worse.

PLASTIC AVENUE

(Right) A couple takes a dry stroll underwater at the Detroit Zoo's Arctic Ring of Life exhibit. The submerged tunnel is made of Lucite, a tough plastic glass substitute known chemically as polymethyl methacrylate resin.

VERSATILE BAGS

(Left) Plastic bags run through a printing process. Ubiquitous and discarded by the billions, plastic bags made of polyethylene from scratch or recycled bags hold groceries, zip up food for the fridge, and haul the trash. Roughly a thousand pounds of plastic material go into making 50,000 bags. Concerns about their role in littering, however—and the fact that they can survive for generations—have resulted in their being targeted by environmentalists as a menace.

SEE ALSO
Synthetic Fibers • 132
Lumber • 172
Glassmaking • 176
Drilling for Oil • 182
Recycling • 186

RECYCLING

JUNK CARS
(Left) In what could pass for contemporary sculpture, crushed cars stacked like cordwood await recycling. The most recycled high-volume consumer throwaway in the U.S., junk cars produce million of tons a year of steel scrap that can be melted down and used again. Even the fluids that are drained out—antifreeze, fuel, and oil—can be reused.

USABLE TRASH
(Right) Sorting and sifting, a recycler gets something back from things thrown away. Recycling reduces the enormous expense of disposal and incineration; it also provides a new resource.

According to waste industry estimates, Americans generate nearly 230 million tons of household trash a year. Included in the massive dumping are billions of batteries, razors, blades, and disposable diapers, along with millions of tons of paper, glass, plastic, leaves, and grass clippings. If we were somehow able to place all our discarded beverage cans one on top of another, they would reach the moon almost a score of times.

Fortunately, recycling helps. It can be done in a number of ways, such as reusing leftovers after a manufacturing process. Steel scrap, for instance, can be used to make stainless steel, or glass scrap may be remelted to make more glass. One Japanese manufacturer, Mitsubishi Rayon, handles thousands of tons of waste plastic a year, using a heating technique to break down acrylic plastic from a complex polymer to a simpler form that becomes the raw material for more acrylic plastic. The process can convert as much as 85 percent of scrap plastic into its reusable, basic form.

Another method reclaims materials from worn-out items, using cardboard boxes and newspapers, for example, to make business cards, paper towels, and more newspapers. Lead batteries can be made into new batteries; steel cans make more steel; discarded glass is remelted to make new bottles. Precious metals such as gold, silver, palladium, and platinum can be recovered from printed circuit boards, plated and inlaid metals, and photographic wastes.

Old rubber can become tennis shoe soles, roofing and road construction materials, running tracks, brake linings, floor coverings, and even TDF (tire derived fuel). To put old tires to good use, they are ground into bits, and then chemically processed to break down their sulfur content and rearrange chemical bonds in the new rubber. Retreading involves stripping away the old tread and bonding a new one on the surface with heat and pressure.

SEE ALSO
Synthetic Fibers • 132
Glassmaking • 176
Steelmaking • 180
Plastics • 184

INDUSTRIAL ROBOTICS

INDUSTRIAL ROBOTS ARE TRUE TO THE ORIGIN OF their name: "Robot" derives from the Czech word *robota,* which means work, or slavery. Task-oriented machines guided by mechanical and electronic means, they perform functions ordinarily done by humans and are especially useful when the job is boringly repetitive, dangerous, or heavy-duty like polishing door handles or helping police to dispose of bombs.

But contrary to robotic images in sci-fi films, these factory work-hands do not physically resemble humans, though they may have mechanical "hands" and "wrists" with segmented gripping "fingers." They run on high-efficiency electric motors, hydraulic systems, compressed air, sensors that respond to physical stimuli, and solenoids, which are current-carrying devices that convert electrical into mechanical energy. Robots do not think for themselves but get their brainpower and directions for what they can do and how to do it from auxiliary computers or embedded microprocessors, and computer programs. They come in a variety of sizes and nondescript shapes, and many configurations that allow for different axes of movement, or "degrees of freedom," as robot experts put it. Robots flex their steel muscular "arms" while ranged along an assembly line, or they do incredibly detailed work while mounted unobtrusively on a table or wall, or hanging from the ceiling on a gantry; some are mobile, like those used in undersea and space exploration, and may be controlled remotely by a distant human operator.

All this "technology of mobility" begins with the actuators—the motors, pneumatic systems, solenoids, and so on that enable the robot to pivot, reach, gyrate, grip, and grab in seemingly limitless ways. The manipulative arm of a robot is capable of moving up and down, in and out, and side to side. It can also perform various "wrist" movements, which include rotating clockwise and counterclockwise.

Robots can be set up to do their designated tasks by a "teach-and-repeat" method, through which an operator, or programmer, employs a portable control device to teach the robot its job manually, which includes, for example, walking the robot's arm through the various positions. The position and functional data obtained are then "memorized" by the robot's computer system for retrieval when needed. Robots also learn by being fed movement data and other essential information prewritten as a computer program; the information to be retained is transferred to the robot either directly from a control room or through storage devices.

To get the robot to go about its business, the programmed instructions are retrieved, and the computer or the embedded microprocessors switch on the motors and other actuators that translate the coded data into robot motion; if it becomes necessary to alter the robot's mission, it is simply reprogrammed using either online or off-line techniques that transfer new data to the computer via storage disks or telephone modems.

ROBOT AT WORK

(Right) A robotic arm tests a piece of equipment, a mechanical chore that can spot, test, and discard defective parts.

OUTSIDE HELP

(Above) Mechanical hands precision-weld thousands of spots on auto bodies, causing sparks to fly over a Honda assembly line in Japan.

SEE ALSO
Medical Automation • 202
Computers • 230
Computer Chips & Transistors • 232

CINCINNAT
ACRON

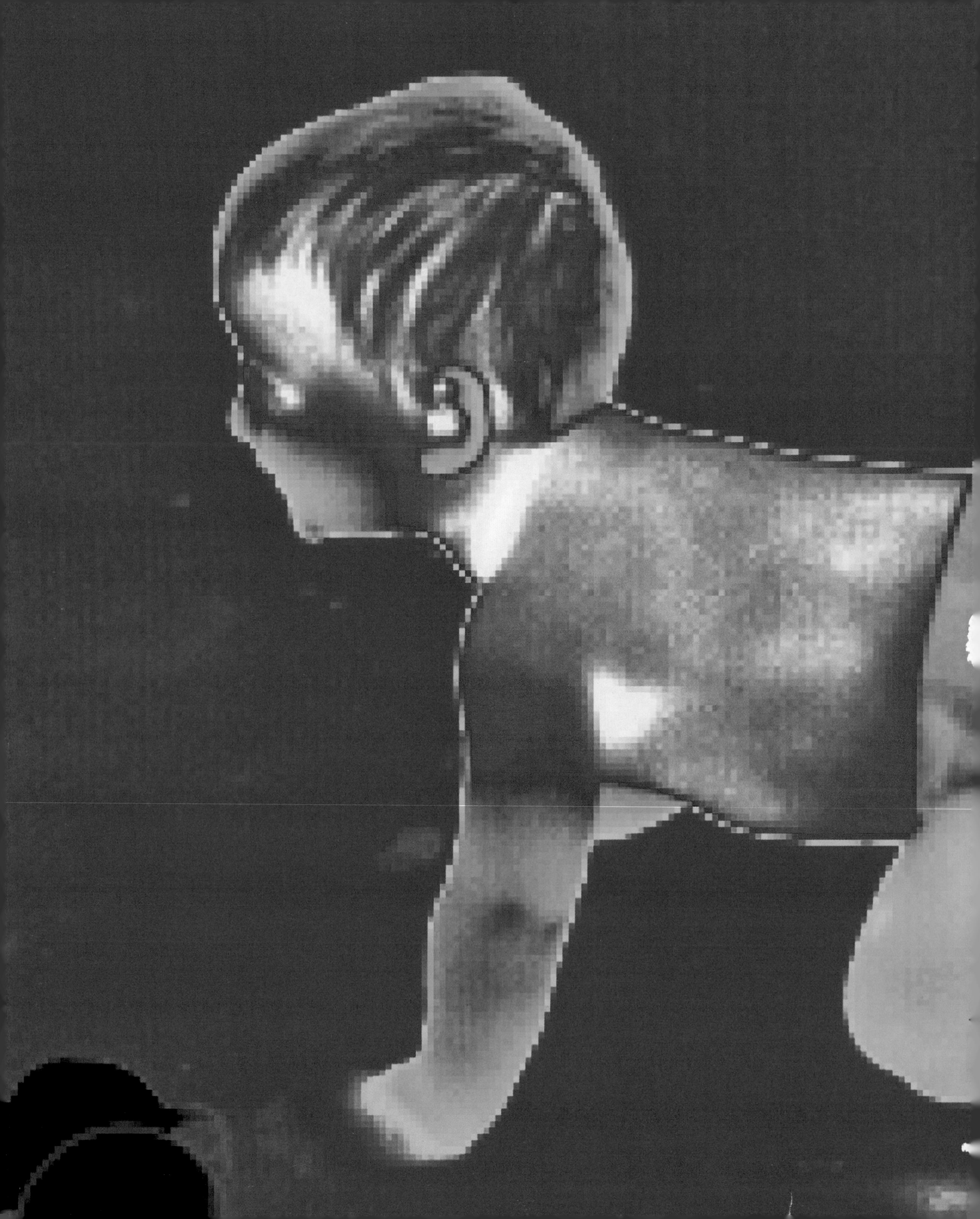

CHAPTER 9

HEALTH & MEDICINE

HEALTH CARE TODAY IS A WORK OF BOTH science and art. Sophisticated imaging systems see structures that eyes alone can never see; ultrasound hears a rush of blood that ears cannot. Satellites beam information to doctors who are far removed from the surgical scene, and robots wield surgeons' tools. Medicines are designed by computers and delivered by patches, sprays, and implants. Surgery is performed through tiny incisions. Reengineered genes are piggybacked onto viruses and steered through a patient's body to repair a defect or replenish a deficiency. Radiation pellets and rays treat cancers. Gerontologists are even exploring ways to tinker with the genetic clock, their aim being to slow it a bit. If they succeed, and if we can eradicate major diseases, we will live out our lives with perhaps a score or more years added, and we will do so in good health.

A diagnostic thermographic image shows the baby's cold feet in mauve and its hotter forehead in yellow.

X-RAY, CT, & MRI

If diagnostic equipment has a venerable grandfather, it is the x-ray machine. Still the quickest way for us to take off our skins and pose in our bones, as someone said many years ago, it images our skeletons and internal organs, using controlled beams of highly energetic electromagnetic rays discovered in 1895 by Wilhelm Conrad Roentgen, a German physicist.

X-rays pass through flesh and thick paper but are stopped and reflected by bones and metal, a characteristic that—when the rays are on their way to a negative photographic plate—results in an image of bones, in white. Various shadows and shadings help identify anomalies, such as breaks or fractures.

Diagnosticians need more refined tools, though, because x-rays can miss many structures and abnormalities, and they project only two dimensions. Computerized axial tomography, called the CT scan, has been hailed as the greatest advance in radiology since Roentgen's discovery. It emerged in 1972 largely through the research of Dr. Allan MacLeod Cormack. A South African native, Dr. Cormack was a professor of physics at Tufts University and shared a Nobel Prize for the rationale behind CT. Linking x-ray and digital technology, a CT scan shows the body in cross sections, from which three-dimensional images are constructed. An x-ray tube revolves about a patient's head, for example, converting the images into a digital code. Differences in density between normal and abnormal tissues are revealed, as well as bone details and the location of tumors and other signs of disease.

A step beyond the CT is magnetic resonance imaging (MRI), which also produces cross-sectional pictures. While a patient lies in a tunnel-like chamber, surrounded by electromagnets creating an intense field, the MRI unit reads the radio signals that return from hydrogen atoms in the water molecules of bodily tissues. These signals are fed into a computer and converted into detailed images of soft tissues, such as those in the brain and spinal cord. Refinements have enabled the MRI to look beyond what is simply structural or anatomic.

A version called fMRI, for functional MRI, is being used by neurologists to look at nerve cell activity and brain function, evaluate patients with Alzheimer's disease and schizophrenia, and assess changes in cerebral blood circulation and volume that accompany various diseases. As with all new medical technology, fMRI has benefited markedly from the development of extremely rapid imaging techniques that use sophisticated hardware configurations and special magnetic "pulse sequences" to provide images in a few seconds as compared with minutes in conventional apparatus. In fact, a development called echoplanar imaging (EPI) allows image acquisition in a fraction of a second.

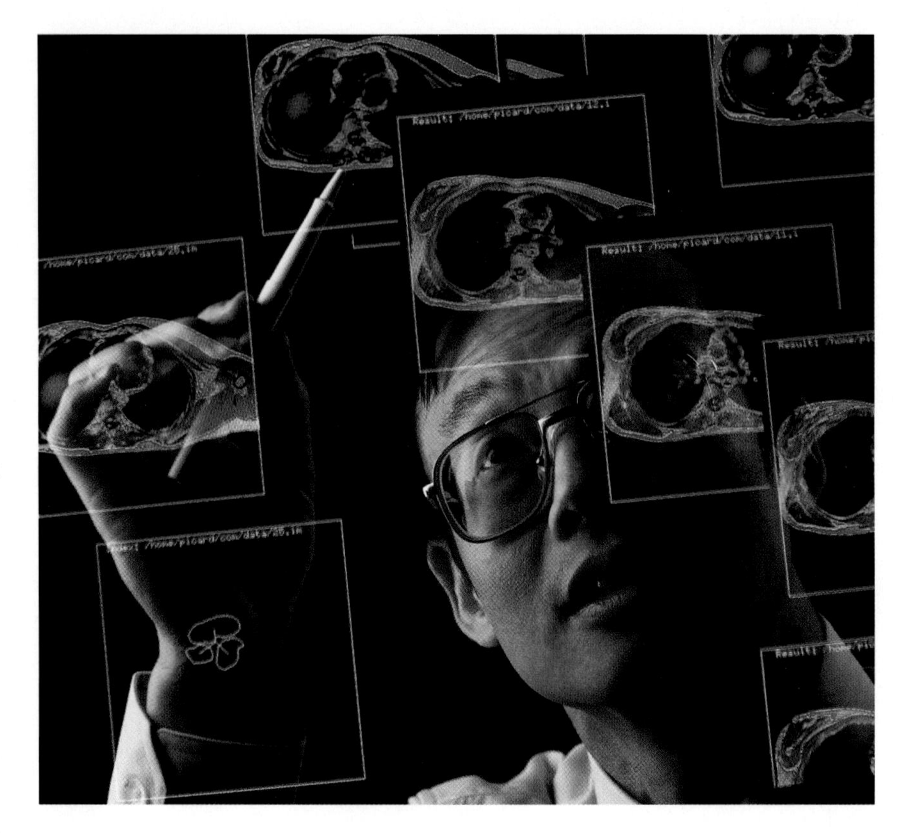

HEART SCANS

Digital scans of a human heart can detect atherosclerotic plaque in arteries and analyze valve function and a variety of conditions that damage or overtax the heart muscle.

SEE ALSO
Endoscopy · 196
Fiber Optic Cables & DSL · 210
Computers · 230

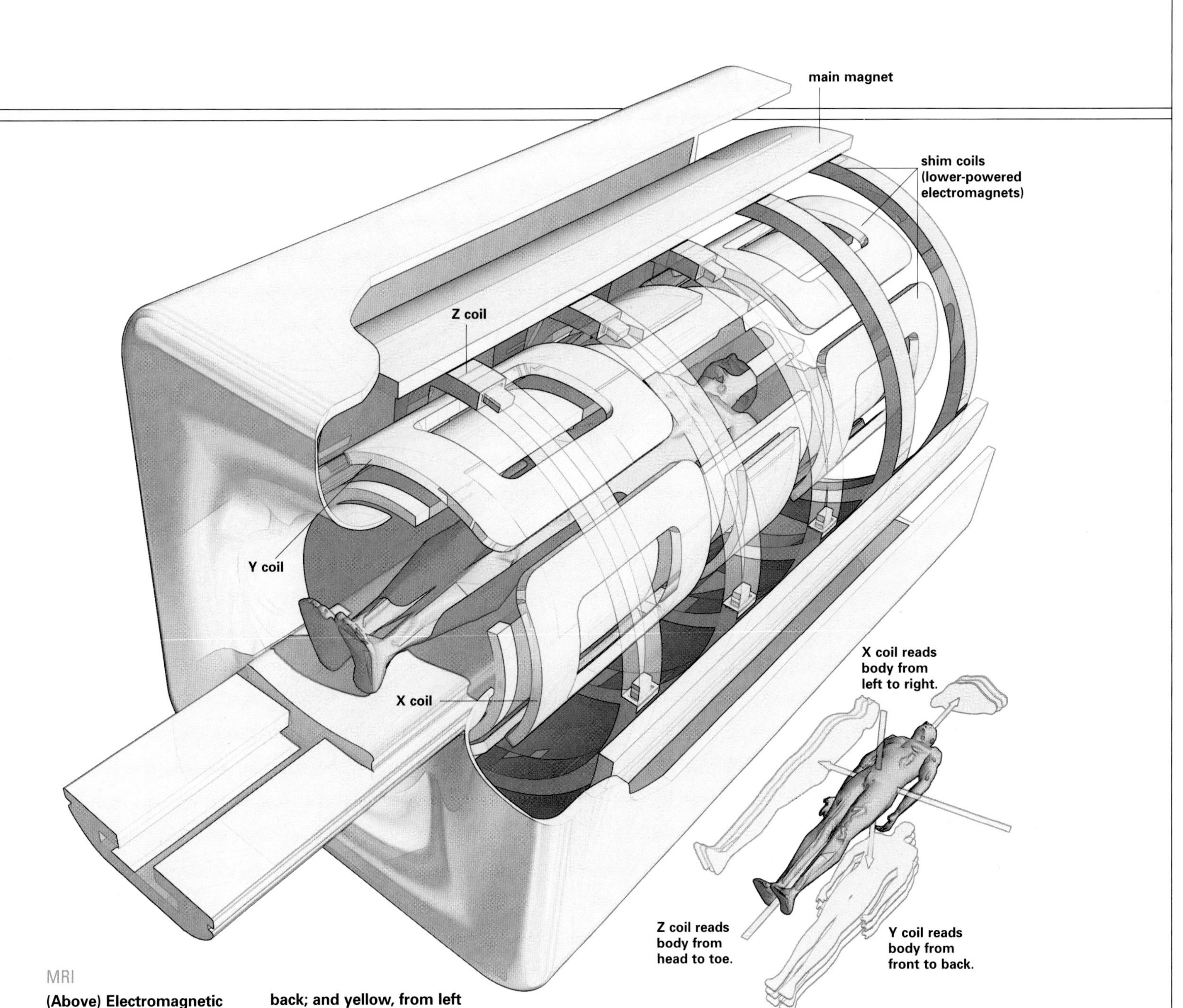

MRI

(Above) Electromagnetic coils in an MRI unit scan a body with radio waves, exciting hydrogen atoms in bodily tissues. The imaging device then reads signals returned by the atoms to create cross-sectional, 3-D views. Low-powered magnets, called shim coils, control the main magnetic field to vary field strength, setting up gradients in different planes. The red coils "read" from head to toe; green, from front to back; and yellow, from left to right. In this way, each portion of the body can be identified with magnetic coordinates and rendered into a computerized, sectional image.

LAID BARE

(Right) Under an MRI, a body resembles a model in a medical exhibit, not only because of its appearance but also because of the vital information it provides.

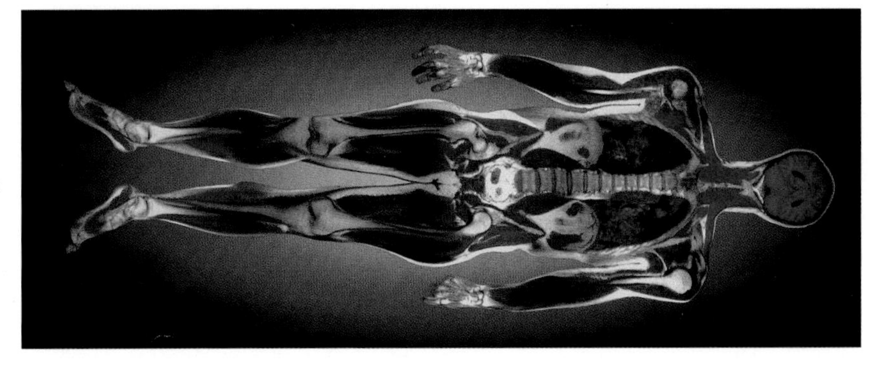

MEDICAL MONITORS

"IT IS QUITE IMPOSSIBLE FOR ANY MAN TO GAIN information respecting acute disease, unless he watch its progress," the eminent 19th-century English physician Richard Bright observed. "Day after day it must be seen; the lapse of eight-and-forty hours will so change the face of disease."

One of the foundations upon which medicine is built is, indeed, observation. A vast store of seemingly infallible medical monitoring systems has contributed enormously to that end. They hum and beep in hospitals' surgical suites, in emergency and patients' rooms, and in the intensive care unit (ICU), the place where the very sick, often with a life-threatening ailment, are carefully watched and treated.

Much of today's monitoring instrumentation is grounded in some of the old standbys. The electrocardiograph still traces the heart's electric current through electrodes pasted to the chest, legs, and arms, charting ventricular contractions and pinpointing anomalies that indicate disease and disorder. The electroencephalograph records electrical activity in the brain, looking for telltale peaks and dips that signal tumors. The pulse oximeter, attached to a finger or earlobe, searches for inadequate amounts of oxygen in the blood by monitoring the percentage of hemoglobin, the complex compound that sends oxygen to other cells.

Yet, medical monitoring instrumentation has become increasingly "smarter," computerized and transformed into either stand-alone apparatus, or networked to connect to a central nursing station where the data can be observed. The oximeter, for example, is a device that for all its simple appearance is a piece of electronic ingenuity. Linked to a computerized unit that displays the amount of oxygen-saturated hemoglobin, it also provides an audible pulse-beat signal, calculates the heart rate, and monitors the rate of blood flow. It does its job by analyzing the wavelengths of a light source that emanates from the probe; the light is absorbed in differing amounts by the hemoglobin depending on saturation or desaturation of oxygen. A computer then measures the absorption and determines the extent of the oxygenation.

Microprocessors, at the heart of the computerized systems, do away with yards of wiring, allowing for more precise measurement of blood pressure, temperature, pulse rate, and the amount of oxygen in the blood. Moreover, while in the past instruments that monitored such essentials were elaborate affairs, today a single computer chip can be programmed to take many measurements.

Other monitoring devices are portable, such as the battery-operated telemetry monitor that checks heart rate and other activity. Carried about in a pocket or holster like a cell phone, it is connected to chest patches; the heart's electrical activity is then transmitted wirelessly, by radio waves, to a monitor screen in the patient's room or at a nursing station. Aside from giving the patient more mobility during tests, wireless medical telemetry can reduce health care costs because it allows several patients to be remotely monitored simultaneously.

MOBILE CARE

(Right) Medical technicians, with an array of monitoring and life-support equipment close at hand, tend to a patient in their speeding ambulance.

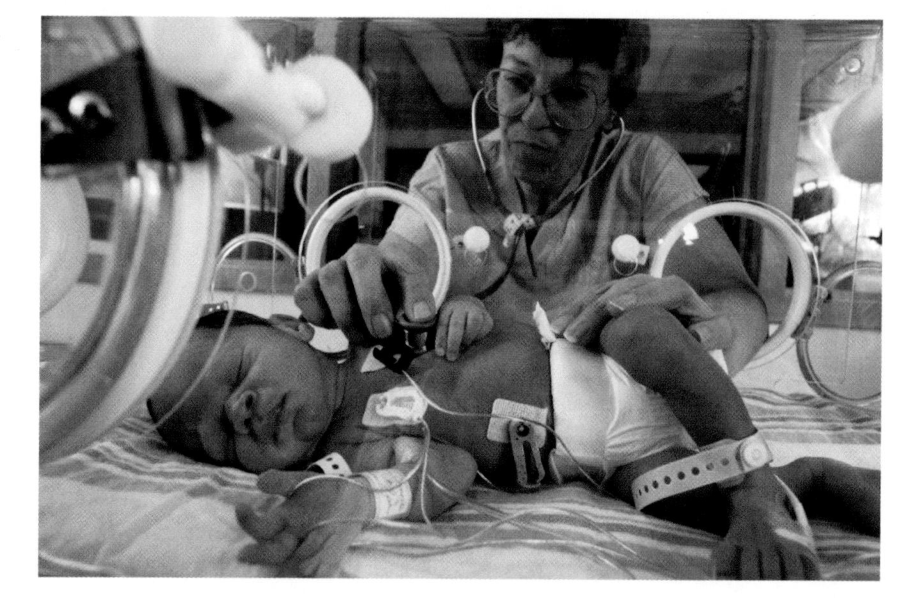

HI-TECH TLC

(Above) As state-of-the-art monitoring equipment keeps an unending vigil, a nurse examines a premature newborn. Years ago, nurses alone, without electronic monitoring equipment, kept babies alive and healthy with basic hands-on care.

SEE ALSO

Sending Signals · 144
Fiber Optic Cables & DSL · 210
Computers · 230
Computer Chips & Transistors · 232

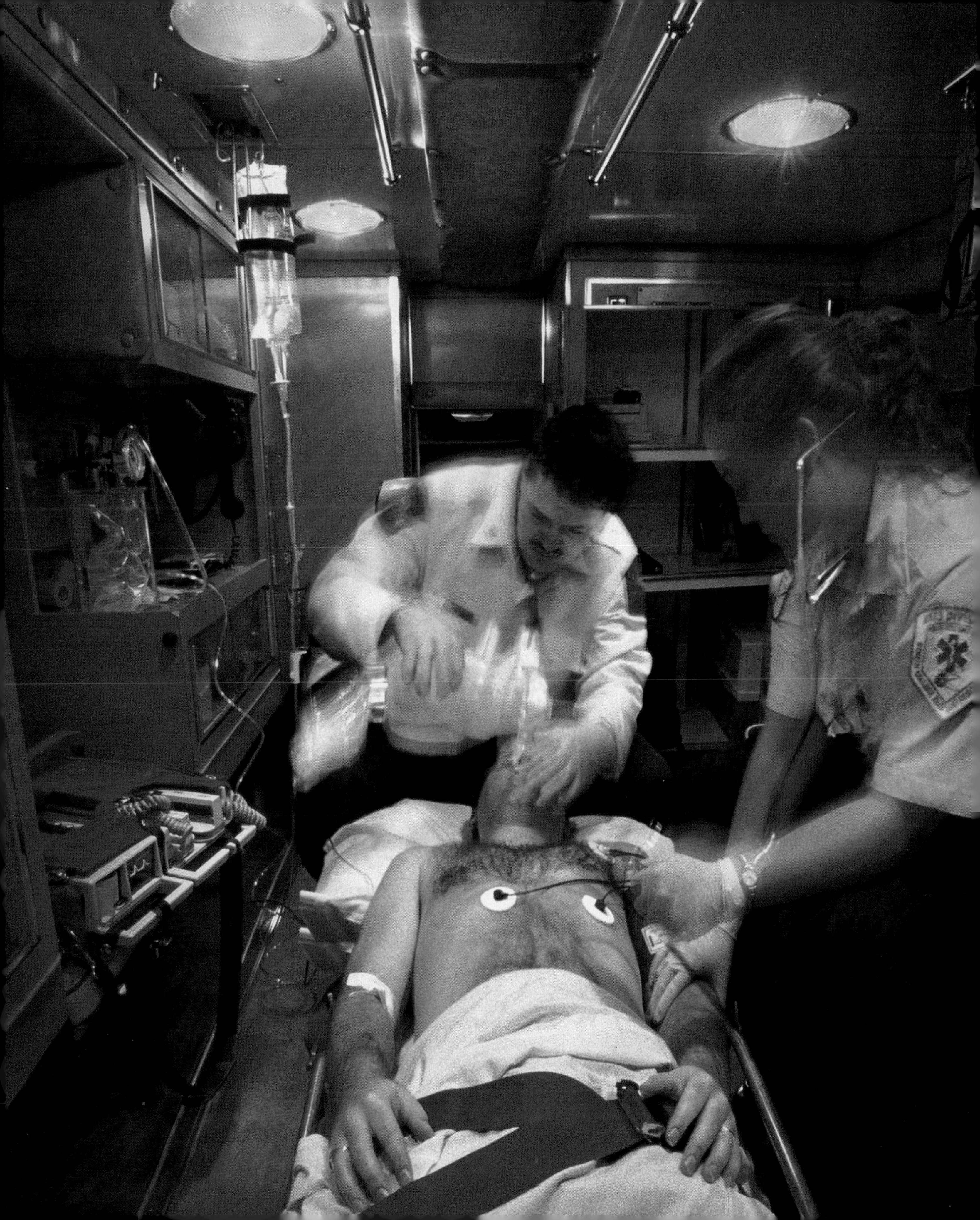

ENDOSCOPY

IN THE CLASSIC SCIENCE-FICTION MOVIE, *Fantastic Voyage*, a group of scientists and their submarine are miniaturized and injected into a patient, where they get a first-hand look at organ systems and the interaction of defending antibodies and invader antigens. Endoscopy affords a somewhat similar view. With flexible fiber-optic tubes inserted through such natural openings as the nose, mouth, anus, urethra, and vagina, doctors can examine a person's internal organs and other structures. One form of endoscopy—a colonoscopy—lets a diagnostician search the large intestine for bowel diseases and bleeding. Another approach involves making a small incision in the abdominal wall and inserting a narrower lighted instrument called a laparoscope. Still another is cystoscopy, which explores the urinary bladder to diagnose bladder stones or cancer. Rhinoscopy examines the nasal passages and the rear of the throat. Special instruments may be attached to such tubes, allowing surgeons to remove lesions and collect samples for biopsy.

Because the procedure is uncomfortable and has a small risk of perforation and infection, doctors have sought a noninvasive way to get a direct look. Enter virtual endoscopy, a blending of CT and MRI scans with high-performance computing. The scans can provide simulated visualizations of specific organs in 3-D animated form. A virtual endoscopist, seated at a workstation, views the inner anatomy while manipulating a computer mouse on a "flight path" through the body. Virtual endoscopy lets diagnosticians change direction, their angle of view, and the scale; they can also shift immediately to new views.

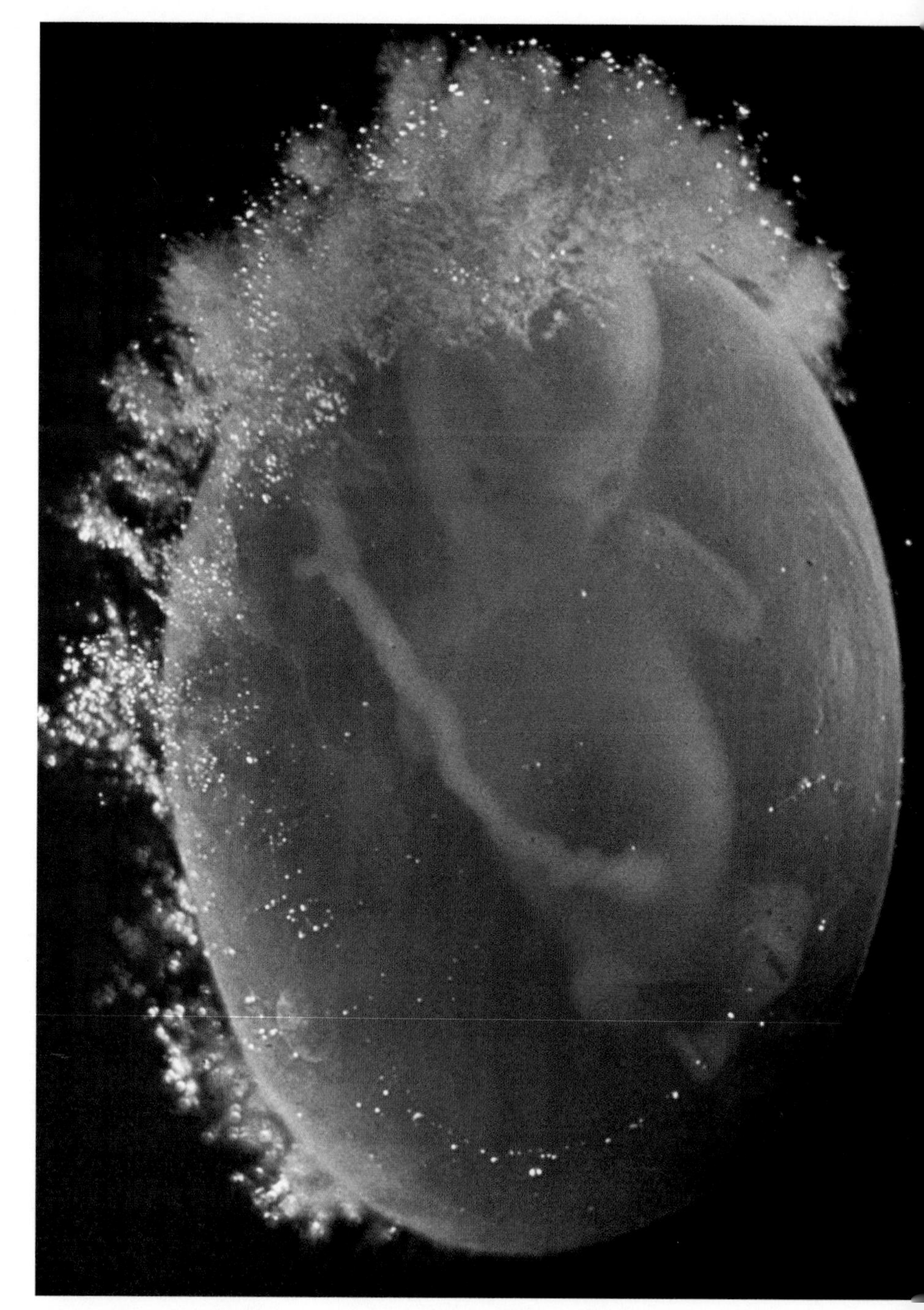

FETAL SURGERY

Even a human fetus can undergo surgery to treat debilitating birth defects after an incision is made in the uterus. Rarely performed, however, because of the risk of inducing preterm labor, the procedure has been refined by fetoscopic surgery, which is minimally invasive.

SEE ALSO
X-ray, CT, & MRI · 192
Laser Surgery · 204
Computers · 230

ENDOSCOPE

Inserted through natural openings or small incisions made in the body, this flexible instrument lights its way with a bundle of optical fibers. Peering through an eyepiece, a physician can manipulate controls to inflate a bowel section and maneuver surgical instruments through a channel to perform a procedure or take a tissue sample. Fitted with the appropriate devices, endoscopes can also cauterize bleeding vessels and suction out tissue debris.

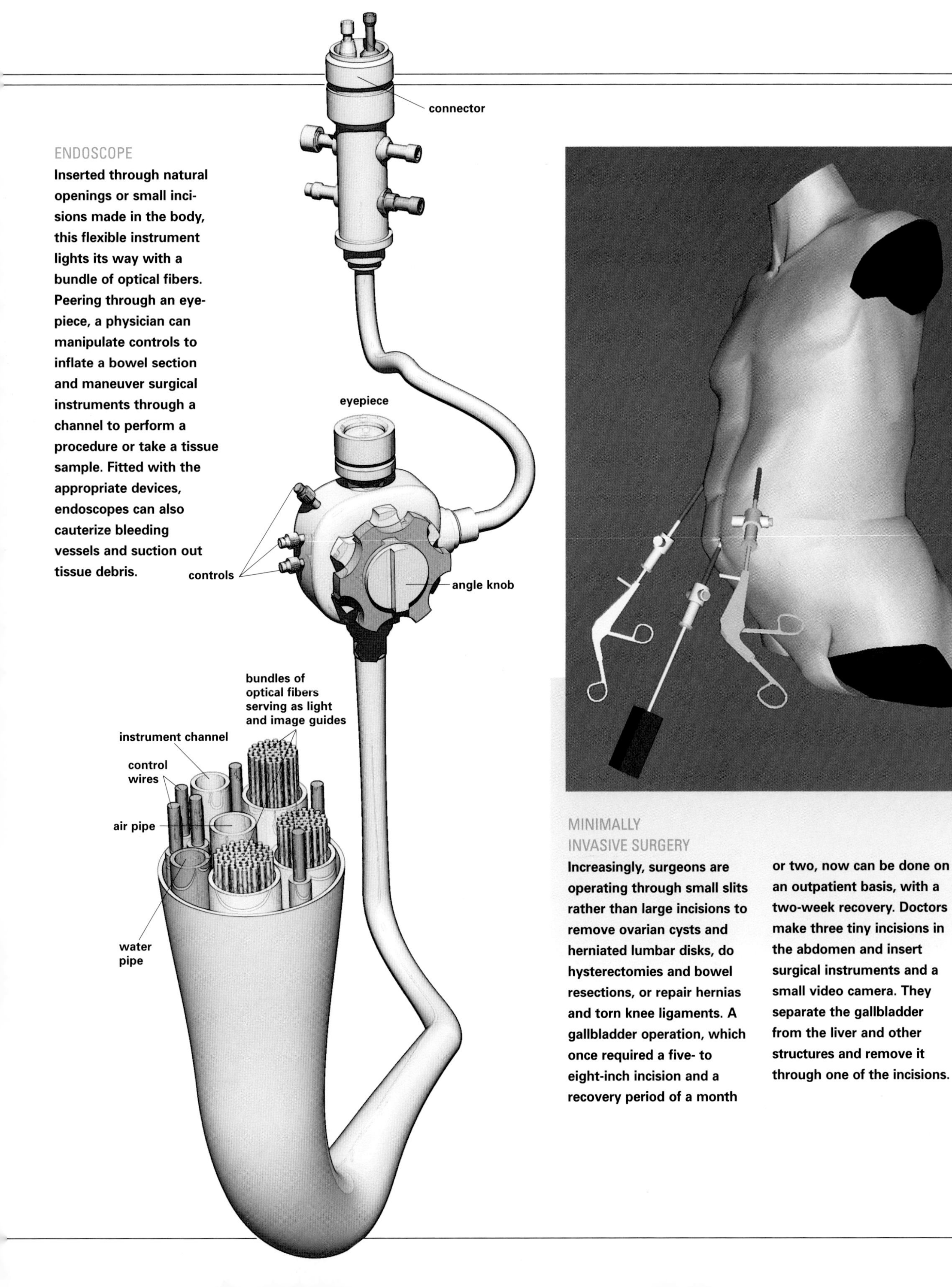

MINIMALLY INVASIVE SURGERY

Increasingly, surgeons are operating through small slits rather than large incisions to remove ovarian cysts and herniated lumbar disks, do hysterectomies and bowel resections, or repair hernias and torn knee ligaments. A gallbladder operation, which once required a five- to eight-inch incision and a recovery period of a month or two, now can be done on an outpatient basis, with a two-week recovery. Doctors make three tiny incisions in the abdomen and insert surgical instruments and a small video camera. They separate the gallbladder from the liver and other structures and remove it through one of the incisions.

KIDNEY DIALYSIS

NESTLED IN THE REAR OF THE ABDOMINAL cavity, the two bean-shaped kidneys function primarily as part of the body's waste-management system, using urine to clear it of urea, the principal breakdown product of proteins and blood plasma salts. Controlled by hormone action, kidneys also maintain water balance and regulate the body's acid-base balance. A person can function with one kidney, but urine production halts and waste products accumulate in the blood when an inflammation or infection causes both organs to fail. Water also accumulates, and chemical concentrations ordinarily regulated by the kidneys are thrown off balance. In advanced form, kidney failure often requires a transplant or a cleansing process known as dialysis.

Peritoneal dialysis takes advantage of the semipermeability of the peritoneum, the delicate membrane lining the abdominal cavity. After an incision is made in the abdominal wall, irrigation fluid is pumped in from a thin plastic tube; waste products enter the solution from the blood and are withdrawn. In hemodialysis, blood from an arm vein is pumped through a tube to a machine that uses an artificial membrane to filter it; the cleansed blood is kept warm and returned to the body in the same vein.

While dialysis is usually done in a hospital on a three-times-a-week schedule, portable machines that can be used at home are being developed; one other advantage is that dialysis can be done at night while a patent is asleep, which, researchers believe, improves its efficacy. Wearable or implanted devices may be on the horizon.

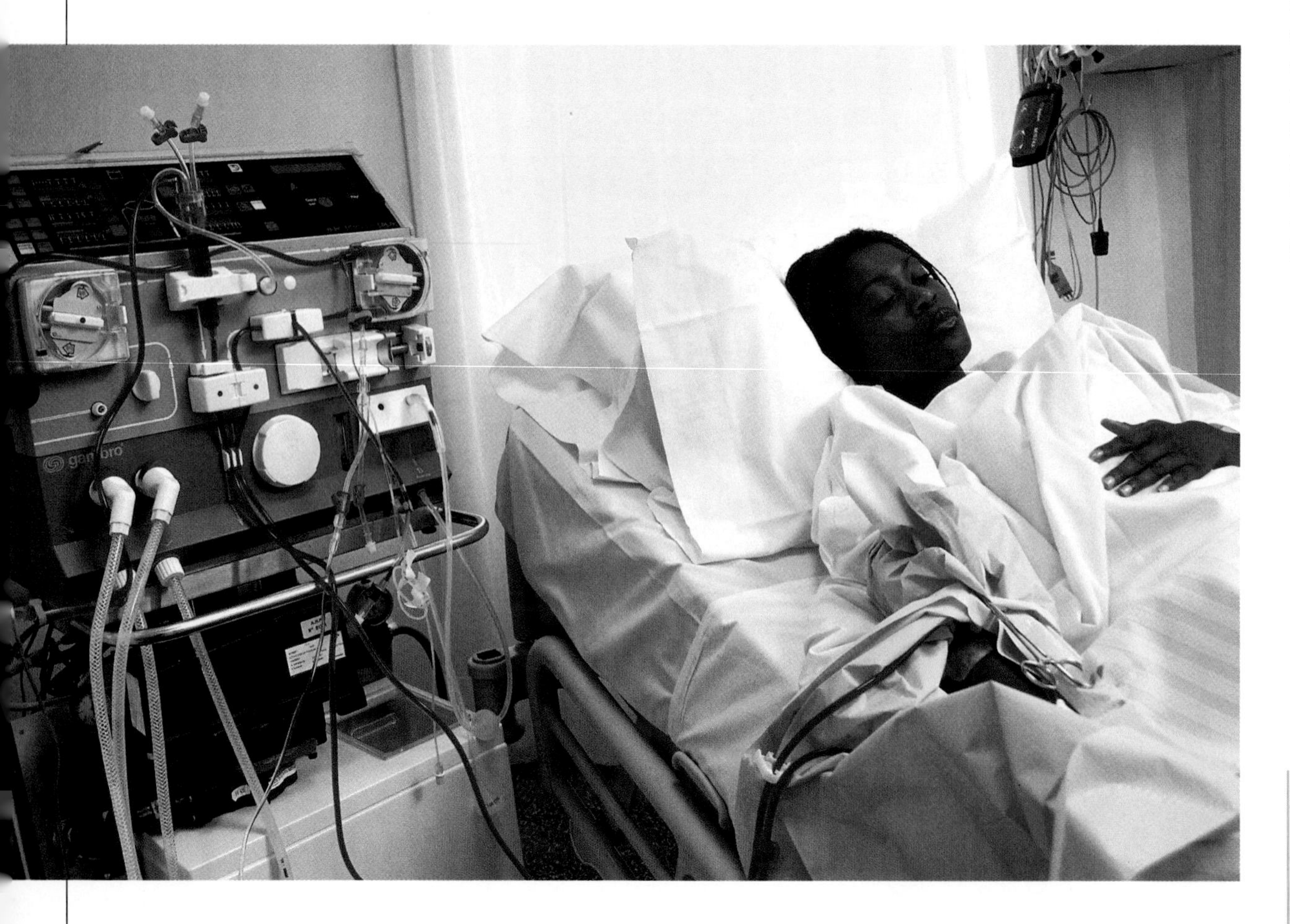

ARTIFICIAL KIDNEY

Conceived of in 1913 as a short-term treatment for reversible acute renal failure, dialysis machines now help transplant patients get by as their bodies accept or reject kidney grafts. In hemodialysis, a tank in the machine holds the dialysate fluid, and a semipermeable membrane filters impurities from the blood. In peritoneal dialysis, the patient's abdominal cavity holds the fluid, and the peritoneal membrane lining the cavity filters out waste products. Either the machine or the patient can exchange used fluid for fresh dialysate.

SEE ALSO
Implants · 200
Designing & Delivering Drugs · 206

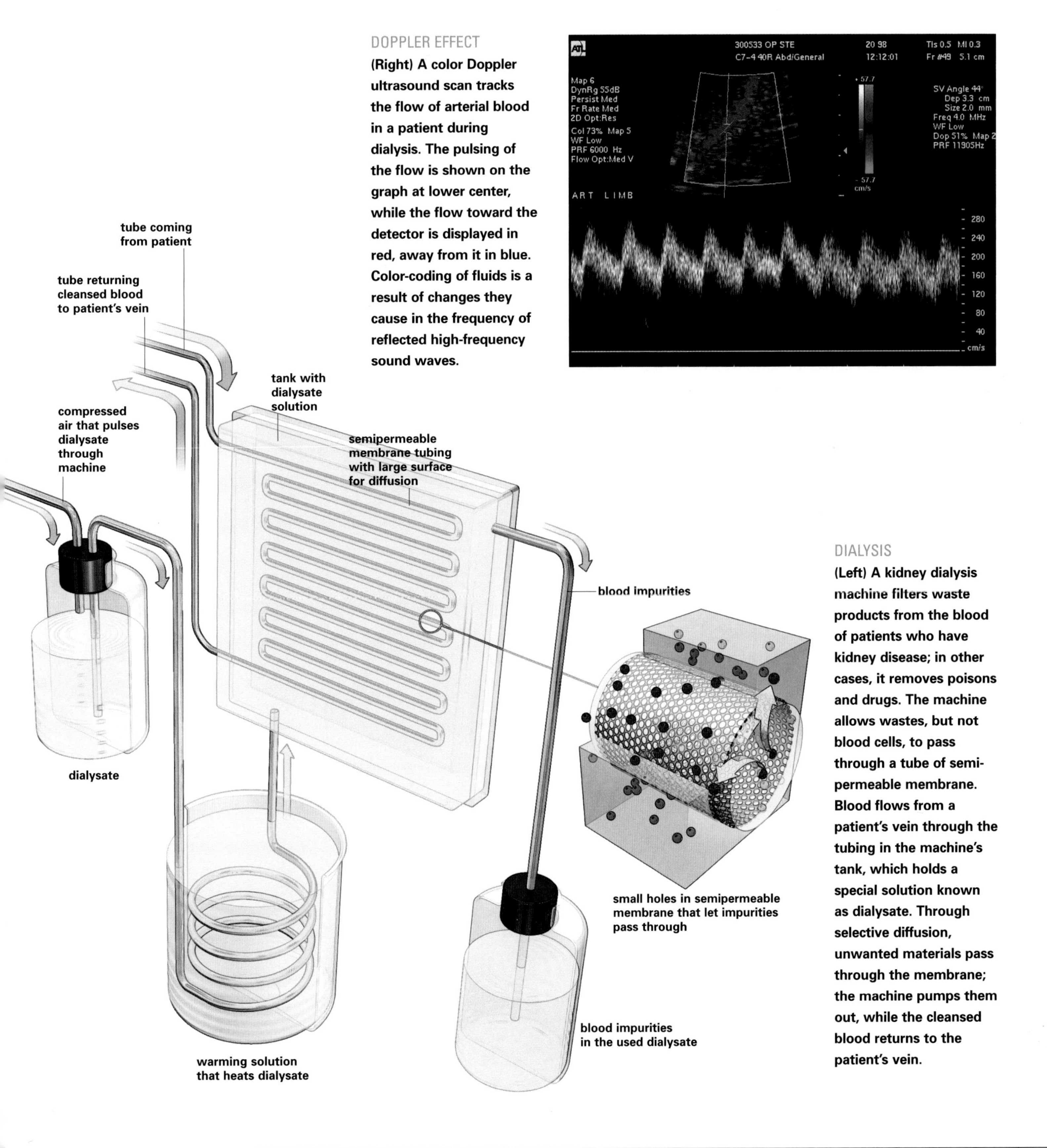

DOPPLER EFFECT

(Right) A color Doppler ultrasound scan tracks the flow of arterial blood in a patient during dialysis. The pulsing of the flow is shown on the graph at lower center, while the flow toward the detector is displayed in red, away from it in blue. Color-coding of fluids is a result of changes they cause in the frequency of reflected high-frequency sound waves.

DIALYSIS

(Left) A kidney dialysis machine filters waste products from the blood of patients who have kidney disease; in other cases, it removes poisons and drugs. The machine allows wastes, but not blood cells, to pass through a tube of semi-permeable membrane. Blood flows from a patient's vein through the tubing in the machine's tank, which holds a special solution known as dialysate. Through selective diffusion, unwanted materials pass through the membrane; the machine pumps them out, while the cleansed blood returns to the patient's vein.

IMPLANTS

MODERN TECHNOLOGY DESERVES ENORMOUS credit for giving patients artificial heart valves and prosthetic limbs, implantable cardiac defibrillators, titanium and ceramic hip joints, and numerous other spare parts. Indeed, according to the U.S. Food and Drug Administration, more than 20,000 firms worldwide produce more than 80,000 brands and models of medical devices for America's market: An implanted pacemaker with a tiny transmitter can automatically send data about the patient's heart to the physician. A wristwatch-like device for diabetics monitors the wearer's glucose level every 20 minutes and sets off an alarm when it gets dangerously high. A capsule-camera can be swallowed and takes pictures of internal bleeding and other abnormalities as it travels through the small intestine.

Suitable materials are the key to a successful implant. Aluminum compounds are used in dental and orthopedic prostheses because their interaction with surrounding tissue is minimal, and they have low levels of friction and wear. Synthetic polymers that resist water and oxidation and can also insulate and lubricate are used to protect electrical implants such as cardiac pacemakers. A number of porous materials and absorbable composites allow for natural tissue growth in and around some implanted devices, and special coatings on metal implants reduce corrosion. A coating of genetically engineered cells may one day prevent an implant from being rejected or from being choked by scar tissue the body makes when invaded.

Prosthetics have also evolved. An amputated arm, for example, can be replaced by a prosthesis powered by a harness and cable attached to the residual limb, or one that uses an external powered device. New imaging devices have ensured near-perfect fits for an artificial limb, and myoelectric devices shift the electrical impulses of a person's muscle contractions to a prosthetic limb, allowing for a natural range of movement. Computerized prosthetics that regulate gait can move joints by controlling a hydraulic and motor-driven system. Implantable electrodes are used to stimulate muscles in spinal cord injury patients. In the future, according to some, brain waves may be used to power prostheses.

BONE IMPLANT

(Right) Metal pins inserted into an ankle stabilize and help repair a fracture. The range of such metallic medical hardware is prodigious and versatile. Nails secure broken bones; hip screws, spinal fixation plates, leg-lengthening apparatus, and locking bolts take care of other structures.

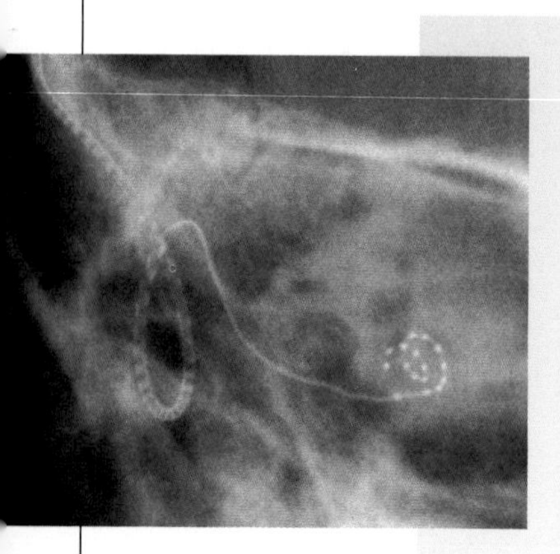

IMPLANTED HEARING

The usual hearing aid is battery-powered, and contains a microphone, an amplifier and an earphone. Fitted behind the earlobe, or nestled in the ear canal, its volume is controlled automatically or by a tiny control. Far different—and often of benefit to people who cannot hear even with a hearing aid—is the cochlear implant, a device that is implanted in the ear's cochlea, a snail-shaped section of the inner ear. We hear because sound waves stimulate the cochlea's sensory cells, and we lose hearing when, among other factors, there is damage to the inner sensory structures. A cochlear implant is a multipart device consisting of an internal and external coil, a microphone to pick up sound, a signal processor that converts the sound into electrical signal, and electrodes that send electrical signals into the cochlea. One coil is surgically implanted in the skull behind and above the ear; the electrodes are inserted into the cochlea and connected to this coil. The external coil is fixed on the skin over the site of the internal one, and its purpose is to transmit the electrical impulses through the skin to the internal coil and thus stimulate the auditory nerve so that it sends neural impulses to the brain; there, they are interpreted as sounds.

SEE ALSO
Synthetic Fibers · 132
Medical Automation · 202

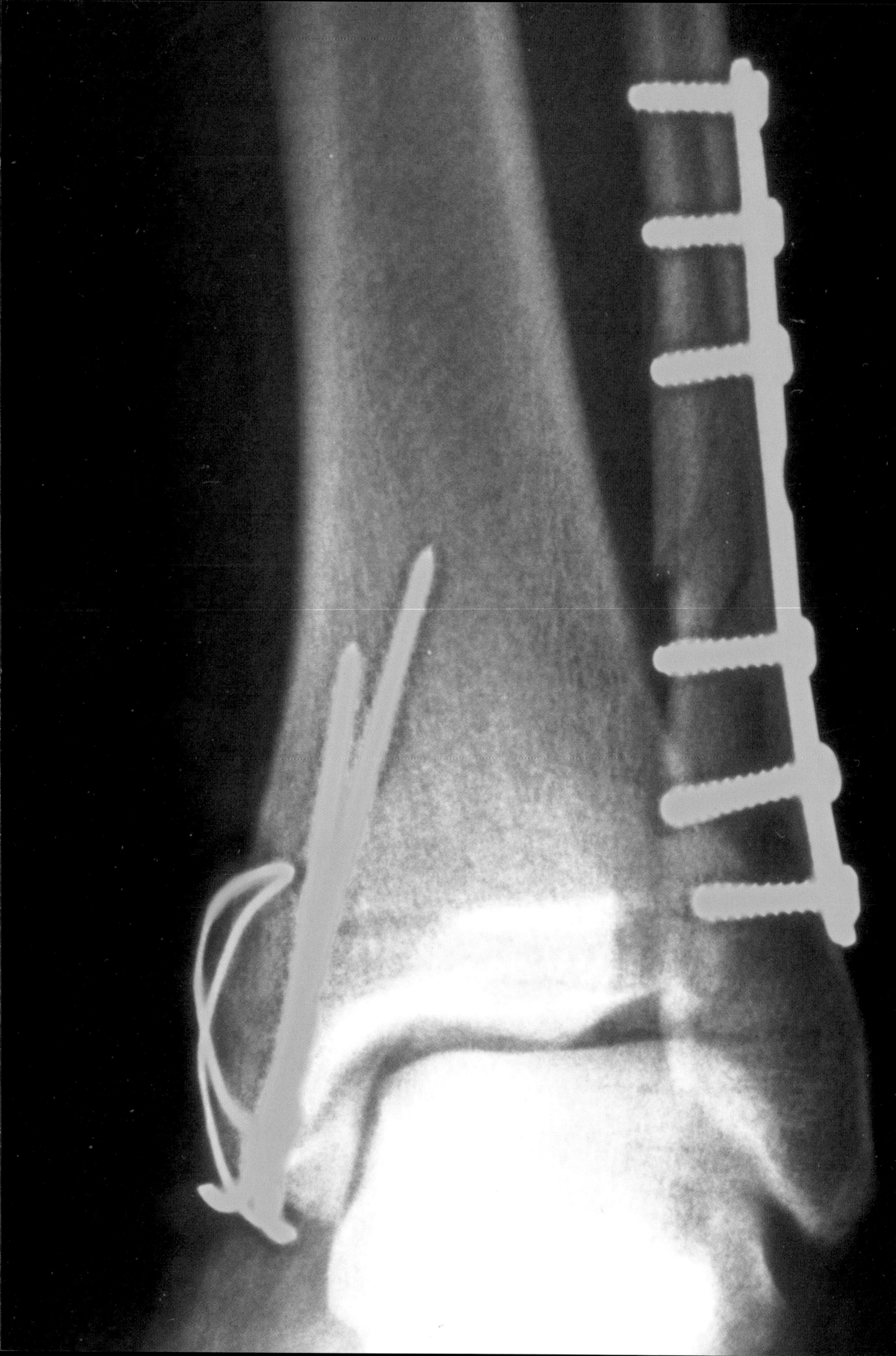

MEDICAL AUTOMATION

PEOPLE ONCE THOUGHT OF ROBOTS ONLY AS amazing creatures in science fiction or as powerful automatons working on industrial assembly lines. No longer. In today's world, automated machines also play important roles in health care: Robotics and other forms of computer-assisted medical intervention make surgery more precise, improve a surgeon's dexterity, reduce complications, and help lower the costs of medical care.

Medical robots are generally assumed to be humanlike in appearance, but most are not. More often, they are fairly shapeless, computer-controlled machine tools. Robots hold and guide laparoscopic cameras or other delicate instruments during surgery, and they perform such intricate tasks as automated precision-milling to ensure a perfect fit of bone and implant during, say, hip-replacement surgery. Some finely tuned robots provide what bioengineers call telepresence—the feeling an operator has when moving scissorslike devices connected to a robot's arms, a sense that the robot's eyes and hands are actually extensions of the operator's own eyes and hands. One robotic arm, ten inches long and developed at the Jet Propulsion Laboratory in Pasadena, California, is agile, steady, and so precise that it can move surgical tools a mere 800 millionths of an inch, an advantage when performing delicate microsurgery in a brain or an eye. Other robotic systems extend human manipulation capabilities. They position and guide needles to do brain biopsies, and they ensure that surgical instruments are in the right place in the nasal cavity during a sinus operation. They help place radiation pellets into diseased organs, locate veins for injection, and they accurately insert screws during spine surgery. Some robots have nonsurgical medical uses as well.

"Patient" robots designed to look like people have been used to train medical students. One classic Japanese model even has vital signs—blood pressure, respirations, and heartbeats—that are programmed into its computer. When a student applies a stethoscope to the robot's chest, the heartbeat and erratic rhythms can be detected. Fingers pressed against the wrist elicit a pulse, and a light shined into the robot's eyes alters the pupils. The robot can also be revived after its heart and respirations are stopped: When artificial respiration is done correctly, the heart resumes beating, blood pressure rises, and the robot's pale lips show color.

But with all this, and despite the fact that robot "hands" can even now go where human hands cannot, or see what human eyes miss, the advice of one physician to "Treat the patient, not the x-ray" should not be forgotten. Medical robots are generally safe, but those who enlist them are also aware that they are only as good as the design of the software that moves them. For this reason, robots will continue to be monitored, as they monitor, to ensure that they follow the same guidelines and safety precautions that physicians and nurses must.

LENDING AN ARM

(Right) A computer-generated image allows a surgeon to guide computer-controlled instruments in an intricate operation on the brain.

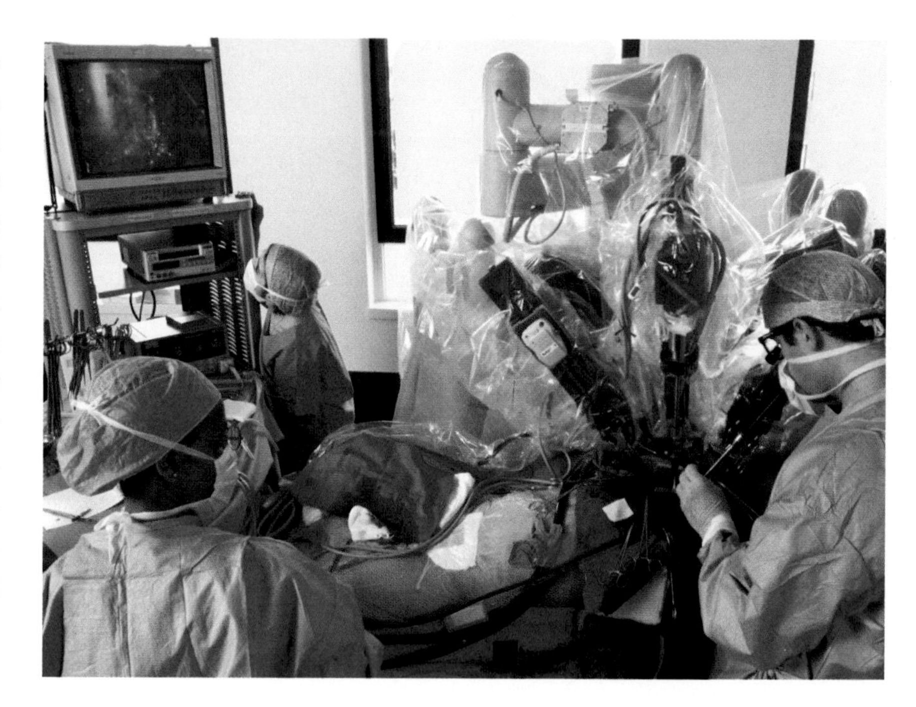

ROBO SURGEON

(Above) A surgical team gets an assist from robotic equipment during a heart procedure. Precise and responsive, robots have been used to perform heart bypass surgery and mitral valve repair, providing a new approach to heart surgery.

SEE ALSO
Medical Monitors • 194
Endoscopy • 196
Implants • 200

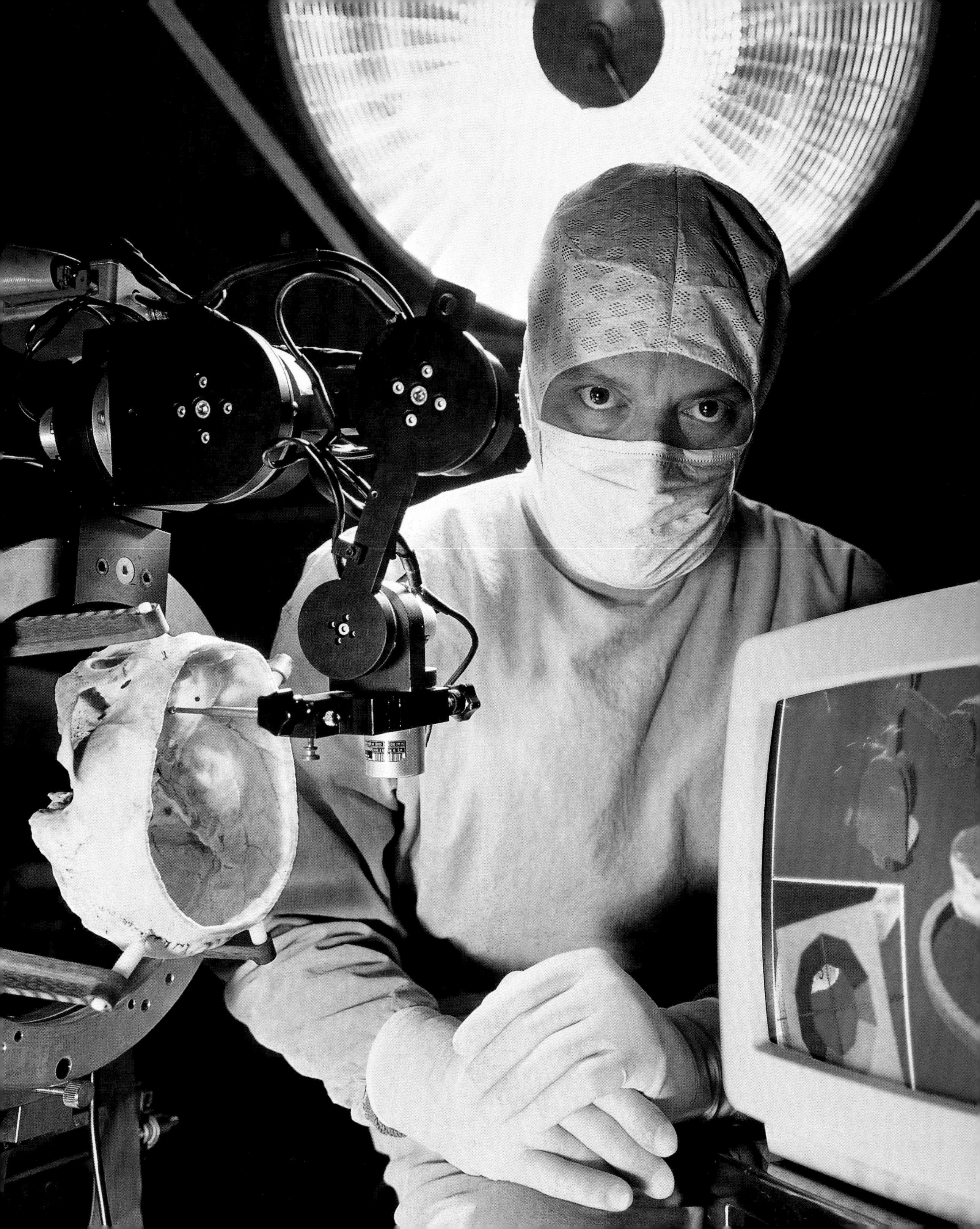

LASER SURGERY

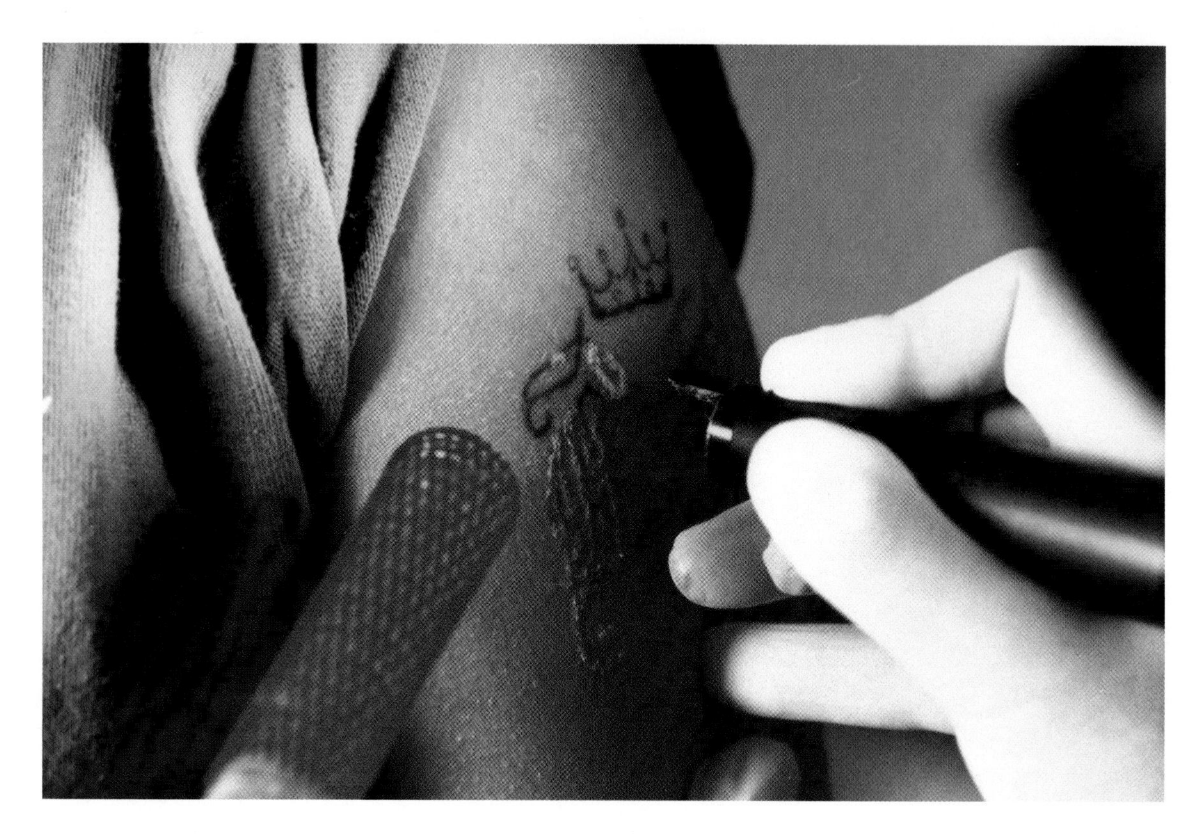

LASER ERASER

(Left) A surgeon uses a laser to remove an identifying gang tattoo from the arm of an ex-member. Laser treatments do not usually require anesthesia and may be performed on an outpatient basis. Able to cut and dissolve tissue, medical lasers differ in their use and power according to their makeup.

HEALING LIGHT

(Right) Aiming the intense beam of a head-mounted laser, a surgeon at Johns Hopkins University's Wilmer Eye Institute repairs a retina. Ophthalmologic lasers can also destroy tumors and arrest abnormal growth of blood vessels.

LASERS CARRY THOUSANDS OF TELEPHONE CALLS simultaneously over fiber optic cables. They play CDs, slice steel beams, guide missiles, measure distances, fashion suits and semiconductor chips, and bore holes in diamonds in a fraction of a second. They also help heal. Medical lasers can vaporize brain tumors, spot-weld tissue grafts, fragment kidney stones, cauterize the lining of a uterus to stop prolonged bleeding, and clear blocked fallopian tubes. In addition, "cosmetic" lasers remove unwanted hair, obliterate tattoos and birthmarks, smooth facial wrinkles, and activate tooth-bleaching solutions.

Used in surgery, a laser's work may be called bloodless or knifeless; it seals blood vessels as it cuts, and sterilizes at the same time. Its precision makes it ideal for microsurgical procedures, such as repairing detached retinas.

Laser stands for "light amplification by stimulated emission of radiation," but its light differs from that of the sun or a bulb, which radiates in every direction. Laser beams are concentrated and narrow, and they move in the same direction. Moreover, they can focus intense heat into small spots: At a width tinier than a pinpoint, a laser can generate a temperature of 10,000°F. And because human tissue is about 80 percent water, a laser can vaporize diseased cells in its target zone.

Wavelength and power are determined by a laser's makeup. Argon gas may be used as a laser medium in eye surgery because the beam's energy can reach the retina without being absorbed by eye fluid. The carbon dioxide laser used in gynecology has a beam that is absorbed by substances containing water. (Argon and carbon dioxide lasers are also used to treat gum disease.) A "biostimulation" laser (also known as a cold laser or a soft laser) provides low-level therapy, such as required in acupuncture, and is used as an adjunctive device for temporary relief of pain.

SEE ALSO
Digitized Music • 146
Fiber Optic Cables & DSL • 210
Computer Printers • 226
Bar Codes & Scanners • 242

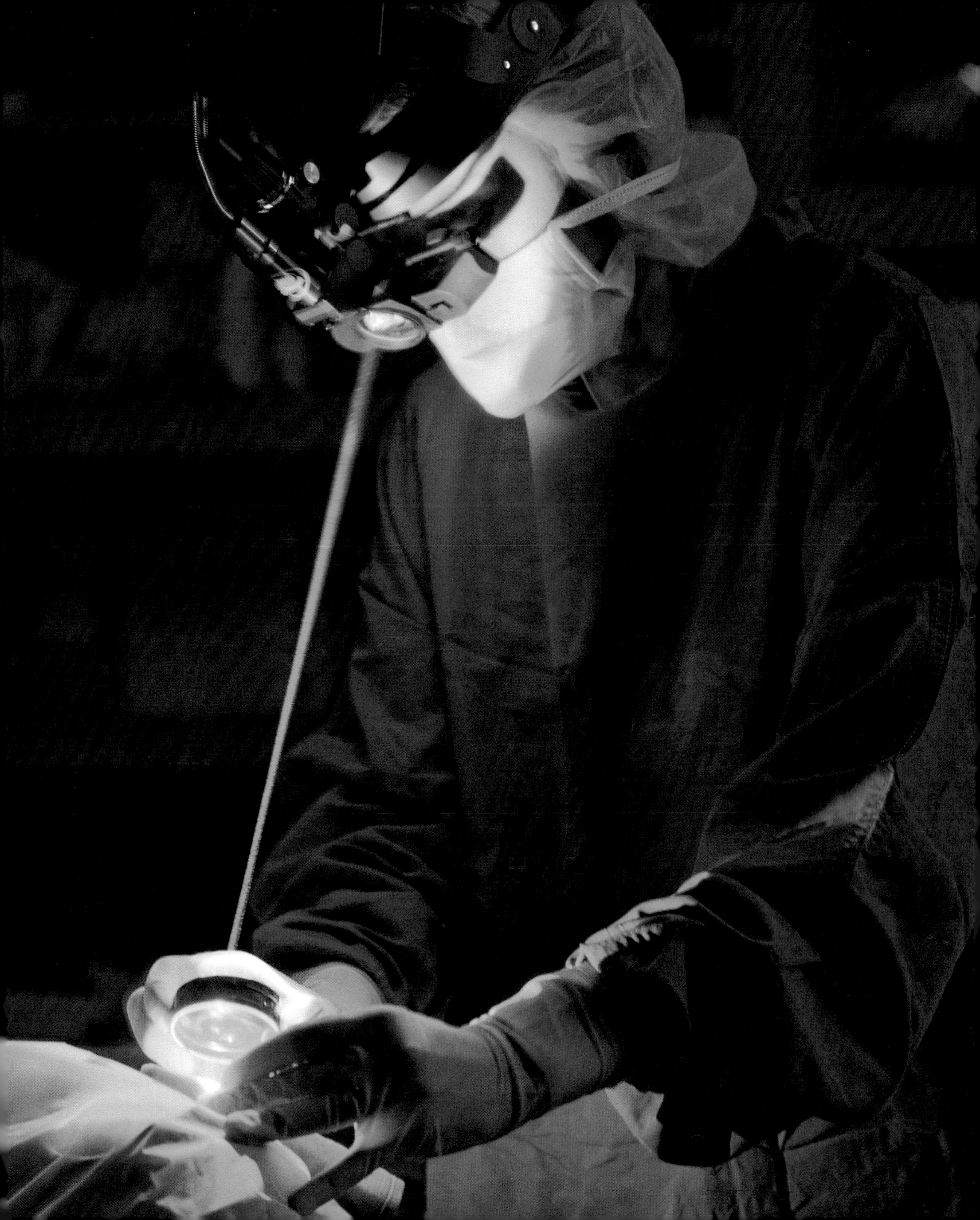

DESIGNING & DELIVERING DRUGS

Canadian physician and educator Sir William Osler had his finger on the pulse of hypochondriacs and quick-fixers when he said that a desire to take medicine is perhaps "the great feature which distinguishes man from other animals."

In just one year, pharmacists in the United States fill some two billion prescriptions. The country's drug companies pour 24 billion dollars annually into research and development, spending a total of 500 million dollars, on average, for each new drug that reaches the market.

Traditionally, creating a drug involved testing many compounds, adding them one at a time to cell cultures and enzymes to determine which had an effect, and then laboriously testing those to see if they needed to be modified. Computers eliminated some of the drudgery by simulating, for example, protein receptor sites on a virus or on a disease-associated enzyme and then providing a model of a potentially useful drug molecule that might fit the receptor. If it fit, a drug would be designed that fooled the receptor into binding with it. Drug design is, thus, a "molecular docking process," a matchup that requires complex steps and intensive screening of enormous numbers of molecules that can be combined. The process becomes even more taxing when one considers that one designed drug may not be appropriate for everyone; thus, researchers are also working to make drugs tailored to individual patients. Another approach to drug design involves inducing animals and crop plants to make therapeutic proteins and antibodies. One technique coaxes goats and cows to secrete a selected protein in their milk, and another gets maize plants to produce disease-fighting antibodies in the nutritive tissues of their seeds.

Taking drugs once meant either swallowing them or receiving an injection. But swallowed drugs may not cross the intestinal lining and enter the bloodstream, or they may be broken down too quickly to have an effect. Injections can be expensive and difficult to administer correctly. Newer delivery methods can compensate: Drug-impregnated, foil-backed skin patches bypass the digestive system and let medication be drawn into the skin via a tiny electric current; nasal sprays take advantage of the nose's permeable mucous membrane, which acts as a portal into the circulatory system. New horizons have been opened by biodegradable drug implants, insulin pumps, and time-release medications.

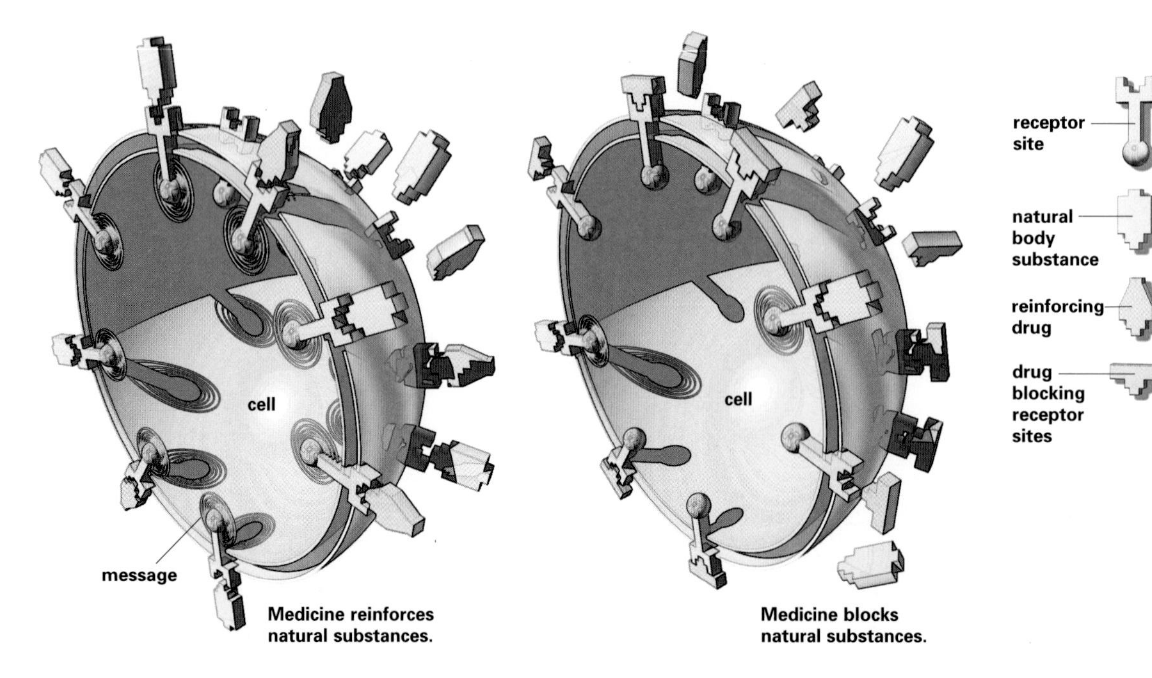

Medicine reinforces natural substances.

Medicine blocks natural substances.

PILL PANTRY

(Right) "For every ill a pill," the British writer and social critic Malcolm Muggeridge observed. He was, of course, absolutely right. The pharmacist here is searching a trove that is a repository of pills for everything from angst to zymosis.

FINDING THE KEY

(Left) A lock-and-key mechanism relying on cell receptors helps drugs work within the cells of the body. In one variation (far left), a drug mimics and reinforces natural messenger molecules that the body produces to call up disease-fighting white blood cells. In another variation (near left), a drug blocks the receptor sites to keep out molecular messages and allow the cell to function normally. Pharmacologists can design drugs with molecules that match protein receptor sites on viruses or disease-related enzymes, fooling the sites into accepting the new drugs.

SEE ALSO
Biotechnology • 118
Incredible Fabrics • 134
Kidney Dialysis • 198
Computers • 230

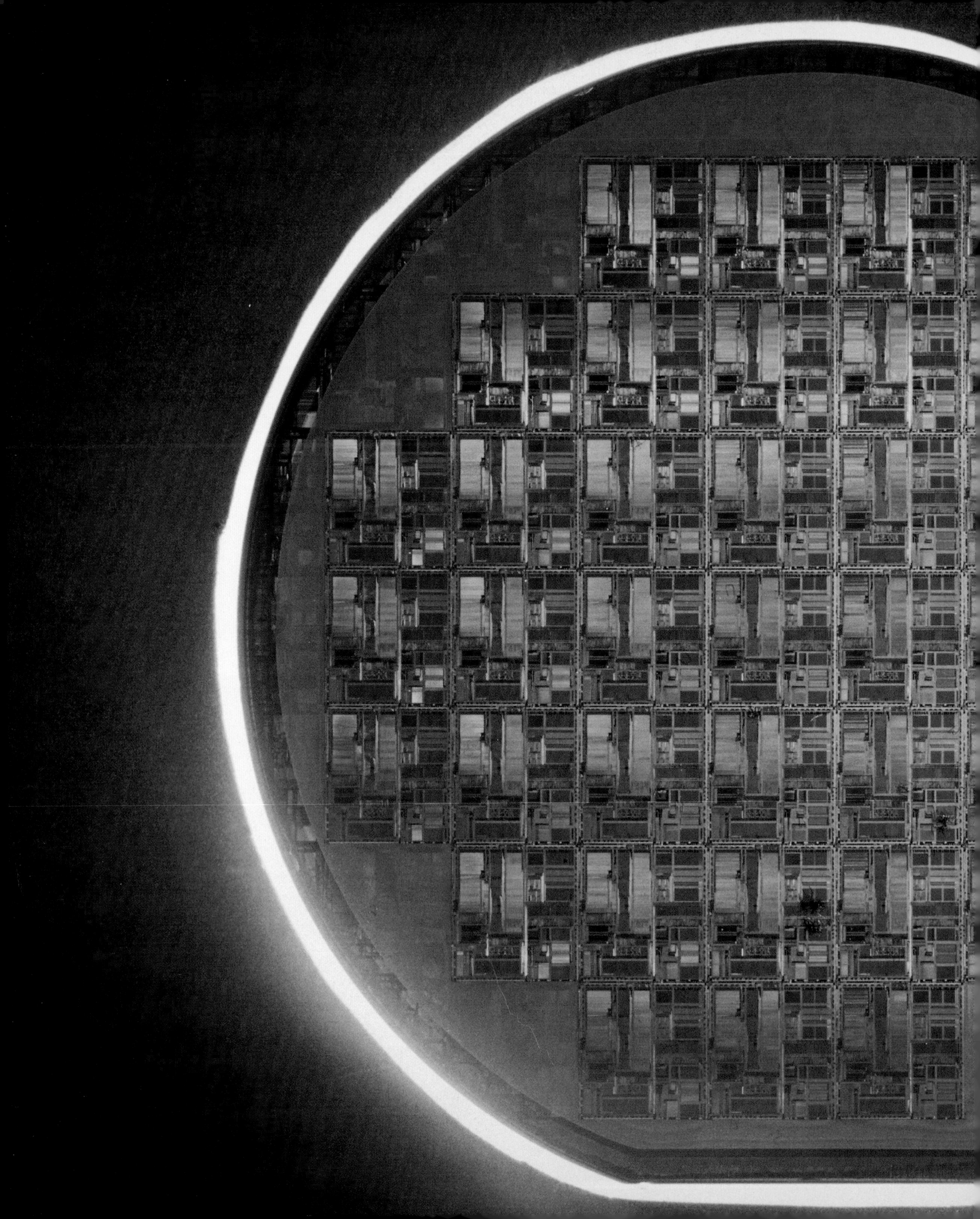

CHAPTER 10

INFORMATION & COMMUNICATION

WE COMMUNICATE INFORMATION IN MANY WAYS, passing it along through our speech and writing, symbols and signs. We also use visual images, gestures, facial expressions, and bodily attitude as we exchange and share facts and notions. Animals, too, have many ways of communicating with one another, and plants may have a system of communication as well. But as far as we know, only we humans have developed the ability to reconstruct and vastly improve our ways of communicating. To do so, we tame and bridle such natural forces as electricity, radio waves, light, and sound, using them alone or in combination. From the cellular telephone to e-mail, from telecommunication devices for the deaf to wearable computers, science and technology have left an indelible imprint on how we get our messages across.

This tiny Intel Pentium chip is capable of controlling a computer's behavior with its millions of transistors and circuitry.

FIBER OPTIC CABLES & DSL

TELECOMMUNICATION—COMMUNICATION AT a distance using telephones or radios, for example—relies on electromagnetic waves. Members of this large energy family include radio and light waves, x-rays, microwaves, and infrared and ultraviolet rays. Called electromagnetic because they consist of electric and magnetic fields (which vibrate at right angles to each other), they are high-velocity energy transmitters that move at the speed of light (186,000 miles a second). Their signals and impulses are capable of carrying sounds, words, images, and data. The higher their frequency, the more energy they have; gamma and x-rays are at the high end; microwaves and radio waves are at the low end.

Radio waves exist naturally as radiant energy from the sun, but when they are generated by electricity in an antenna tower, they transport sound and images. In the form of radar, radio waves help detect distant objects and determine their positions; radar operators measure the time the waves take to travel to an object, reflect off it, and return. The visible light segment of the electromagnetic field has also been used to communicate: In laser form it beams through fine strands of pure glass, its digitally coded pulses translated into thousands of telephone conversations that can be handled simultaneously.

Sound also comes in waves, but they are pressure waves, not electromagnetic ones. Best defined as a measurable, mechanical disturbance, sound is amenable to translation into electricity. This is a valuable quality when transmitting voices, whose words, whether flowery or vernacular, must still be reduced to the same workmanlike electrical signals before they can be sent coursing over conventional copper wire or through tubes of glass.

But all these data-carrying signals can sometimes get in the way of one another, as is the case when trying to make a phone call while someone is accessing the Internet on the same line. Enter broadband, a special telephone line that now enables one to make a phone call and work on the Net simultaneously and on one line. Known also as DSL, for digital subscriber line, a broadband connection lets one surf the Net at a much faster speed than is possible over a regular phone circuit. To accomplish this, it splits the signals into two channels, one for voice and the other for the high-speed data. Bandwith, which indicates how much data can be transmitted through a connection, is the key here, and is generally measured at bits per second (bps). A conventional computer modem— an electronic device that converts digital data into analog signals that can be sent over phone lines—can move data at more than 57,000 bps; but because a DSL modem squeezes more out of a phone line, the speed at which information is exchanged is greatly increased.

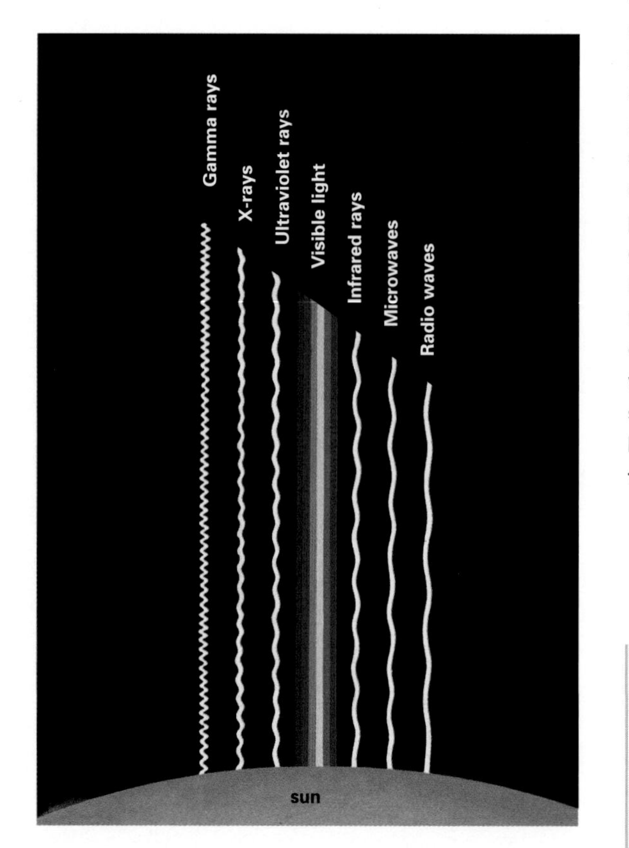

WAVELENGTHS

(Left) Electromagnetic radiation ranges from high-frequency gamma rays, at the top of the electromagnetic spectrum, to low-frequency radio waves.

SHEATHED OPTICAL FIBERS

(Right) Wrapped around a steel strengthening wire, these fibers transmit digital data in the form of laser pulses. Each fiber's core and cladding are made of silicon glass and semiconducting materials, and the plastic sheathing prevents stray light from escaping to other fibers. Narrow-core fibers (far right) carry clear signals over long distances, while wide-core fibers allow the signal to spread out and blur, limiting data's transmission rate.

SEE ALSO
Radio • 142
Sending Signals • 144
Laser Surgery • 204
Telephones • 212
Cell Phones • 214

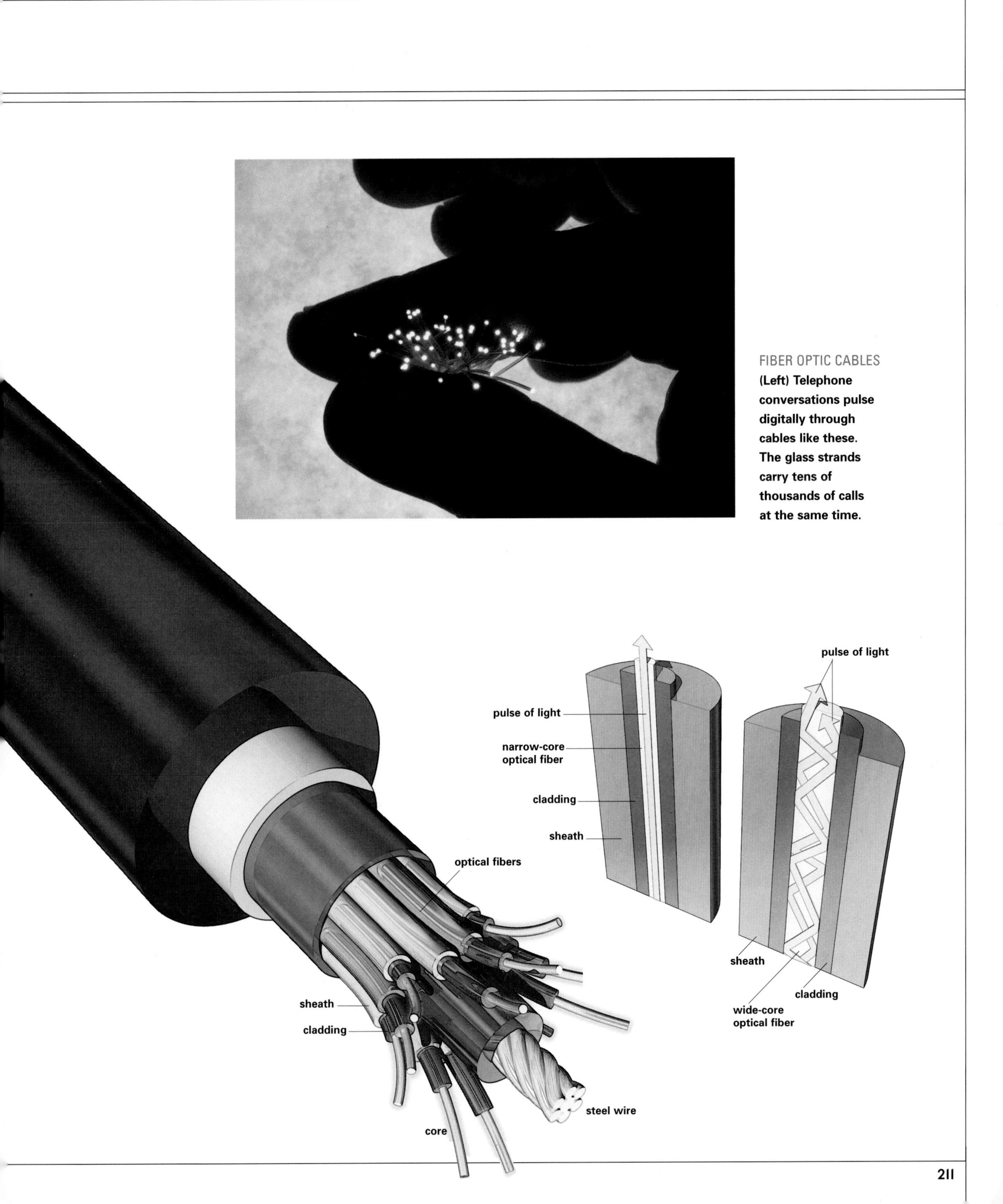

FIBER OPTIC CABLES

(Left) Telephone conversations pulse digitally through cables like these. The glass strands carry tens of thousands of calls at the same time.

TELEPHONES

ALEXANDER GRAHAM BELL'S FIRST TELEPHONE bore little resemblance to the sleek instruments we use today. Instead, it was a rather odd-looking "gallows frame" rig consisting of vibrating metal diaphragms and a bar magnet wrapped in wire.

The principles behind the phone's function are essentially unchanged, but the way the device processes and transmits voices—and looks—is now vastly different. Basically, a telephone is a mouthpiece with a tiny microphone and an earpiece with a receiver. When a person speaks into the mouthpiece, sound waves cause the microphone's diaphragm to vibrate.

Speech is converted to electrical signals that can be transmitted over cable, radio, and microwave connections, or to pulses of laser light carried over fiber-optic equipment. Automatic switching equipment identifies the number dialed and makes the connection. For years, the electrical signals in vocal transmissions were analogs—they were analogous to the vibrations in conversation—but modern systems convert the sound waves into digital information at the local exchange, hiding it in a binary code of 1s and 0s. On the receiving end, the process is reversed by a vibrating diaphragm that picks up the reconverted analog signals and recreates the speaker's voice.

Today's telephones can be cordless and portable, operating as radio transmitters and receivers. Cordless versions work on the same principles as those plugged into a wall jack. Indeed, their bases are connected to an outlet through which they receive calls. But the difference is that a cordless phone converts the electrical signal from a caller into an FM radio signal, and sends it through the air over various radio frequencies instead of from the base and the handset. When one speaks into the cordless phone, voice is sent back to the base via another FM signal, where it is converted into an electrical signal which is sent back to the caller over the phone line.

So-called speaker phones, with built-in microphones and speakers, allow the phone to be used without a handset and enable several people to participate in the call.

OLD FAITHFUL

(Right) The pink, dial-style analog telephone is now a collectible, or a retro fixture that has been updated to accommodate today's digital requirements.

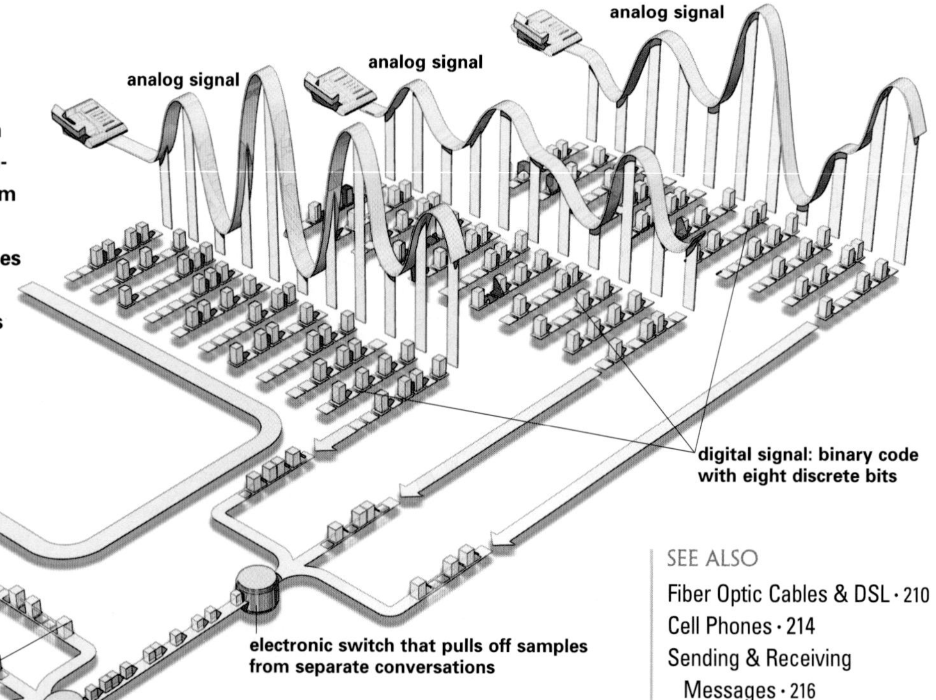

TELEPHONE CALLS

In the United States, where 100 million households have telephones, a vast network of some 150 million main telephone lines carries long-distance and local phone calls. The process of making and receiving a call involves converting the voice into an analog signal of electrical current and sending it to a local exchange that converts it to a digital code of eight electrical "on" or "off" pulses. Electronic switches combine the binary data from many different conversations and send them over a single wire. The receiving end reassembles the components of each conversation, reconverts them to analog, and sends them to the appropriate phones.

SEE ALSO
Fiber Optic Cables & DSL · 210
Cell Phones · 214
Sending & Receiving Messages · 216

1
ABC
2
DEF
GHI
JKL
5
MNO
6
PRS
7
TUV
8
WXY
9
0
OPERATOR

CELL PHONES

THE WEDDING OF THE TELEPHONE AND THE radio was perhaps the most significant development in the history of telephony. Roaming from room to room with a cordless phone, or smartly snapping open a compact cellular model while on a stroll in a park, a subscriber to a phone service now has an expansive degree of mobility.

Cellular phones—so-called because they cover compartmentalized, cell-like areas—are more impressive than the now standard cordless ones. Like cordless phones, they transmit over radio waves, but via an antenna located in a base station in each cell, or by way of satellites. Antennas are connected by phone lines to exchanges, which link cell phone users to one another or connect them to parties using conventional phones. Eventually, extensive satellite telephone systems will connect directly with each subscriber's phone, enabling callers to reach out to all corners of the world. Another generation of cell phones will be able to instruct a global positioning system satellite—alerted by a 911-like message—to beam a signal to the receiver, thereby determining the exact location of, say, a driver in distress.

At the center of it all is, of course, wireless technology, the same through-the-air transmission system that uses radio frequencies and other types of wave phenomena to make robot rovers scour the surface of Mars, direct a "remote" to control a TV set, open garage doors, power pocket-personal computers, and let people talk to one another over two-way radios.

While copper wire and fiber optics still handle our communications systems, the dream of a wireless world is not far-fetched. Many people now rely on palm-size PDAs, personal digital assistants that serve as minicomputers, and manage information such as appointments and addresses. Text is entered on a small keyboard (or on a full-size collapsible one) or with a stylus, and when the PDA is attached to a wireless modem they can surf Web sites and send and receive e-mail.

WALK AND TALK

(Right) A familiar sight all over the world, a cell phone to the ear has become not only a vital means of communication, but, for some, a status symbol.

CELLULAR NET

(Left) Cellular phones, with their own radio receivers and transmitters, pick up signals from antennas on base stations located in cell-like areas or from receiver dishes tuned to satellites. The frequencies in a cell group differ from one another, but beyond a group the same frequencies may be reused throughout a network. As a person moves from cell to cell, the call automatically switches to the appropriate frequency.

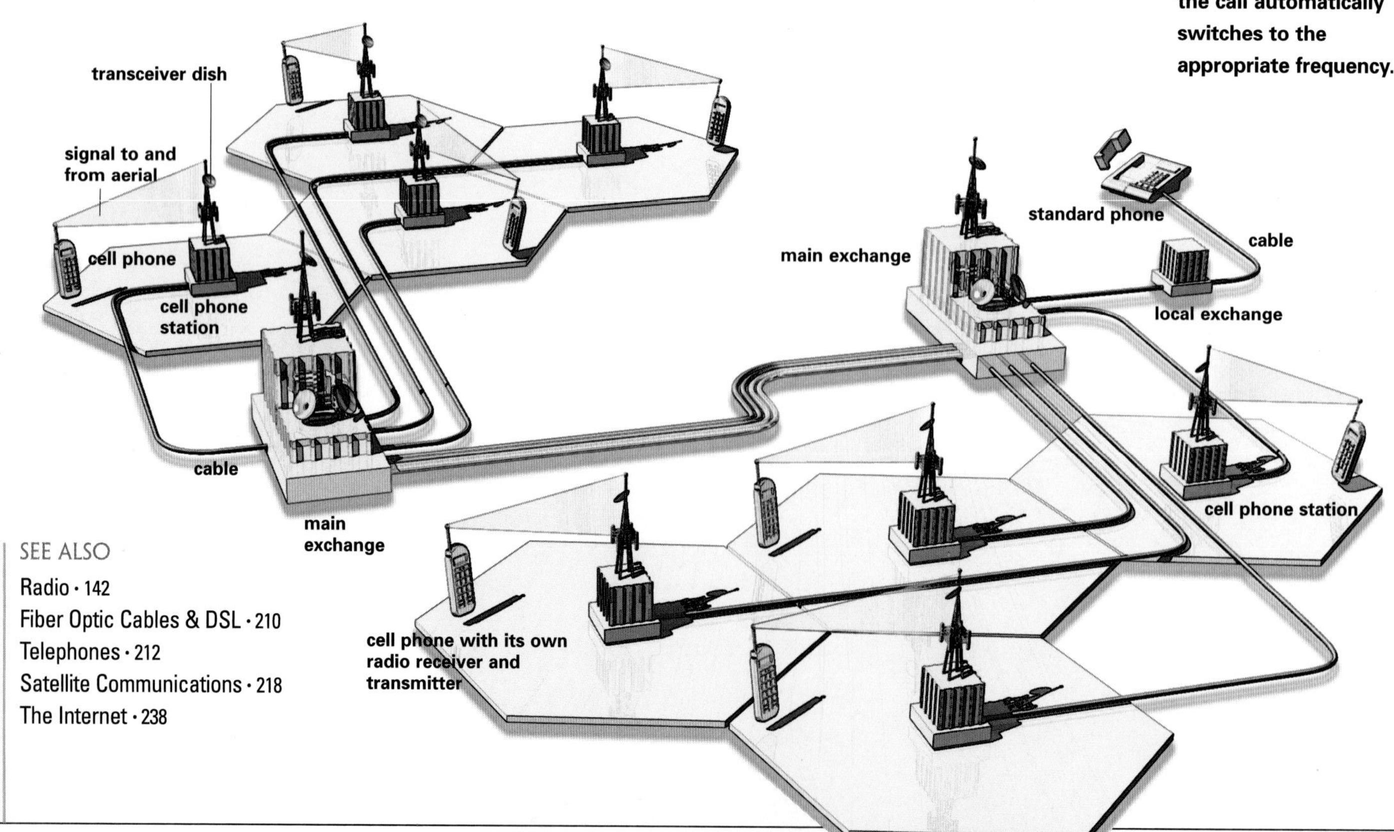

SEE ALSO
Radio • 142
Fiber Optic Cables & DSL • 210
Telephones • 212
Satellite Communications • 218
The Internet • 238

TALK

SENDING & RECEIVING MESSAGES

THE FACSIMILE, OR FAX, DATES BACK TO 1843, when English clockmaker Alexander Bain was given the first patent for facsimile transmission. Fax machines send exact copies of documents or pictures. All it takes is a phone call to connect the recipient and the sender, who feeds the material to be sent into the machine. Inside, a light beam plays over the text or photograph and reflects an image of the light and dark portions into an arrangement of photoelectric cells that, in turn, convert the information into electrical current. Amplified, the current is transmitted over the open telephone circuit to the recipient's machine, where a printer reassembles the imagery on paper.

Pagers and telephone answering machines take messages for people who are temporarily away from their phones. The answering machine, which is essentially a telephone with a recording device inside, detects an incoming call and, after a preset number of rings, plays a prerecorded message inviting the caller to leave a message. A call made to a pager, or beeper, is converted into a code that can be accessed only by a device that can read it.

A desktop scanner, like the most commonly used flatbed model, is a versatile piece of optical technology that captures images and text, and saves them in a computer; while stored, they can be altered, and then printed, or sent out over the Internet. When documents or pictures are placed under a cover on the scanner's glass plate—much like documents are placed on a copy machine—light from a lamp illuminates them. Then a motorized system of mirrors, sensors, and an array of extremely light-sensitive diodes that convert light into electrical charge moves over the objects. When the light hitting the scanned object is reflected back, the scanner analyzes it and reconstructs it electronically. Computer software then allows the scanned material to be transferred to the computer in a format the computer understands.

INSTANT TRANSMISSION

(Left) A Modern fax machine uses digitizing technology not found in early facsimile machines, which sent news photographs via phone, or in teletypes, typewriter-like machines linked to phone lines that transmitted news letter by letter to receiving machines.

FAXING

(Right) When someone feeds a document or a picture into a fax machine, a light-scanner reads the data, and image sensors convert it to electrical pulses. Next, the machine converts the electrical stream to digital form, and a modulator combines it with a carrier wave, allowing the machine to send the information out over an open phone circuit to the recipient's machine. In receiving mode, the demodulator converts the arriving waves into digitized information that prints by the same process employed in a photocopier.

SEE ALSO

Radio • 142
Fiber Optic Cables & DSL • 210
Telephones • 212
Photocopiers • 224

VOICEMAIL

Unlike the telephone answering machine it is fast replacing, a voice mail system is a personal call mailbox that requires no additional equipment other than a Touch-Tone phone set up to accommodate it. No tapes are needed to record messages, which are answered even when the phone is in use. An electronic system with a prerecorded answering message, voice mail operates through computerized equipment at the telephone switching facility. To retrieve a call, a person enters specific security codes, a feature that offers more privacy than an answering machine that more often than not broadcasts its message to all within earshot.

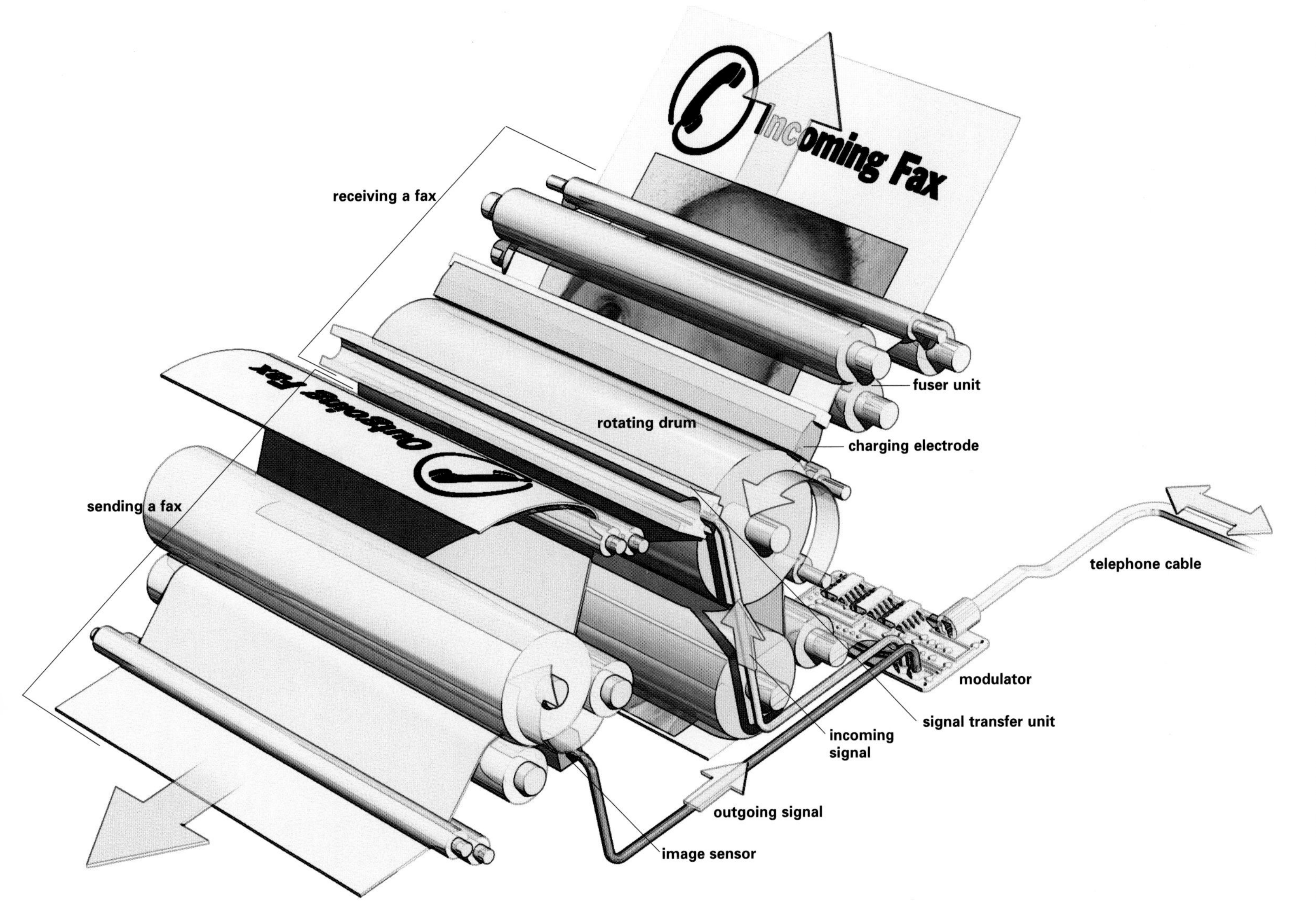

SATELLITE COMMUNICATIONS

Sputnik I, the first artificial object to orbit the Earth, was launched in 1957 by the former Soviet Union. Today, communication by radio, telephone, and television is unthinkable without these miniature moons. Carried into orbit by rockets or space shuttles, satellites handle far more than TV, news, and phone calls. They also track weather systems and monitor soil moisture, map Earth's features, broadcast navigational signals, help plan battlefield strategy, and even carry up-to-the-moment financial information to Wall Street traders or allow us to swipe our credit cards through machines at gas stations.

Satellites can range over the planet, but the so-called geostationary orbiters used for communications and weather observations are synchronized with the Earth's rotation. This "parks" them above the same point on the surface, a position that allows for uninterrupted contact between ground stations in a satellite's line of sight. Three relay satellites uniformly spaced at such a height can easily cover the entire surface of the planet, receiving television and hundreds of thousands of telephone messages from one continent, amplifying them, and relaying them via other satellites to different parts of the globe. U.S. Iridium satellites, for example, are launched into polar orbits and provide voice circuits, data, and paging. Their circular "spot" beams cover "cell" areas some 90 miles across, and they can communicate with each other as well as with Earth stations. They are valuable in delivering communications to and from remote areas and, along with serving the U.S. Defense Department, are used to aid heavy construction projects, emergency services, and mining industries.

GLOBAL INTERCONNECTIVITY
(Left) A geologist examines the laser positioning system in a California crater. GPS measurements indicated that the spot moved up 38 centimeters and northwest 21 centimeters after the Northridge earthquake near Los Angeles.

GPS
(Right) U.S. global positioning system satellites, in pink, and Russia's Global Navigation satellites, in yellow, continuously send signals retrievable from any place on Earth. Used to help troops in war, the timed digital transmissions from GPS can determine a position to within 20 yards.

SEE ALSO
Radio · 142
Television · 148
Cell Phones · 214
Bar Codes & Scanners · 242
Managing Money Electronically · 244

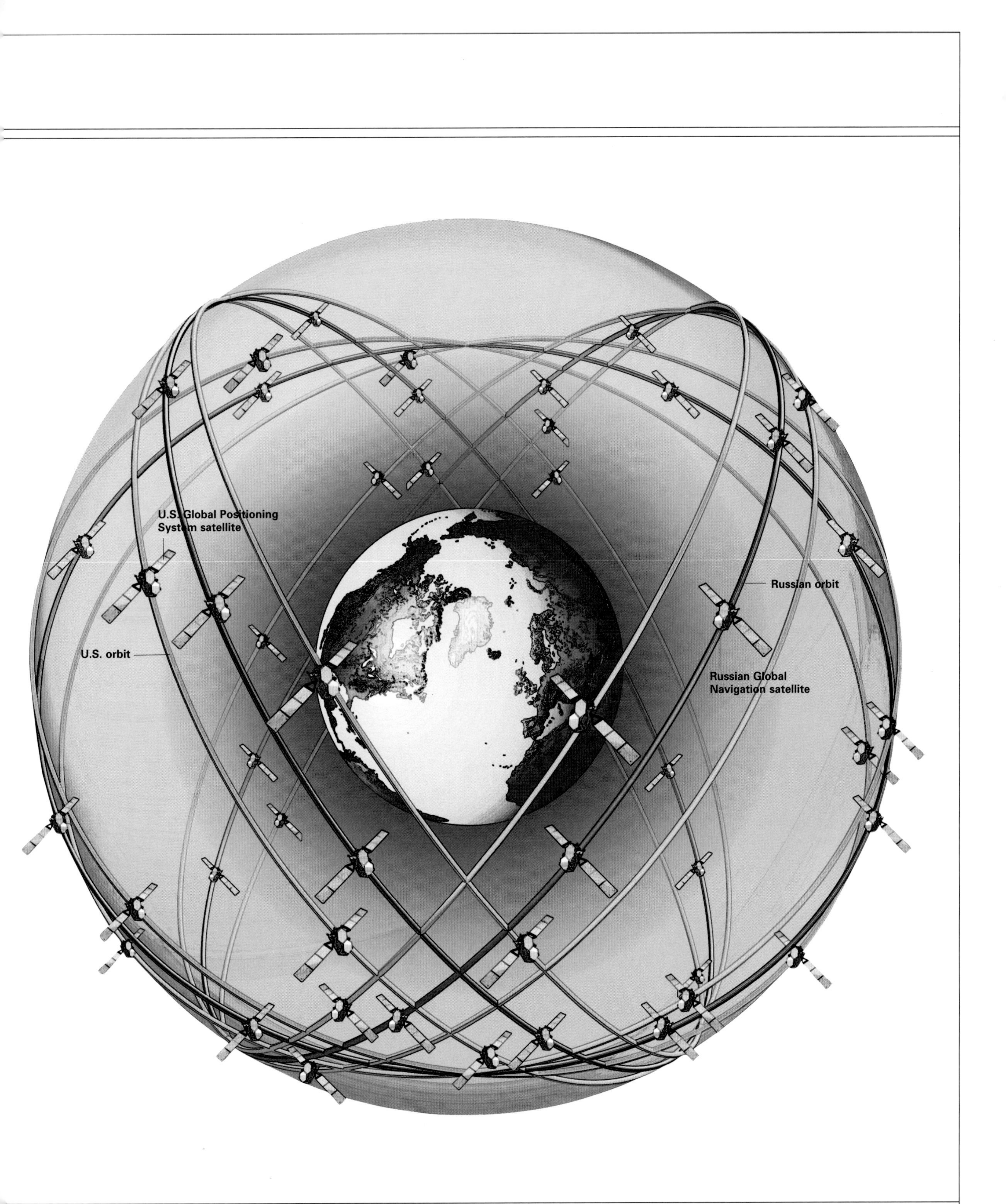
U.S. Global Positioning
System satellite
U.S. orbit
Russian orbit
Russian Global
Navigation satellite

TYPEWRITERS

PATENTED IN 1868 BY CHRISTOPHER LATHAM Sholes of Wisconsin, the first practical typewriter had keys that worked on a "pianoforte" action: When fingers pressed the keys, a series of connected levers raised the type bars; these struck the paper curved around a roller. A ribbon of inked fabric ran across the top from one end of the machine to the other; each type bar had upper- and lowercase letters, and a row of keys held numerals and punctuation marks. Pressing a shift key lowered the type bar, enabling the uppercase letters to strike the ribbon.

The electric typewriter came into use in the 1920s. Early versions had an electric motor that lifted the type bar and powered the typing stroke. The motor also returned the carriage, turned the roller, and made the keys jump at a typist's touch. The type bar was replaced in some later models by a spherical typing head, an "element" with 88 characters that moved along with an inked ribbon holder as keys were struck.

An improved variation is the electronic typewriter, which has the added feature of a "memory" that can store text either internally or in an external cartridge or diskette. Some replace the type ball of the electric models with a daisy wheel, a disk with letters and numbers stamped on protrusions around the outer edges.

Today, the typewriter's functions have largely been taken over by the word processor—except for specific controls, a computer's keyboard resembles a typewriter's, keeping the so-called QWERTY configuration, after the first six letters below the numbers. The computer's shift key, however, does not lower a type bar. It sends a digital signal to a word-processing program.

Typewriters generally see minimal use these days, except in underdeveloped countries that have little access to computers, or among individuals who are not computer literate or who still prefer one for addressing envelopes and filling out forms.

MANUAL TYPEWRITER

(Right) Typewriters do not write; they only type, as some people say. Still, the machines have churned out prose and poetry for generations of novices and professionals, many of whom could get 70 words a minute and more out of the old manual models.

TYPEWRITING

A mechanical typewriter (right) has type bars linked to key levers. When a key is pressed, its type bar strikes paper through an inked ribbon. The electric typewriter (bottom) uses an electric motor to lift the type bar and power the stroke.

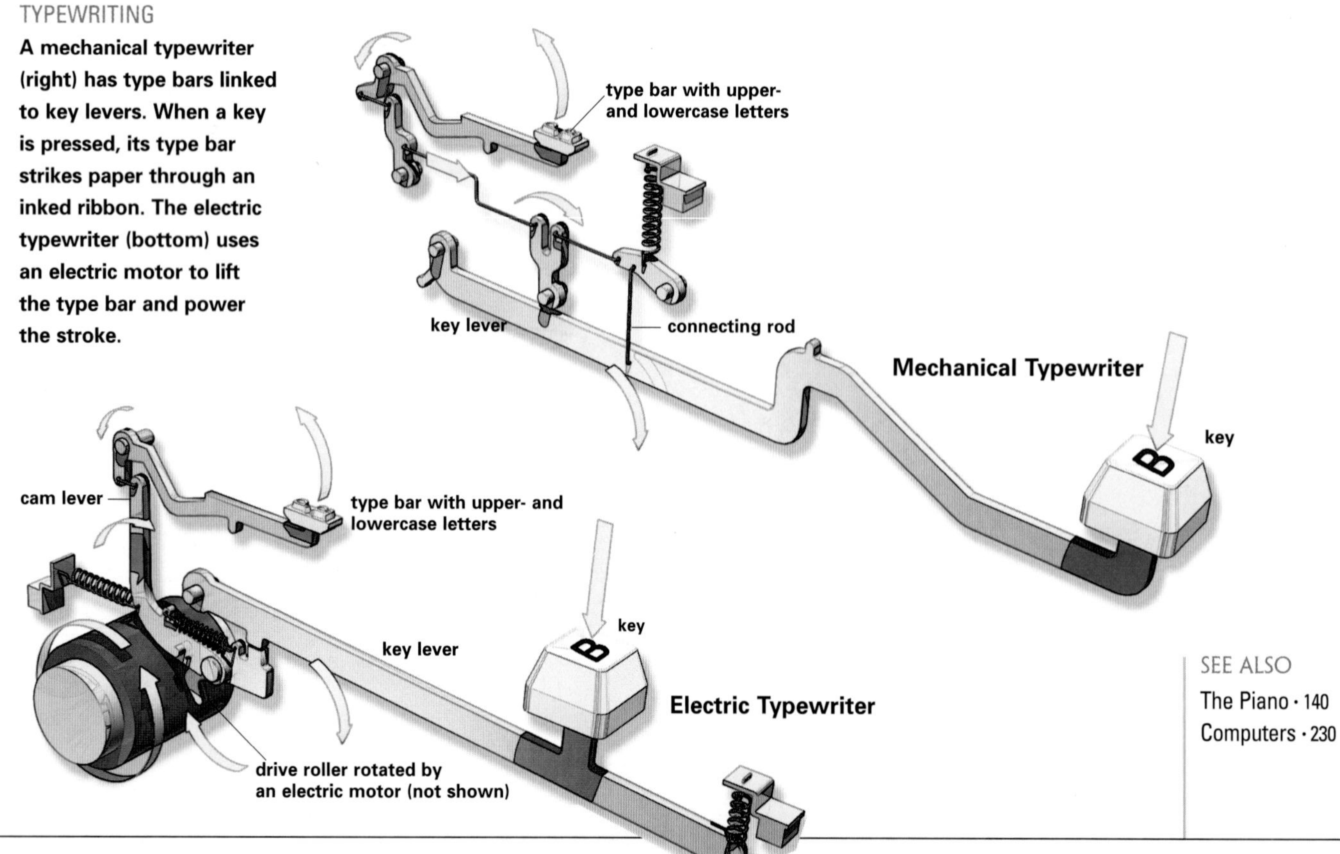

SEE ALSO
The Piano • 140
Computers • 230

BACK SPACE
RIBBON REVERSE
Q W E R T Y U I O P
A S D F G H J K L
SHIFT KEY
Z X C V B N M
SHIFT KEY

PRINTING PRESSES

Undisputed masters of invention, the ancient Chinese are credited with giving the world some of the earliest examples of printing. Their creations, which took the form of wood-block images on paper and silk, perhaps drew on even earlier techniques developed by the Babylonians and Sumerians to make individual name seals.

The need for mass distribution drives modern printing, and while Gutenberg's invention of printing by movable type, around 1446, revolutionized the way information is disseminated, the advent of so-called web presses and offset lithography put virtually everything fit to print into the hands of millions of readers. A newspaper's web press—a machine that prints a continuous roll of paper—can run as fast as 3,000 feet a minute.

Offset lithography is essential to modern printing. This versatile process transfers photographed images and text to a metal plate through photochemical action. The word "offset" means that the ink used to coat the plate does not print directly onto paper. Instead, the inked plate prints on a rubber "blanket" cylinder, and from there images are offset firmly onto paper.

Another system typesets with computer-generated text; a laser beam forms characters at the direction of the computer's electrical pulses, then transfers them to a film or light-sensitive paper. Computerized typesetting systems, which can set thousands of characters a second, also can circumvent paper by storing text and layouts as electronic files before sending them to be printed.

PRESSWORK

(Right) A six-color high-speed press rumbles along, blurring the action but not the quality of the print and images it leaves behind on paper.

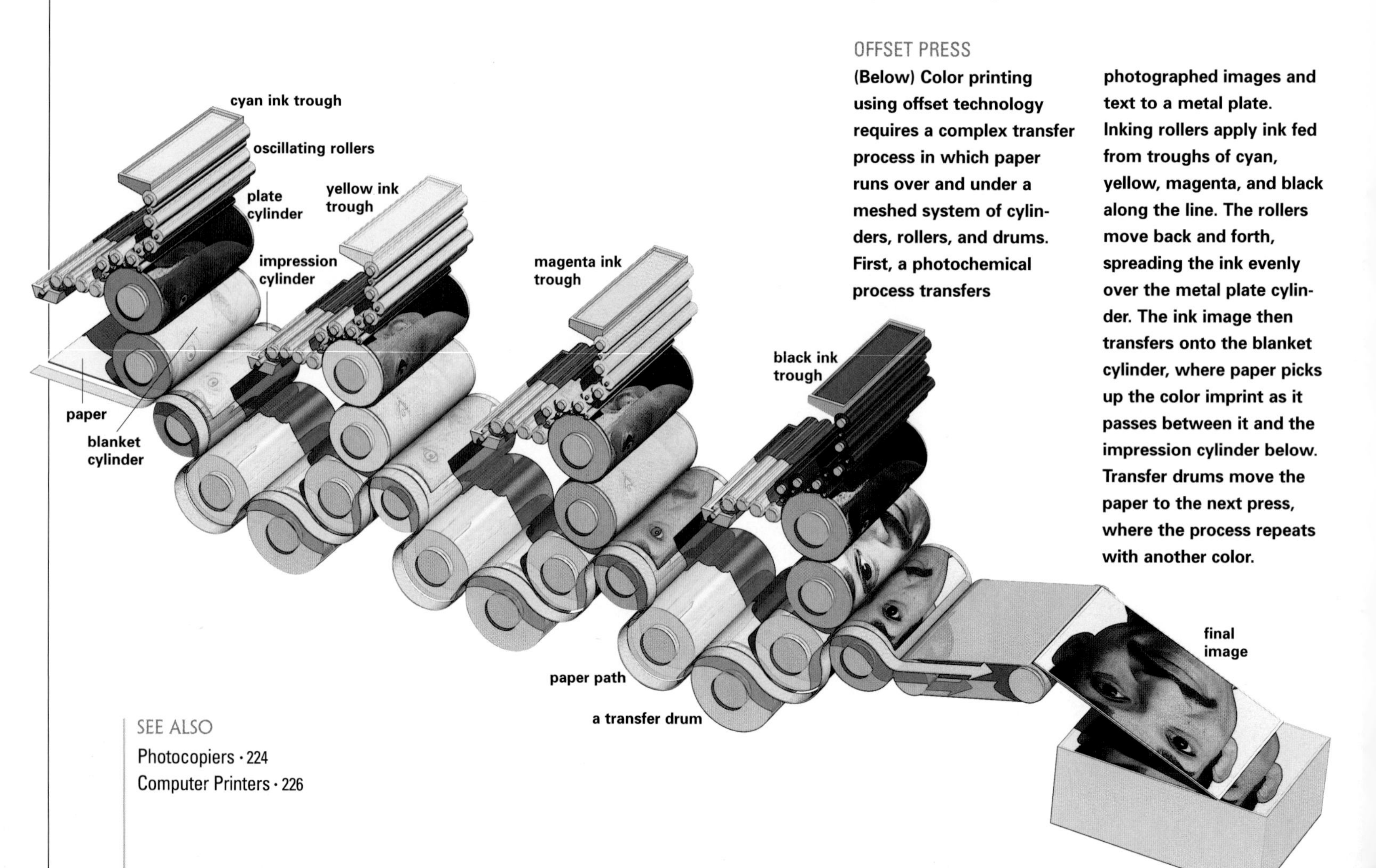

OFFSET PRESS

(Below) Color printing using offset technology requires a complex transfer process in which paper runs over and under a meshed system of cylinders, rollers, and drums. First, a photochemical process transfers photographed images and text to a metal plate. Inking rollers apply ink fed from troughs of cyan, yellow, magenta, and black along the line. The rollers move back and forth, spreading the ink evenly over the metal plate cylinder. The ink image then transfers onto the blanket cylinder, where paper picks up the color imprint as it passes between it and the impression cylinder below. Transfer drums move the paper to the next press, where the process repeats with another color.

SEE ALSO
Photocopiers · 224
Computer Printers · 226

PHOTOCOPIERS

USING WAXY, FINGER-SMUDGING CARBON paper was once the only way to make duplicate copies of printed material. By the late 1800s, this tedious process was replaced by the mimeograph, which relied on stencils cut into coated fiber sheets by a typewriter; a stencil, wrapped around an ink-soaked drum, made copies on the paper as ink seeped through the cut outlines of the characters. Offices also had the hectograph, a messy contraption loaded with a large roll of gelatin-coated fabric stretched across the top and fixed at the other end. An ink "master" typed on special, coated paper was pressed onto the gelatin sheet and then peeled off, leaving its imprint in a purple, smudgy, reverse image. To make a copy of the imprint, a sheet of transfer paper was placed on it and run over by a roller.

In 1959, the Haloid Xerox 914 copier emerged. A relatively cumbersome apparatus, it made copies on ordinary, untreated paper. Despite problems such as paper scorching, it gave rise to the modern Xerox Corporation and transformed an industry. Today, standard copying is a high-speed process called xerography (from the Greek for "dry writing.") It works on the principle of photoconductivity, by which certain materials conduct electricity under the influence of light. A key element in the copy machine is selenium, a by-product of copper refining and a poor conductor of electricity except when bathed in light. Selenium "reads" the electrical difference between a document's white and dark areas; the white areas, which reflect light, lose their electrical charge; the dark areas do not. Then it converts the pattern into an electrical version of the image. Charged ink powder, called toner, is dusted on; this material is attracted only to the charged image and is then fused onto the paper by heat.

Quality of the copy depends on the paper used. Standard copy machines generally accommodate lightweight moisture-balanced paper while computer printers, which are related to the photocopier, may sometimes require special paper depending on their function. However, paper is now sold as "multipurpose" or "dual-purpose."

MAGIC TOUCH

An office worker has only to push a button or two to produce quick and fairly inexpensive copies of a document. Centuries ago, scribes such as learned monks carefully copied manuscripts with quill pens.

SEE ALSO
Patterning Fabrics • 130
Film Camera • 156
Computer Printers • 226

COPY MACHINE

Light from a halogen lamp illuminates a document fed into a photocopier (below). White areas reflect a lot of light, while print reflects little. A mirror and lens send the light onto a revolving drum with a photosensitive coating, which converts the light into a latent electrical image. Positively charged toner adheres to the negatively charged areas on the photosensitive drum that correspond to dark areas on the original document (top right). A charging electrode sends a negative charge to paper on the drum, transferring toner and the translated image onto the sheet. Finally, heated rollers fuse the image to paper.

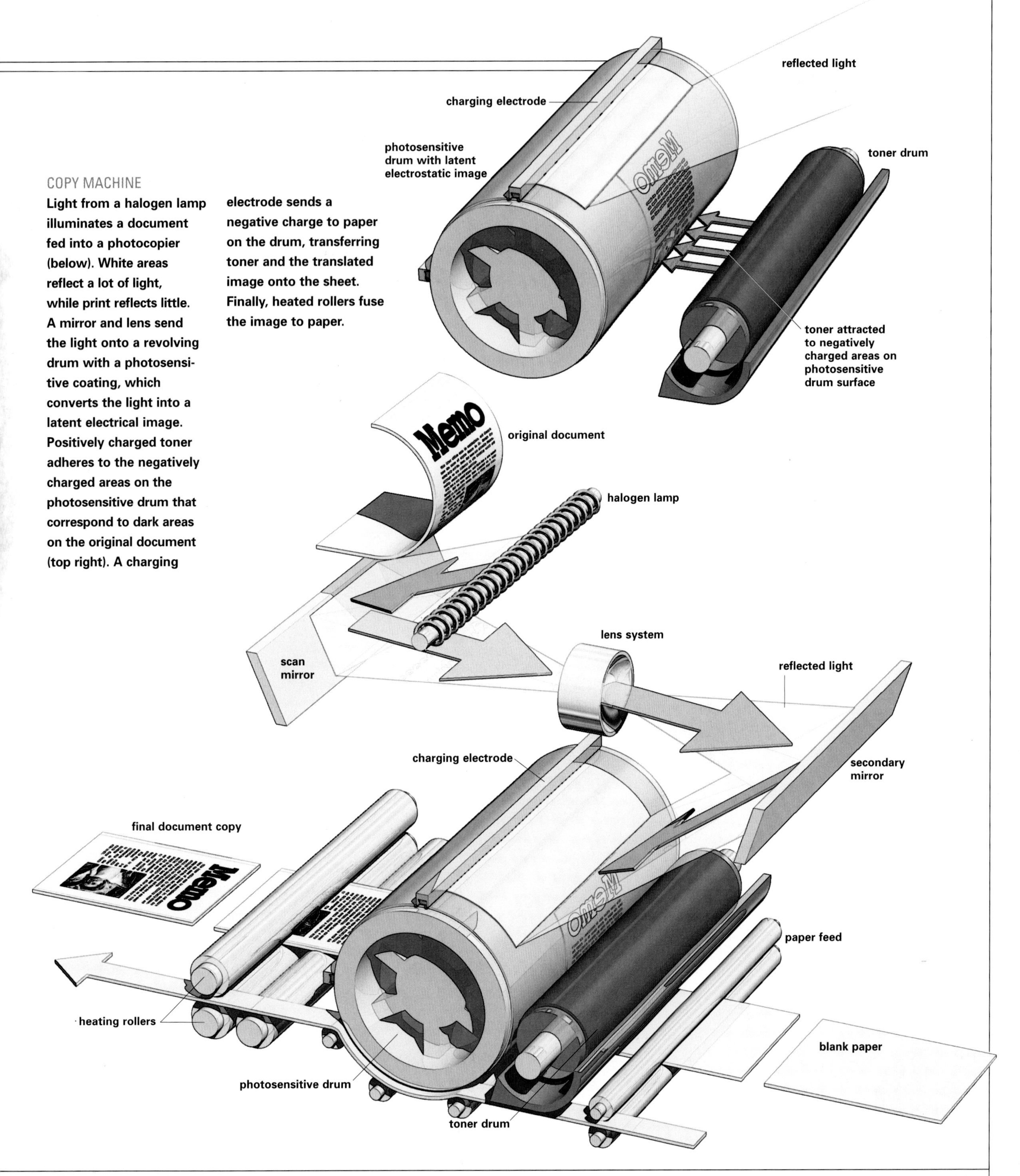

COMPUTER PRINTERS

EVEN IN A DIGITAL WORLD, ANY THREAT THE computer presents to the paper industry may well be a paper tiger. True, a computer can create and share data without a printer, but people who regularly use computers still want and need printers because in the last analysis paper is, well, paper—convenient, cheap, easy to use, familiar, and crash-proof.

The computer printer is an offshoot of the photocopier, and indeed the laser printer uses somewhat similar technology, including a drum, toner, and a paper feed. After the printer receives digital information from the computer about what is to be printed, its laser beam "draws" an image of the data on a photosensitive drum. The drum loses its charge wherever the beam strikes. A roller gathers toner and applies it to the drum. The toner particles stick to the charged areas, as in a copy machine, and are transferred to sheets of paper passed through a fuser. There, heat and pressure fix the image-carrying toner to the surface.

An ink-jet printer works by spraying precise dots of ink on sheets of paper through tiny nozzles; the dots become letters, numbers, and punctuation marks. The more nozzles the printer has, the faster it will pass over a page to produce a copy. The size and shape of the nozzles affect the size and quality of the ink drops. Early jet-printers frequently produced fuzzy and smeared images, but new printheads offer incredibly high resolution (measured in dpi, or dots per inch); cartridges with special inks have resolved the smearing problem. As with all computer-related technology, printer development is in perpetual motion: new generations scan, fax, and copy as well as print, while others can print on ink-jet printable CD-ROMS.

Perhaps a little-known wrinkle in computer printers is the Braille embosser, a device that punches thousands of Braille characters on sheets of paper. Embossers, which "print" on heavier than ordinary paper, are connected to a computer that needs special software to translate text from a word processing program into Braille files. The software that does this has a formidable task since Braille follows unique spelling, punctuation, and word-reducing rules.

A LASER PRINTER

(Right) A laser printer receives digital information from a computer. Like a photocopier, it has a drum, toner, and paper feeder; however, it relies on a laser fired at a spinning mirror. The mirror focuses a data image onto the negatively charged drum, which loses charge where the light strikes. Dark areas stay charged, attracting the toner that creates the image on paper.

IMAGE MAKER

(Left) A designer prepares material that will be fed into an offset printer. Creating such images has become relatively easy with software that allows a designer to resize, brighten, and make them printer-ready.

SEE ALSO
Patterning Fabrics · 130
Laser Surgery · 204
Photocopiers · 224
Computers · 230

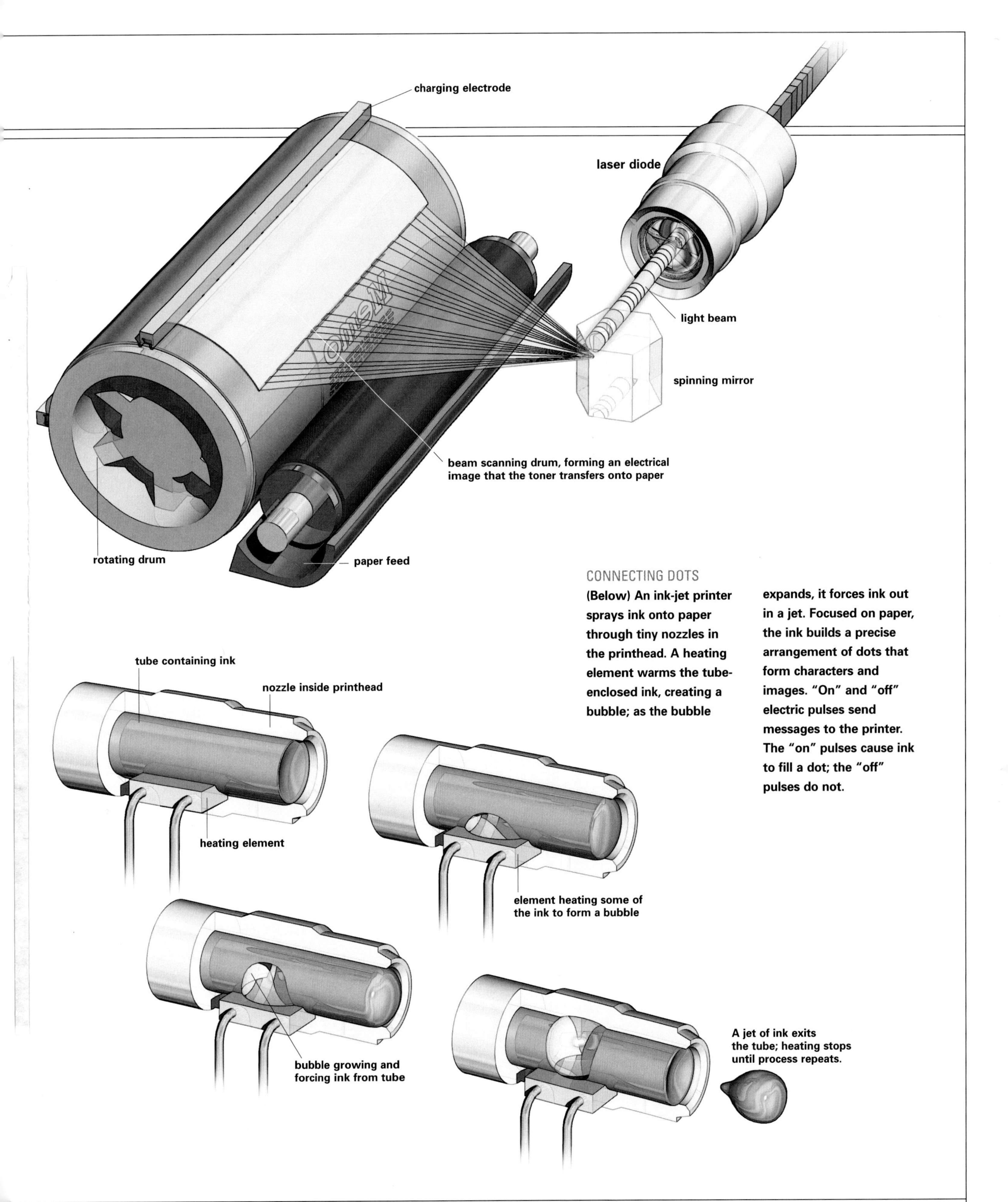

CONNECTING DOTS

(Below) An ink-jet printer sprays ink onto paper through tiny nozzles in the printhead. A heating element warms the tube-enclosed ink, creating a bubble; as the bubble expands, it forces ink out in a jet. Focused on paper, the ink builds a precise arrangement of dots that form characters and images. "On" and "off" electric pulses send messages to the printer. The "on" pulses cause ink to fill a dot; the "off" pulses do not.

CALCULATORS

WITH BRAINS MADE OF SILICON CHIPS carrying numbers coded into electronic pulses, and powered by minuscule batteries or solar cells, modern calculators have astonishing speed and accuracy. They also do infinitely more than basic arithmetic. Apart from handling complicated mathematics and scientific equations, the memory banks of special calculators do cooking conversions, and bar drink and carpentry measurements; they can crunch home financing and interest numbers, and do metric and currency conversions.

A number of interrelated electronic processes are at work inside calculators, and two of the most important are the binary number system and the principles of Boolean algebra. Instead of using the based-on-ten decimal system, the binary system counts by twos (1s and 0s only).

Inside a calculator, decimal numbers and a host of functions offered on the keyboard are converted to a sequence of coded binary numbers that are stored on a memory chip. Next, Boolean algebra, named after English mathematician George Boole in the 1840s, comes into play. Boolean principles are behind thousands of linked microscopic "switches," or logic gates, on a chip. Labeled with such Boolean nomenclatures as OR, AND, and NOT, the gates (made of transistors) evaluate all the steps of calculating by switching off for 0 and on for 1—a judgment that is the binary equivalent of false and true—and sending these digital signals to appropriate locations. In units called half-adders that connect to form full-adders, the logic gates process signals so quickly that when the equal (=) key is pressed, the answer appears instantaneously.

DOING THE MATH

A far cry from the pencil and paper computations once required to solve mathematical problems, calculators now do it all and in far less time. Here, a Nigerian student in Lagos punches the keyboard to find her answer.

SEE ALSO
Computers • 230
Computer Chips & Transistors • 232
Computer Memory • 234

BINARY					DECIMAL
8	4	2	1		
0	0	0	0	=	0
0	0	0	1	=	1
0	0	1	0	=	2
0	0	1	1	=	3
0	1	0	0	=	4
0	1	0	1	=	5
0	1	1	0	=	6
0	1	1	1	=	7
1	0	0	0	=	8
1	0	0	1	=	9
1	0	1	0	=	10

1 x 8 + 0 x 4 + 1 x 2 + 0 x 1

BINARY CODE

(Left) A computer or a calculator converts information to a so-called base-2 system, a binary code that relies on 1s and 0s instead of the based-on-10 decimal system. A number's binary code equivalent uses blocks, or positions, arranged from right to left and worth 1, 2, and powers of 2 (such as 4 and 8). Building the binary 10 requires a 2-block, an 8-block, and no 4- or 1-blocks.

INSIDE A CALCULATOR

(Below) A calculator operates on the "yes" and "no," "on" and "off" principle that translates the binary code into matching voltage states. For example, when the binary system writes decimal 10 as 1010, it says the following: "Blocks of 2 and 8, yes; blocks of 1 and 4, no." In a calculator, switches see that 0s, which mean "no," or "off," are low voltage and that 1s, which mean "yes," or "on," are high voltage. When a person presses decimal numbers on the keypad, a decoder produces the binary equivalents held in the storage cells.

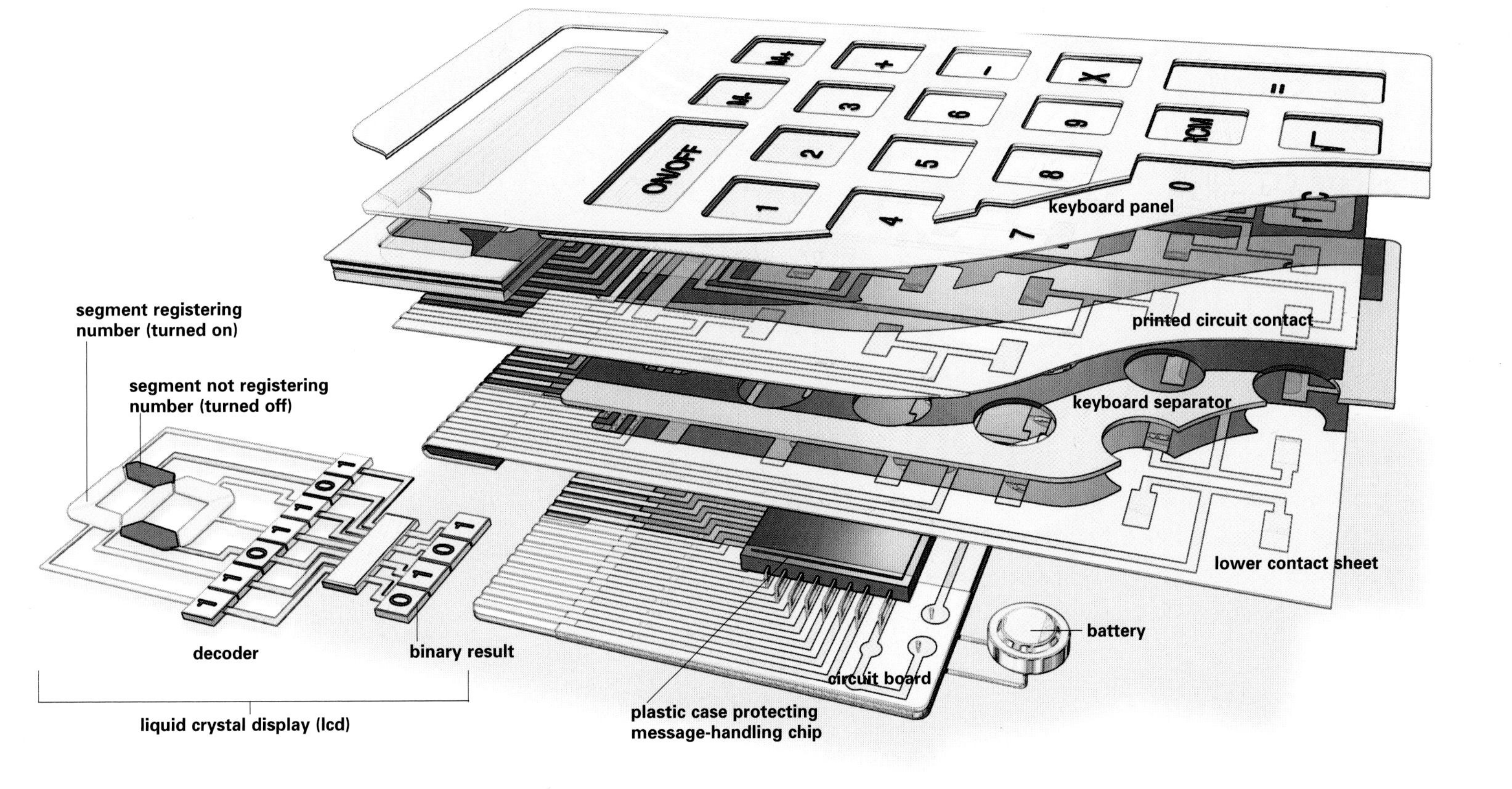

COMPUTERS

IN 1834, CHARLES BABBAGE, AN ENGLISH BANKER'S son and a mathematician, conceived what he called an "analytical engine," a conglomeration of levers, gears, cogs, and wheels designed to run on steam power. Answers would be printed automatically, while the machine controlled itself internally by punched cards, an idea suggested by early mechanized looms that relied on such cards to run the pattern-weaving apparatus. While Babbage's idea never caught on, he had, incredibly, mapped out most of the key elements found in modern computers. These included an arithmetic/logic unit, memory, input and output mechanisms, and automatic sequencing of the instructions without human prodding.

Today, just as the TV has become a multimedia phenomenon, so too has the computer. It does far more than solve math problems, it also cracks codes, translates languages, stores reams of information in small spaces, and lets us compose a novel, surf the Internet and transmit e-mail. PCs broadcast video news, handle electronic banking, make our travel arrangements, and buy and sell real estate for us. An ever growing list of peripherals, some of them wireless, enable them to accept and scan photos from a digital camera, perhaps signaling the demise of "chemical" photography; "burn" CDs and "write" DVDs; download music; play games; and link micro-printers to pocket PCs.

Behind such multiplicity is the enormous capacity now built into a computer's memory bank, such as in RAM, the random-access memory storage system that contains programs and the systems that run the computer; the more RAM, the more efficient and faster the computer. Improvements in the CPU, the central processing unit that communicates with peripherals and performs a host of mathematical calculations and data-moving functions, are also key. Nowadays, processors manage around a thousand MIPS (millions of instructions per second), compared with earlier single-digit MIPS performance. Peripherals-expanders, known as SCSI (small computer systems interface), have also turned computers into jacks-of-all-trades; installed into a computer, a high-bandwidth SCSI can support several devices on a single connector.

EVER CYBER-READY

(Left) Wearing a 5-Gig hard drive computer boosted by a powerful video chip, a user experiences the view of a 15-inch monitor.

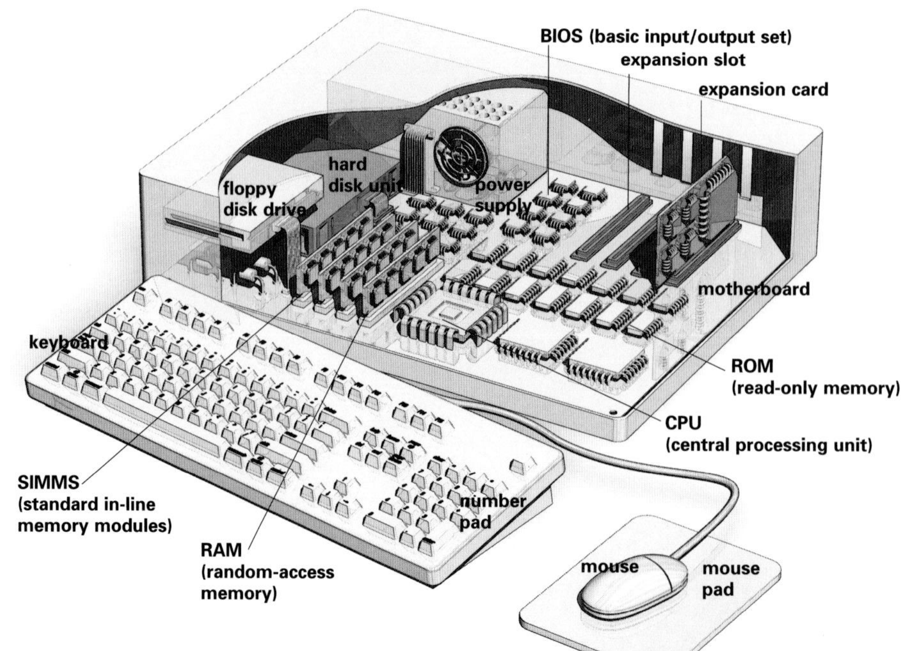

INSIDE A COMPUTER

(Below) The electronic circuitry of the central processing unit (CPU) contains arithmetic/logic and control units for calculating, transferring data, and executing instructions. Two memory units store program instructions and data for the CPU: RAM (random-access memory) holds data only while the current is on; ROM (read-only memory)—a permanent strongbox—stores digital information.

SEE ALSO
The Loom · 126
Typewriters · 220
Computer Chips & Transistors · 232
Computer Memory · 234
Backup Storage · 236

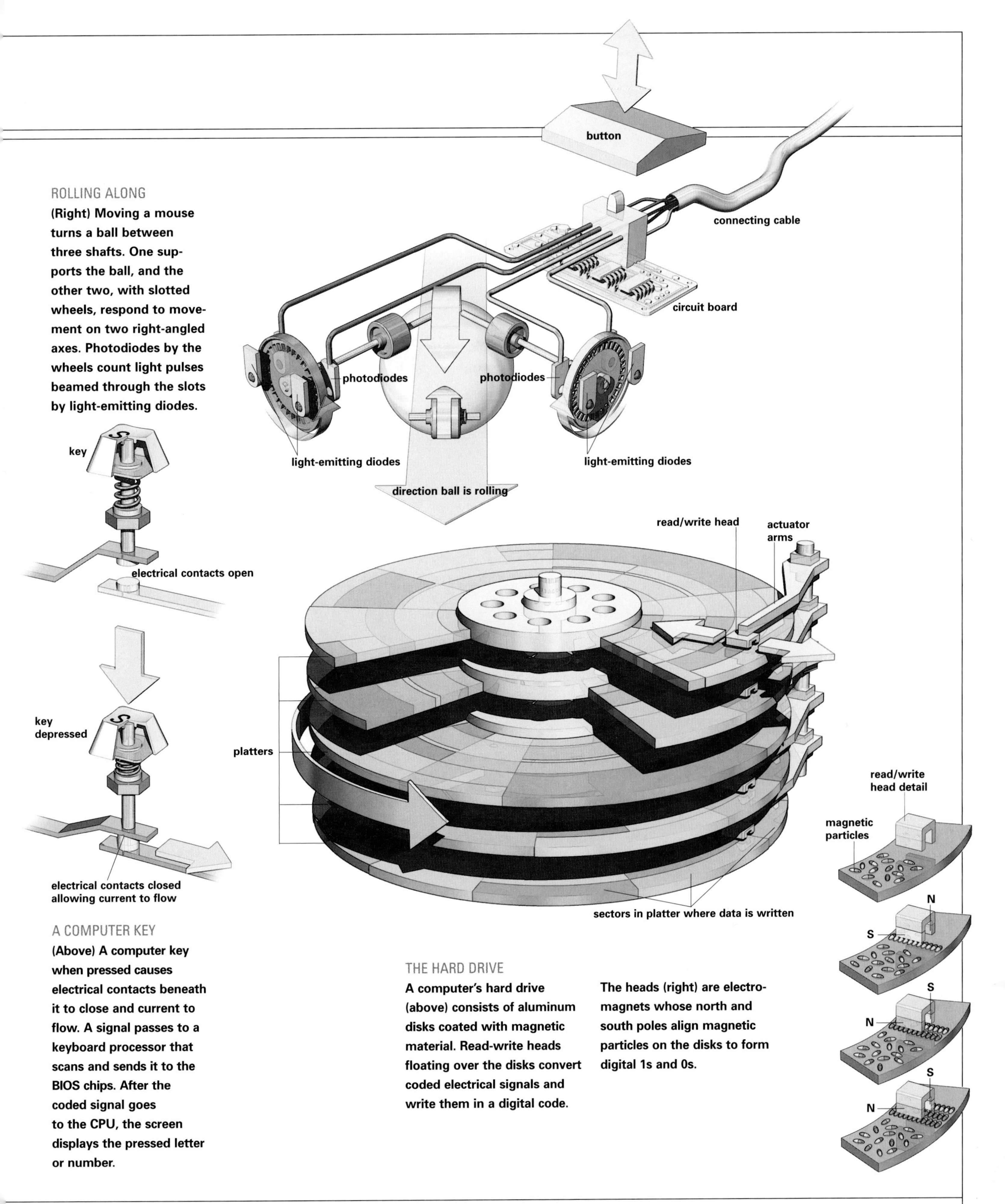

ROLLING ALONG

(Right) Moving a mouse turns a ball between three shafts. One supports the ball, and the other two, with slotted wheels, respond to movement on two right-angled axes. Photodiodes by the wheels count light pulses beamed through the slots by light-emitting diodes.

A COMPUTER KEY

(Above) A computer key when pressed causes electrical contacts beneath it to close and current to flow. A signal passes to a keyboard processor that scans and sends it to the BIOS chips. After the coded signal goes to the CPU, the screen displays the pressed letter or number.

THE HARD DRIVE

A computer's hard drive (above) consists of aluminum disks coated with magnetic material. Read-write heads floating over the disks convert coded electrical signals and write them in a digital code.

The heads (right) are electromagnets whose north and south poles align magnetic particles on the disks to form digital 1s and 0s.

COMPUTER CHIPS & TRANSISTORS

EARLY COMPUTERS RELIED ON VACUUM TUBES—the airless glass "bottles" in which electric and magnetic fields control the movement of electrons—to switch electrical signals and to add, multiply, store, and compare data. Developed for the radio industry, the tubes permitted machines to calculate several thousand times faster than earlier electromechanical relays. The transistor, infinitely smaller and far more frugal with power, would put that speed to shame. Invented in 1947 at Bell Laboratories, the transistor is an electronic device made of semiconducting material. This material carries current only under certain conditions, such as when a tiny amount of voltage is applied. Computer microchips, the tiny flakes of silicon that control all of a computer's behavior, can each contain millions of transistors linked by fine connections to form integrated circuits.

Transistors serve as electrical switches that quickly turn current off and on; this action precisely controls the flow of binary numbers, the digital data the computer uses to perform its multitude of chores. Depending on their arrangement, and purpose of the chip—memory, central processing, logic, or timing that synchronizes all the signal activity of the chip—transistors, with their ability to interpret and channel specific digital information, govern virtually everything a computer does. They control moving an icon with a mouse or deleting a letter with a keystroke; they also perform complex arithmetic and store programs and data.

In a recent example of what's referred to as "grid-computing," Harvard physicist John Huth linked 2,700 central processing units from computers at 27 different institutions; the resulting "supercomputer" ran an average of 500 jobs simultaneously, and sent more than 100 terabytes (a million million bytes) of information over the Internet. Grid technology, still a fledgling science, is promising.

Perhaps even more awe-inspiring is the amazing carbon nanotube, a tiny cylinder of carbon atoms that is 500 times smaller than a silicon-based transistor and 1,000 times stronger than steel. IBM researchers built the world's first array of transistors out of the tubes, a major step in constructing molecular-scale electronic devices. The more transistors there are on a chip, the faster the processing speed. Thus miniaturization will undoubtedly have a monumental impact on chip performance.

Other research has focused on actually getting transistors to assemble themselves by bonding nanotubes to strands of DNA, the biochemical code that determines life. Transistor designers are also researching so-called quantum mechanical versions, which have been likened to traveling through a tunnel in a mountain that would be impossible to climb. The devices essentially enable electrons to "tunnel" under energy barriers to get to places that would appear impossible to reach. The shortcut means speed to destination, perhaps more than ten times the speed of some of the fastest transistor circuits.

The other advantage is that quantum transistors would run on far lower power than conventional transistors, and far fewer of them would be required in an integrated circuit.

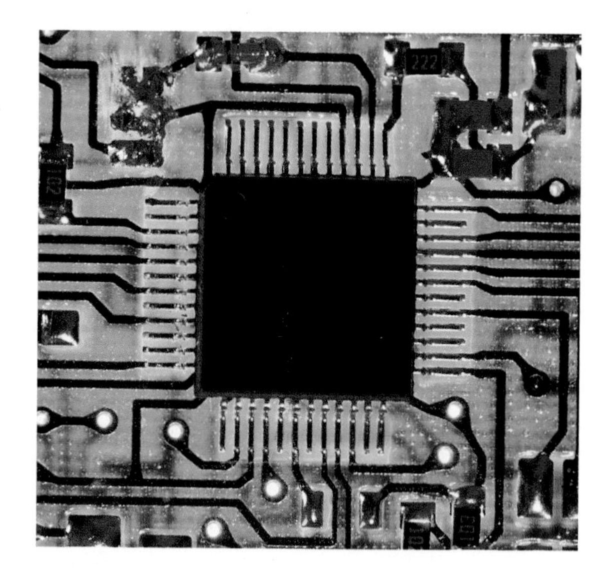

MICROCHIP

(Left) Engraved with millions of circuits, wafer-thin bits of silicon contain tiny transistors that control current flow with their on-off switching capability. Microchips have different functions and a variety of arrangements on a motherboard, the printed circuitry holding components of a microcomputer system.

MICROPROCESSOR

(Right) Complex integrated circuitry—packed with miniature diodes, transistors, capacitors, resistors, and conductors—handles a computer's central processing. The Pentium processor shown here has millions of transistors.

SEE ALSO
Radio • 142
Calculators • 228
Computers • 230
Computer Memory • 234
Backup Storage • 236

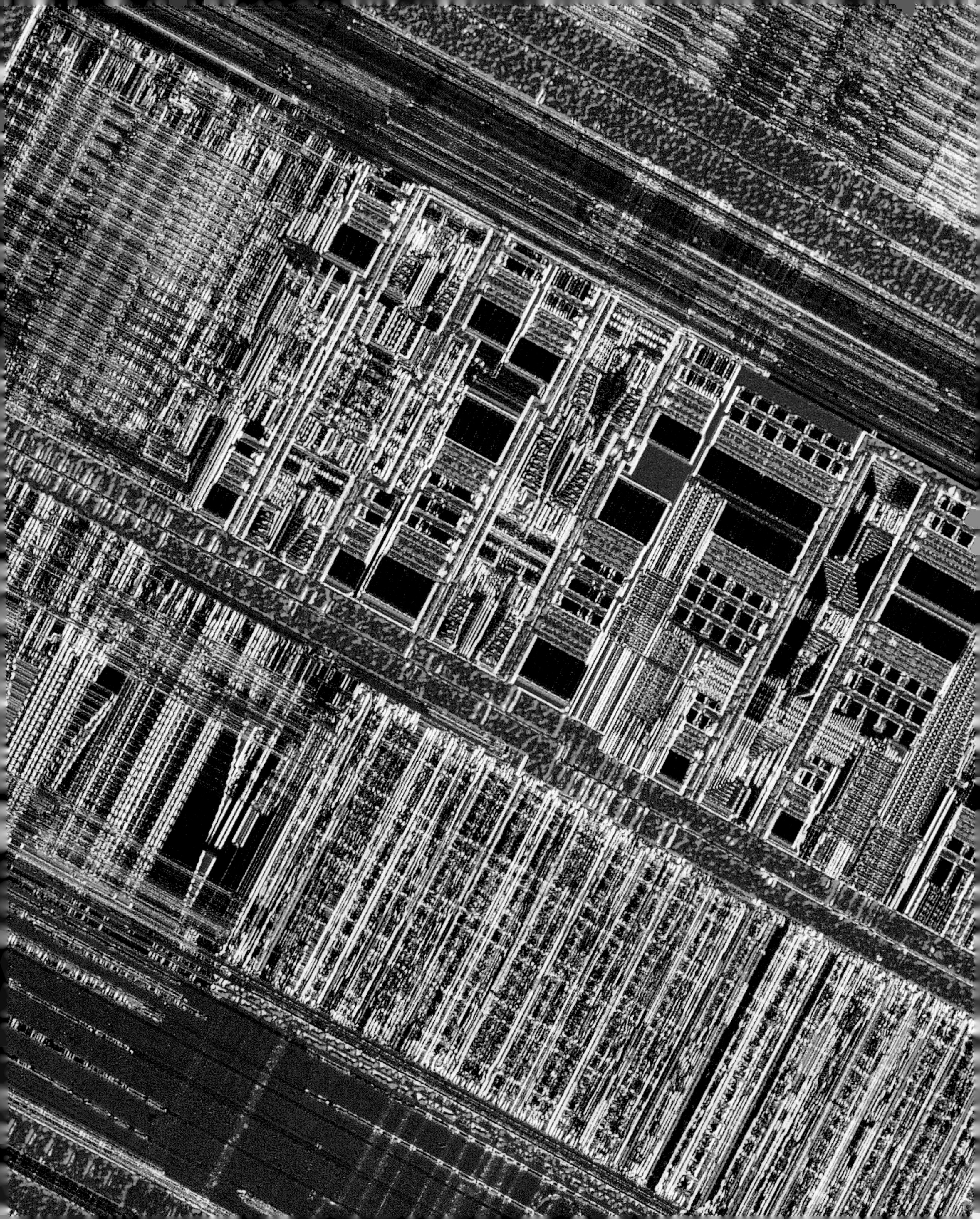

COMPUTER MEMORY

A computer program is a set of coded instructions that tell the machine which operations to perform and when and how to perform them. A program is what enables the computer to process words, play games or music, or store data.

Instructions and other information must be remembered if they are to be followed, however, and computers use different kinds of memory to do that within their circuitry. Random-access memory (RAM) is where program instructions and data are stored until the central processing unit—a microprocessor on a chip packed with computational controls that can move data—can access them. RAM is dependent on chips with memory capacity measured in accumulations of bytes, the basic units of data. (A byte is a series of eight consecutive binary digits that can represent one from hundreds of alphanumeric characters.) Typically, a home PC might have a minimum of 64 megabytes of RAM, a prodigious capacity, considering that one megabyte can store a million characters. However, RAM can be far larger, going to 256 or 512 megabytes, depending on how much is required to run various applications, like graphics. Additional RAM can be installed in a computer by simply inserting a new chip. (RAM storage capacity should not be confused with what is available in the computer's hard drive, the permanent storage areas also referred to as disc space that can hold billions of bytes.)

RAM, which comes in several types (DDR RAM, which is faster than the ordinary, comes in many new computers) is the computer's short-term memory, concentrating on the moment. While more of it means faster computing, it holds data only when the current is switched on; hence, the reason for the cardinal rule of computing: save, save, save.

ROM, for read-only memory (meaning it can be read but not altered), is the other key memory bank. It is a permanent strongbox that stores digital information, such as start-up and operating programs and various computer systems, even when the computer is switched off. Another type of computer memory is EPROM (erasable programmable read-only memory), which, as its name signifies, is erasable and reusable. It may be used to store a computer program called BIOS (basic input/output system) that the central processing unit draws on to operate start-up functions.

CIRCUIT BOARD

(Right) Repository for computer chips and other interconnected electronic components, a circuit board is a model of intricacy. Computers have one or more boards that are known as cards, chief among them the motherboard, or system board, which contains the central processing unit (CPU), memory and other elements.

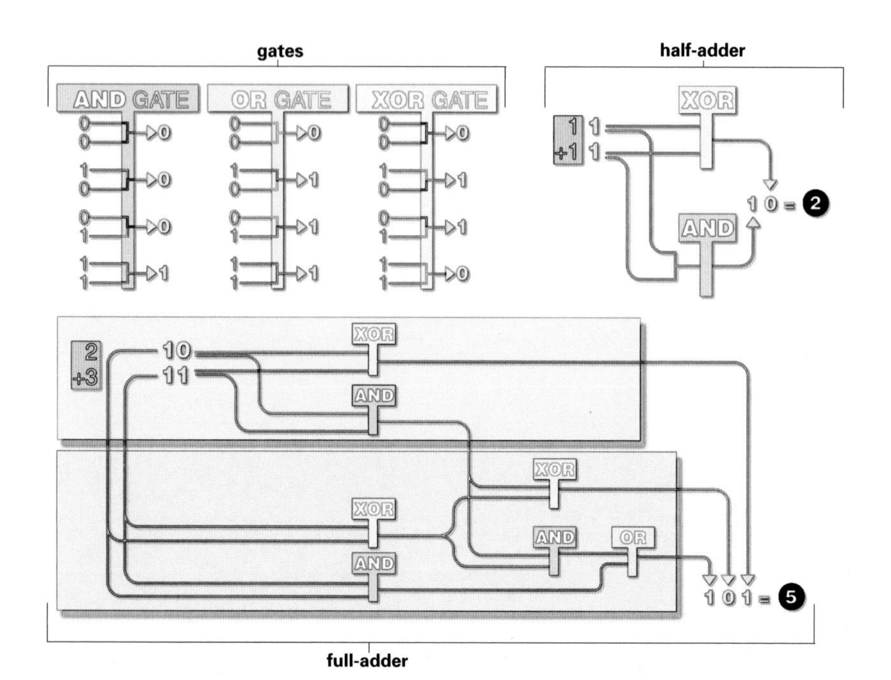

NUMBER CRUNCHER

(Left) Logic gates, or transistors on chips, control binary signal flow. Gates AND, OR, and XOR switch voltage off for 0 and on for 1 (binary equivalents of false and true). Adder circuits do math: Half-adders add two binary digits; full-adders process two digits and a carry. The switching steps below add 2 and 3 in binary code to get 5.

SEE ALSO
Computers · 230
Computer Chips & Transistors · 232
Backup Storage · 236

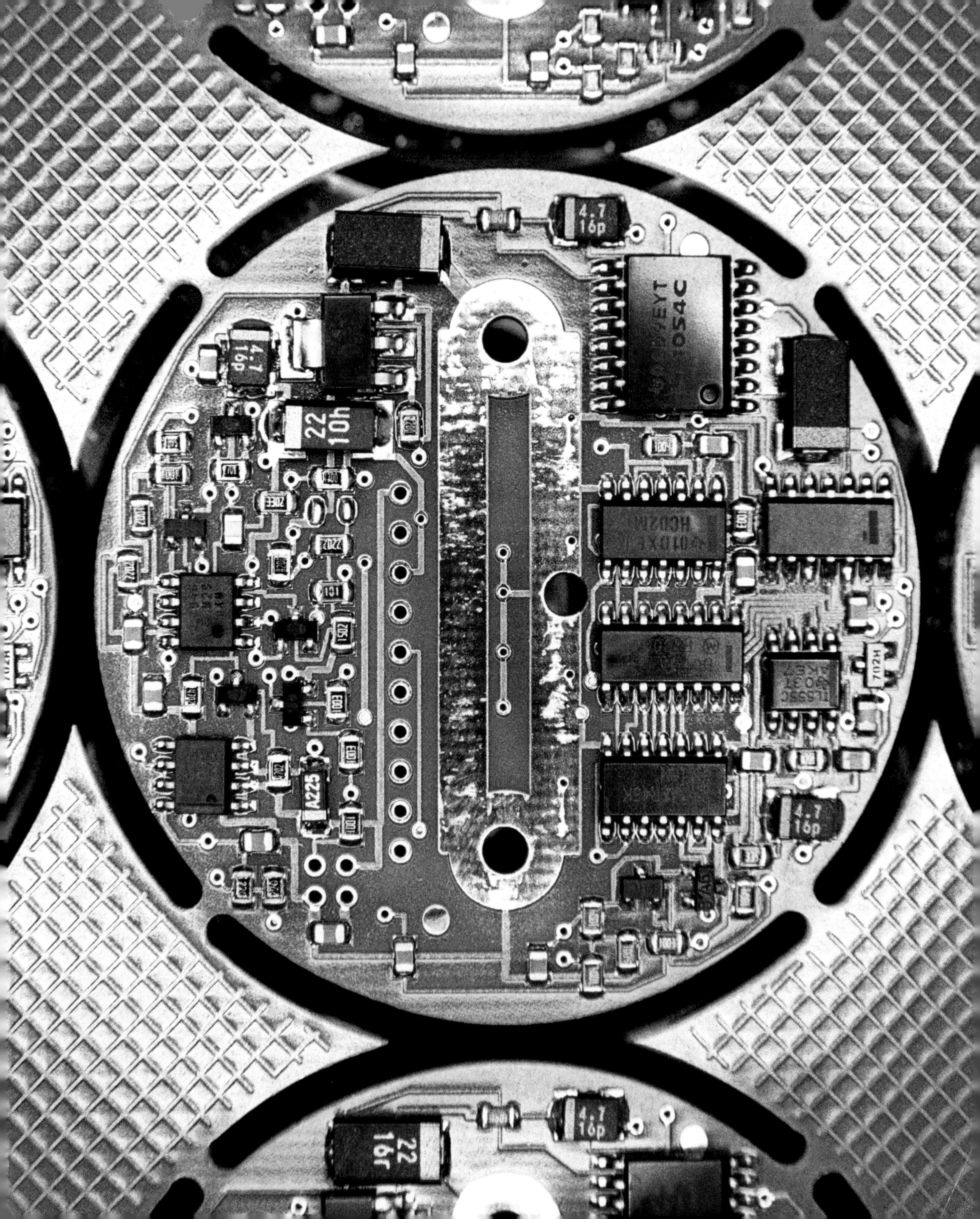
4.7
16p
22
10h
A225
22
16r

BACKUP STORAGE

BECAUSE A COMPUTER GENERATES VAST quantities of information, storage is essential. The depositories in general use are built-in hard drives that store volumes of digital code, typically measured nowadays in gigabytes (a gig, as it is called, stores a billion characters, or a thousand megabytes) and removable disks of all sorts.

A hard drive is a storage medium that can hold many gigabytes of information. It comprises a stack of hard disks that spin about in a magnetic enclosure. Information sent to the drive is written on the disk faces in a magnetic pattern of digital 1s and 0s by electromagnetic heads. These heads are similar to the ones used in a tape recorder and can read as well as write, much as the tape recorder both records and plays back.

The old standby, the 3.5-inch floppy disk, is also magnetic, but its heads touch the spinning disk surface while that of a hard drive's float millionths of an inch over it. Inserted into the computer, the floppy still does its job well, but its drive is a slower reader than a hard drive, and it holds far less data, somewhere around two megabytes. A new generation of removable storage devices that can hold enormous amounts of data in the multigigabyte range has nudged floppies aside. Some devices, with storage equal to that of 70 or more floppy disks, are inserted directly into a slot in the computer and can store and read with magnetic heads that mimic those of a hard drive. Others, notably those with gigs of space, may require both an external drive unit that is plugged into the computer and an insertable storage disk.

CD-ROMs, compact laser-read disks that are just like the audio version, can also digitally store computer text as well as music and images. Some, like the CD-R (for CD-Recordable), which holds about 700 megabytes, can be written to only once. All test and audio that it accepts can be presented, but not erased, and they are best used to store data for long periods. The CD-RW (for CD-ReWritable), on the other hand, can be erased and may be reused repeatedly. (Fast computers with so-called DVD-ROM drives can read both types of CDs.)

On a broader scale, storage-area networks (known as SAN) are local or far-flung networks of storage devices connected to one another and to servers and clients. High-speed operations that use fiber optic connectors, their storage apparatus is shared. SANs can also accommodate subnetworks with backup storage devices called NAS (network-attached storage) that can be plugged into the system and have a hundred and more gigabyte capacity.

SHINY CONVEYORS

(Right) CD-ROMs store a wealth of entertainment and information on digitally coded, laser-read, and big-byte-capacity disks.

DISK-MAKER

(Below) Whether floppy, hard-drive, or CDs, computer disks are digital storage mediums that encode and decode data.

SEE ALSO
Computers · 230
Computer Chips & Transistors · 232
Computer Memory · 234

CAPITOL

THE INTERNET

When the Internet was created in 1969 as a U.S. Defense Department program known as ARPANET (Advanced Research Projects Agency), its primary goal was to establish a secure communications system that would connect computers between different locations and survive in the event of hostilities. Soon, researchers and academics began to use the system. Today's Internet is many networks arranged like a gigantic spider's web throughout the world. Any one corporation or governmental agency does not own "it," and it is publicly accessible, with many millions of people plugged into its World Wide Web, the www. While Internet "traffic" still travels over standard telephone lines, it now also moves through cable-television lines, and via satellites, fiber optic links, and radio signals.

At the heart of the system are the Internet Service Providers (ISP), computerized links to the Internet. Chains of ISPs communicate through public network access points (NAPs, which cover a broad geographic region) and local access points, called LAPs. A phone line or one of the other transmission modes grants access to an ISP, while a browser, a software program that connects with a server, allows the user to view and navigate (surf) the Internet. To help locate information there are search engines, which are enormous databases.

Data available on the Internet are broken down electronically into "packets," which may contain coded address information, and are sent out simultaneously over a variety of routes from server to server, moving through exchange or access points. But in order to accommodate a request for information, servers must send the electronic information to the right "computer client"; moreover, once they get there all the data contained within have to be collected and reassembled. All of this is the job of the communications "protocols," which are sets of rules and regulations established for the servers. The most important of these are the TCP, the Transmission Control Protocol, which collects and reassembles the fragments of data, and the IP, the Internet Protocol, which handles the proper routing. Protocols employ various forms of cyberspeak to get their work done, like the familiar http://, for "hypertext transfer protocol," which is one of the most important "languages" used in addresses and to move hypertext files on the World Wide Web. (Hypertext refers to any text with connections to words and phrases in a document that can be selected by a client, enabling another document to be retrieved.) Other protocols translate electronic mail (e-mail) and direct file transfers, identify the origin of information, and allow one to log onto, or determine who's logged onto, a remote computer.

Web addresses have another essential component: a specific "domain" name, and the "dot parts" that follow it, such as .com, for commercial; .org, organization; .edu, education; .gov, government; .mil, military; .rec, recreation; and .net for network business. When complete, a Web address is called a URL (Universal Resource Locator). Typing in a combination of, say, the protocol http:// with a unique domain name on the computer alerts the servers online to the nature of the request, and where to send it.

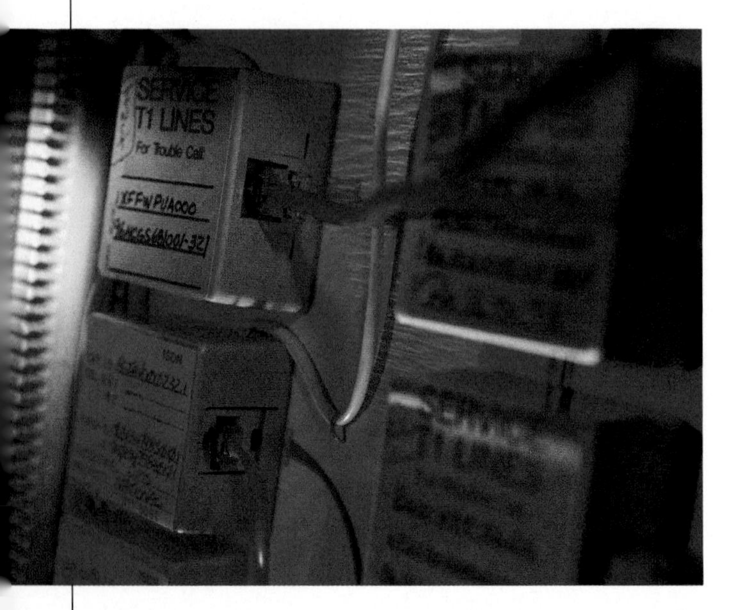

PLUG IN TO THE NET
(Left) Easily accessible with simple telephone jacks and cord, the World Wide Web's more than 92 million gigabytes of data (enough to fill a two-billion-volume encyclopedia) inform, educate, and entertain countless users. Perhaps the most valuable tools of the information age, the Web and the Internet it draws on have dramatically changed the way the world obtains, compares, and keeps track of knowledge and news.

WI-FI USERS
(Right) Customers at the Canvas Café in San Francisco take advantage of free wireless Wi-Fi (Wireless Fidelity) Internet access. Wi-Fi enables one to connect Wi-Fi-configured computers to the Net from hotel rooms, homes, restaurants, coffeeshops, libraries, and malls.

SEE ALSO
Fiber Optic Cables & DSL · 210
Satellite Communications · 218
Computers · 230
E-mail · 240

E-MAIL

When the fast riders of the Pony Express were carrying the mail, they needed eight to ten days aboard numerous horses to get a letter from Missouri to the Pacific coast. Today, even overnight mail by truck and plane seems slow, now that we have electronic mail.

E-mail is mail sent from one computer to another via a telephone network, while the Internet over which it routinely runs is a World Wide Web. Sending e-mail or raising an Internet site seems easy, but the technology that gets it from here to there involves an intricate process that begins at the keyboard and includes more station stops than a Pony Express route. E-mail requires a software program that allows one to compose, send, receive, forward, and reply to text files. Like letter mail, e-mail uses addresses, most of which include the symbol "@," which separates the recipient's (and the sender's) name from his or her mail service name.

When a message is sent through the network, from computer to computer, it follows a routing program called SMTP, for Standard Mail Transfer Protocol. Messages carry an elaborate letter and number code that tells each computer station what it must do with the message so that it can reach its destination in an electronic mailbox. Messages can be delivered in seconds or minutes to computers anywhere in the world, and usually only for the price of a local telephone call. You can also attach documents and digital imagery to e-mail. Personal e-mail can be accessed from anywhere with the appropriate equipment and software. Aided by a voice encoder, e-mail can compress a message and deliver it orally; or get it read by an animated character when the recipient is occupied elsewhere. Encryption add-ons enable the user to code the message for only the eyes of the recipient, and wireless devices allow users to send and receive e-mail with a single small battery. Like regular mail, e-mail is often plagued by junk mail, unwanted material known as SPAM. Filtering engines generally take care of the intrusion.

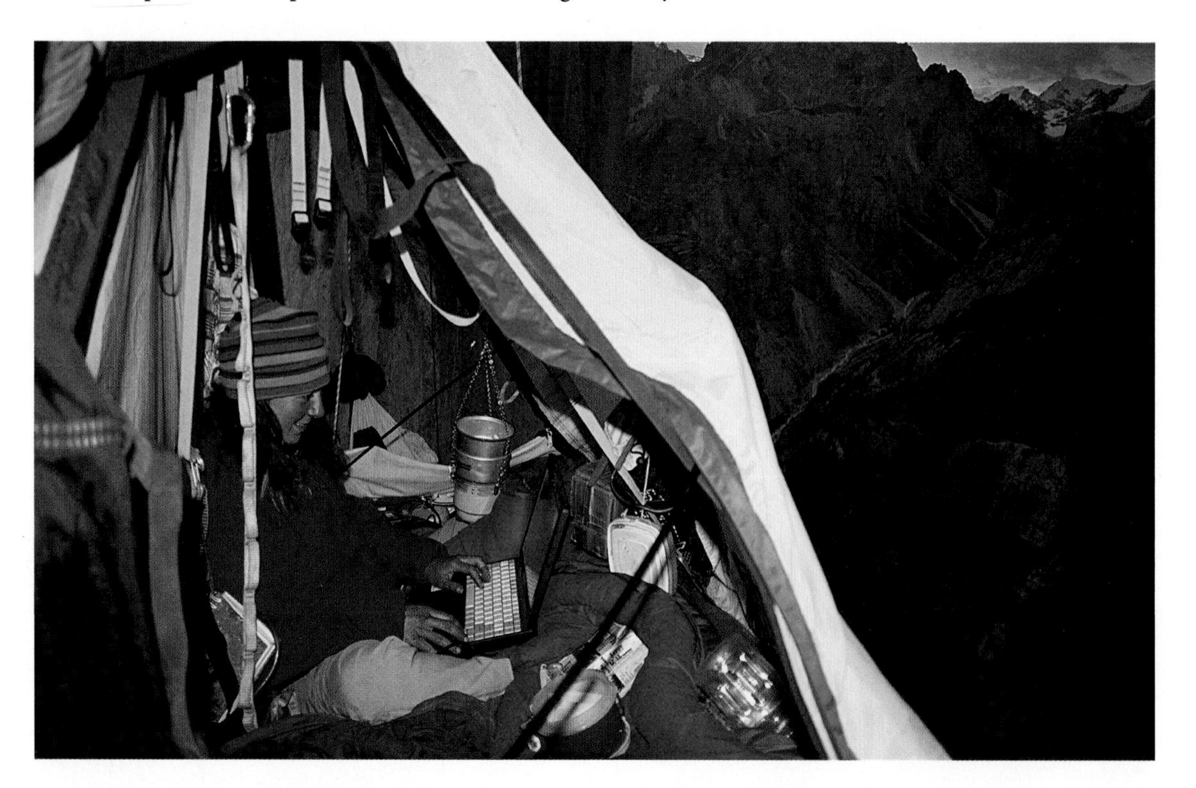

ACCESS

(Right) Satellites in five positions of the Astrolink system may soon provide easier, cheaper, and additional Internet access. Placed in geosynchronous orbit, each satellite will remain above the same spot on the Earth's surface. Internet users will be able to access the system with simpler, nontracking terminals.

ALWAYS IN TOUCH

(Left) Seated in a tent atop Tahir Tower in Karakoram, Pakistan, a woman reads and delivers her e-mail as casually and efficiently as she would in an office.

SEE ALSO
Sending Signals • 144
Cell Phones • 214
Satellite Communications • 218
Computers • 230
The Internet • 238

NETWORKING

(Below) A typical network connects businesses, homes, universities, and government agencies, allowing access to information. Powerful gateway computers operated by service providers connect different wide-area networks to one another, providing Internet access and a host of services, relaying digitized information and making each network's computers able to communicate with one another. Satellites and phone lines relay data and messages between computers, each one equipped with a modem or other device for transmitting and receiving data. Routing computers along the way decode instructions on the transmissions that tell them how and where to send the messages. Conceived in 1969 by the Advanced Research Projects Agency at the U.S. Defense Department, the first nationwide network linked computers at four universities.

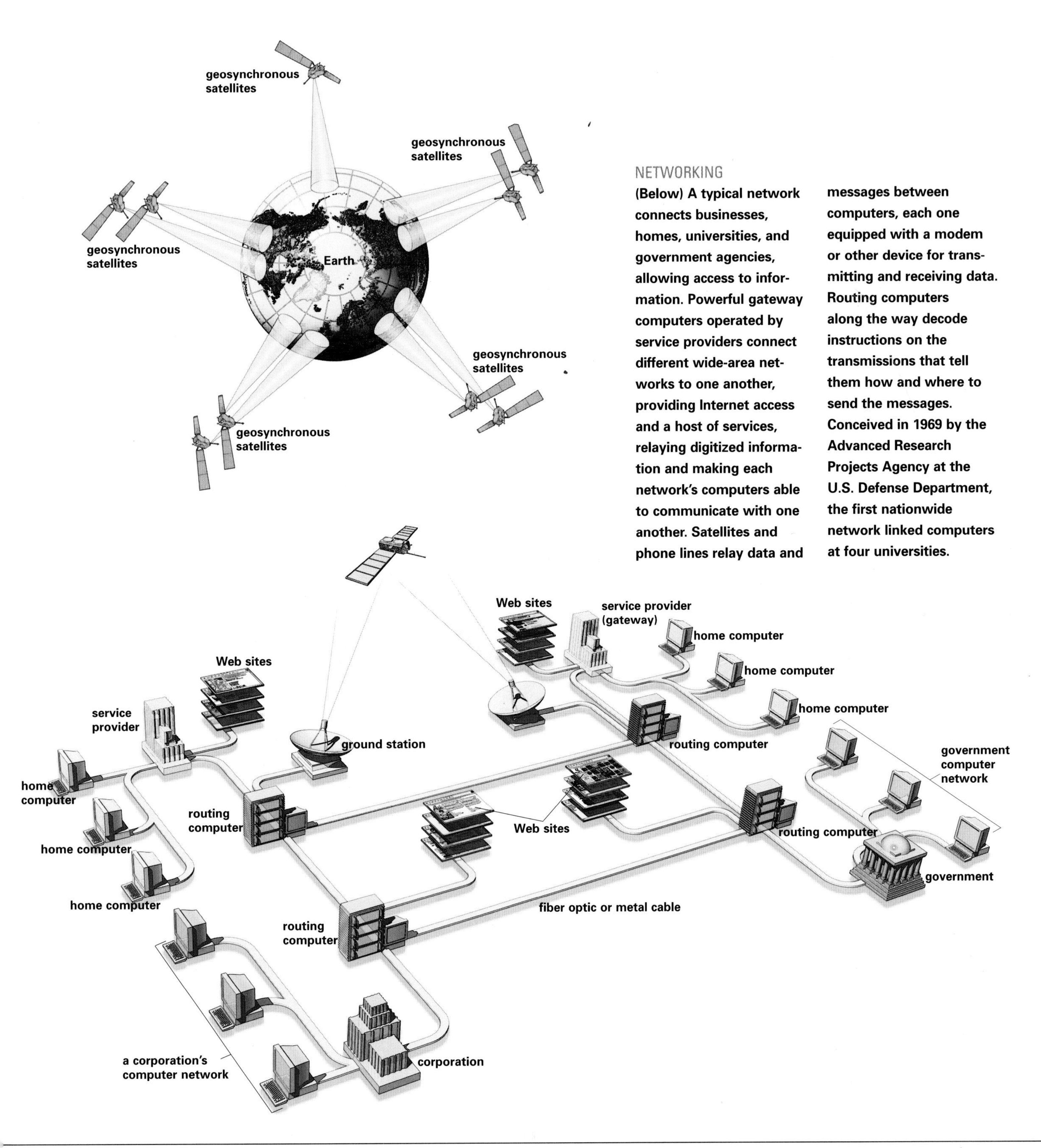

BAR CODES & SCANNERS

USING COMPUTER TECHNOLOGY'S OMNIPRESENT servant, the binary system, bar codes represent the decimal price and other details about an item in a series of parallel vertical lines and white spaces. Such codes also are used to track documents and packages, gene sequences in databases, and books in libraries and bookstores, and they are standardized. For example, bar codes at supermarkets and retail stores come under the Uniform Product Code, which assigns a unique numerical signature to grocery products; the UPC identifies the manufacturer and the product. Bookstores use codes set by the International Standard Book Number (ISBN) system; they identify a book's price and the country that published it.

Bar codes are not meant to carry information like addresses or Social Security numbers; rather, they are a set of numbers a computer uses to access specific data from files. A drugstore item, for example, might have a bar code that is a product number, not a price or description of the item. When a cashier passes the item over a scanner, or under a handheld one, a laser beam set to a specific frequency reads the binary code in the bars; it is transmitted to a decoder and a computer, and the computer goes to the files and locates the item and its price. The information may then be sent to the store's main computer for processing before being flashed onto the cash register display where it and other items purchased are subtotaled and totaled.

BAR CODES

(Right) No cash register bells ring as a scanner tracks a package via its bar code. Introduced in the 1970s, a bar code identifies a product and its price. Businesses that must track inventory rely on bar-coding; so do makers of identification cards and researchers who monitor gene banks. Scanners that read the codes come in various models: handheld or built into counters.

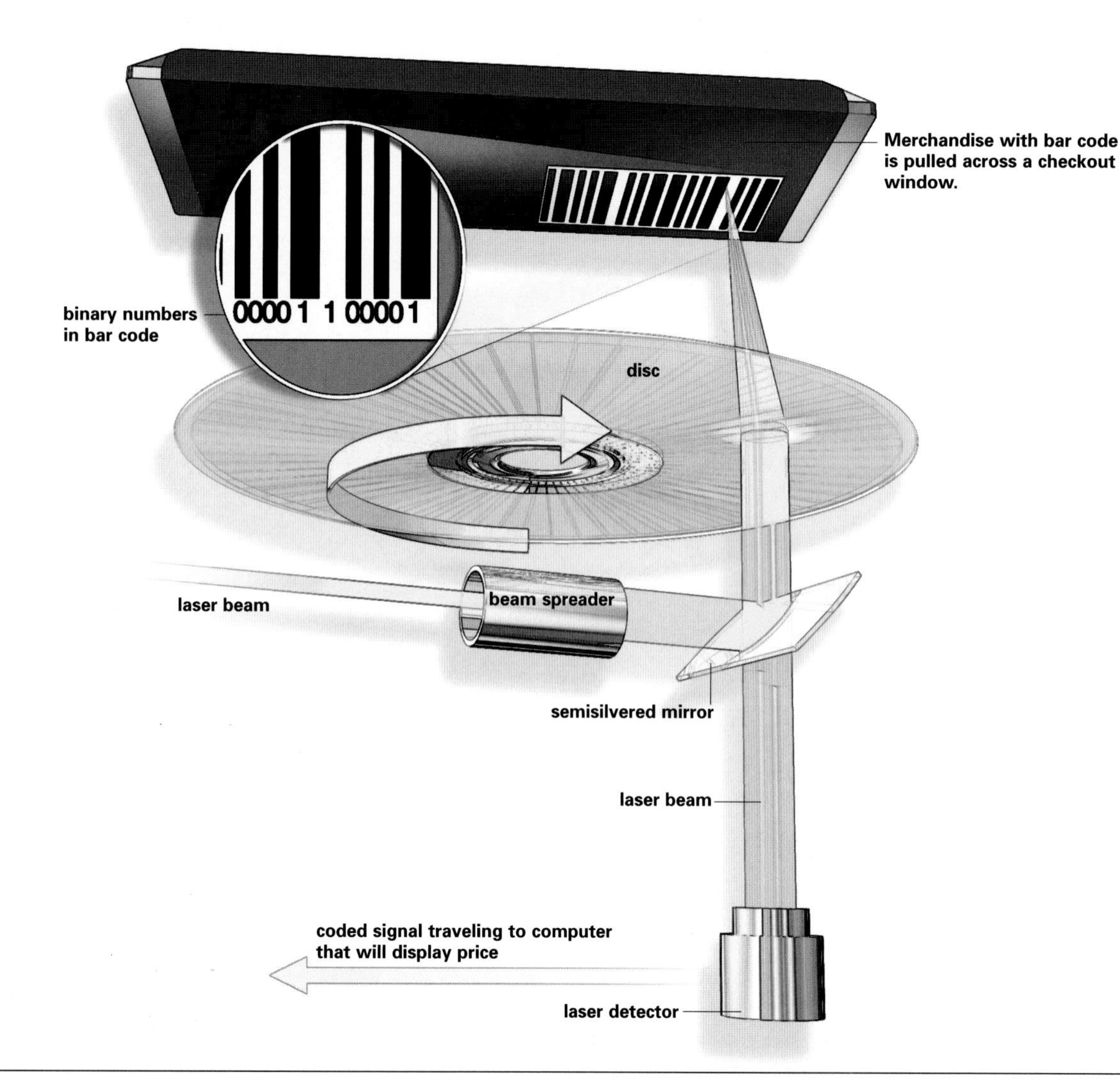

SCANNING

(Left) When a clerk draws a bar-coded item over a scanner in a checkout counter, a laser beam enters a spreader and reflects off a mirror to a disc and up through the opening. The beam reads the encoded information and bounces back to the scanner's laser detector; the signal then goes to a computer that displays the price. The bars of the code represent the binary digits 0 (in white), which reflect more strongly, and 1 (in black); combinations of the units identify the item.

SEE ALSO
Laser Surgery · 204
Calculators · 228
Computers · 230
Sending & Receiving Messages · 216

SCAN
DECODE
LS 2000

MANAGING MONEY ELECTRONICALLY

DRAWING CASH

(Left) Able stand-ins for the human tellers inside the bank, ATMs dispense and manage cash for these Malaysian women.

ATM

(Right) The automated teller machine—perhaps the most visible symbol of electronic banking—provides fast cash. The machines not only dispense bills but also take deposits, allow transfers of funds, and answer balance inquiries. To activate an ATM, the user inserts a plastic card with a data-rich magnetic strip into a slot and then punches a PIN (personal identification number) on a keypad. A computer verifies the identification number and cash balance, making note of the transaction requested. When customers withdraw cash, electronically counted money comes out of built-in strongboxes, or cassettes. Many Americans use ATMs about eight times a month.

"Remember that time is money," Benjamin Franklin advised a young tradesman. If by that he meant time wasted stood in the way of gain, he would have appreciated the ease and speed with which money gets into our hands, pays bills, and is moved around in bank transactions.

Technology now allows us to withdraw cash from an ATM (automated teller machine), buy and sell stocks online, apply for a loan, check bank balances, and shift funds from one account to another without waiting in a teller's line. The ATM is a prime example. Activated by a plastic card, a bank's ATM first determines whether you qualify for service. If you have enough money in your account, you get money from the machine. The key to it all is the dark magnetic strip on the back of the card. Its invisible tracks hold around 200 bytes of information in letters and numbers, and when the card is inserted into an ATM, the data and transaction request are sent electronically to a computer at the financial institution that operates the ATM. There, the account number, the balance, and the user's personal identification number (PIN) are verified. If everything checks out, the computer subtracts the money withdrawn, and a vending machine-like mechanism dispenses the cash, all in a matter of seconds.

Such cards already are being replaced by so-called Smart Cards that do their own computing. Armed with a processing power that greatly widens their capability, these cards contain electricity-carrying silicon chips that store thousands of bytes of information.

Shopping online and banking transactions are, of course, computer-driven. When an online customer, for example, sends a desired item to a "shopping cart," a number of steps are set in motion. Credit card information is first encrypted so it cannot be accessed indiscriminately while it is transferred over the Internet. A program on the seller's computer processes the order and credit information, and transmits it to a card processor, which in turn sends it to the credit card company for verification; the card company then transmits a yes or a no back to the processor, which relays the information to the shopping card program. If the sale is approved, the credit card processor transfers funds to the merchant.

SEE ALSO
Plastic • 184
Computers • 230
Computer Chips & Transistors • 232

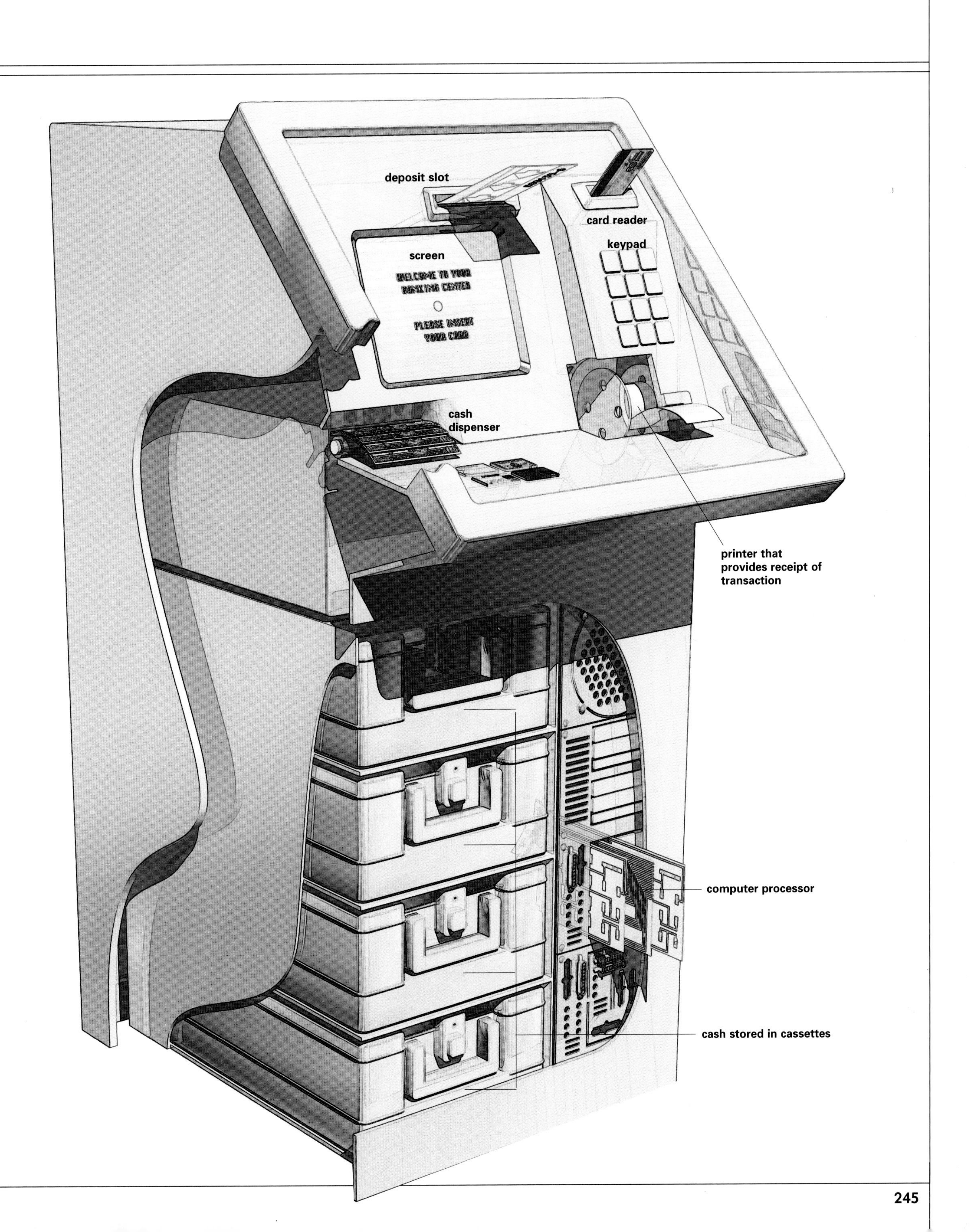
deposit slot
card reader
keypad
screen
WELCOME TO YOUR
BANKING CENTER
PLEASE INSERT
YOUR CARD
cash
dispenser
printer that
provides receipt of
transaction
computer processor
cash stored in cassettes

THE STOCK EXCHANGE

A STOCK EXCHANGE IS AN OFTEN CHAOTIC place—an agency auction market where the securities of corporations and municipalities, like stocks and bonds, are bought and sold daily. The world's largest exchange is New York's, with roots going back to 1792 on Manhattan's Wall Street. Some 3,000 companies are listed on the Big Board, and an equal number of individuals—brokers, support staff, and various specialists—man the trading floor, handling a daily volume of one billion shares.

Once, the only technology that graced a trading floor or the offices of brokers and corporate executives was the stock ticker, a glass-enclosed wooden pedestal through which stock quotes chattered, printed on streams of paper tape transmitted by the Western Union Telegraph Company. Today, electronics so rule the New York Stock Exchange that what was a whole day of trading ten years ago is now handled in the opening half hour, sometimes even in the first 15 minutes; a central computer can fill orders at the astonishing rate of more than 200 transactions a second.

The typical trading order goes through a series of computer-assisted steps from the time a customer places it with a broker by phone or through computer access, to when it reaches the floor of the Exchange. (Some customers may skip a broker and trade directly on the Internet.) On the floor, an electronic order-routing system stores the order, then sends it to a screen at a broker's booth—or directly to a trading post where specialists for particular stocks work—for speedy electronic processing and execution of the deal.

Computers also track the value of groups of stocks, and allow the broker to compare it with stock prices on other exchanges and compete for the best price for the customer; automation also identifies the most active stocks and sends changing stock values to the familiar scrolling "ticker" that crawls across the bottom of a television screen or on the big boards of brokerage offices. Computers also play a role in the market's "circuit-breaking" or "trading collar" strategy: When the market is threatened with decline because of panic selling, trading can be automatically halted for a period of time when stocks drop significantly in value.

Adding in the Exchange's own cellular telephone network, a wireless data system that enables brokers to get orders and send reports and market "looks" with mobile handheld devices from anywhere on the floor, and 3-D data displays that can be accessed on the Web to bring the Exchange's floor action to investors, it is easy to understand why 99 percent of all orders are delivered to the point of sale electronically.

THE BOARD

(Left) Riveted by illuminated stacks of numbers, investors focus their attention on the electronic display at the Shanghai Stock Exchange, founded in 1990. The lighted board reflects market fluctuations, and is a familiar fixture in exchanges and brokerage offices throughout the world. Organized markets differ by country in their connection with the government. The New York Stock Exchange, for example, is not operated by the government but is regulated by U.S. law; the Shanghai Exchange is governed by the China Securities Regulatory Commission.

TRADING FLOOR

(Right) A mecca for brokers and investors, the New York Stock Exchange is the world's largest and most technologically advanced equities market. The Exchange helps financially shrewd individuals invest for the short and long term.

SEE ALSO
Cell Phones · 214
Computers · 230
The Internet · 238

10:30:33
10

CHAPTER 11

OTHER WORLDS: NEAR & FAR

TOOLS THAT LET US OBSERVE REALMS TOO SMALL or too distant to be seen by the naked eye are often baffling. Even the familiar optical devices—contact lenses, bifocals, and binoculars, for example—work in ways we may only vaguely understand. Employing lenses, mirrors, and prisms, the precision instruments we use to view microscopic or macroscopic worlds may have direct or indirect links to laws of reflection and refraction (which deal with changes in the direction of light as it passes from one medium to another). Some devices follow different laws, using lenses made from electromagnets to focus or change magnification, and beaming electrons rather than light to study objects. Others have no lenses at all and look nothing like conventional telescopes. Called radio telescopes, the dish-shaped devices receive radio waves from space, allowing us to "see" other worlds, near and far, in yet another way.

Made transparent as its opening turned during the camera exposure, the Keck Telescope dome rises on Hawaii's Mauna Kea.

OPTICAL LENSES

Optics is the science of light and vision, and optical lenses are devices that aid vision by focusing, bending, and spreading light rays emitted or reflected by an object. Unlike the lens in a human eye, which can shift its focal length through muscular contraction and relaxation, optical lenses have focal length and power built in, depending on their shape. They generally consist of two curved surfaces, or one flat and one curved; these curves may be concave (inward curving) or convex (outward curving). Various surfaces and thicknesses determine a lens's focal power and function, while combinations of different lenses—the compound lenses cemented together in a basic light microscope, for example—prevent the blurring, distortion, and other anomalies that can occur with single, thin lenses.

Eye specialists have many techniques available to them for measuring the eye's refractive power and for determining a person's need for corrective lenses in eyeglasses or in implants like those used after cataract surgery. To compensate for nearsightedness, lenses are ground in concave shapes; for farsightedness, lenses are convex. Cylindrical lenses are used for astigmatism, a condition in which light does not focus properly on the retina because of a defect in the curvature of the natural lens. Prisms, which bend, spread, and reflect light, are used for other defects.

Combinations may be required. In bifocals, the upper and lower parts of a lens are ground differently to correct for both close and distant vision. Trifocals are ground with a center lens for intermediate distance. Intraocular cataract lenses that replace clouded natural lenses are made of a variety of synthetic elastic materials and may be monofocal, which provide vision at a fixed distance, or multifocal, which broaden viewing distance near to far. Foldable inserts actually unfold after surgery.

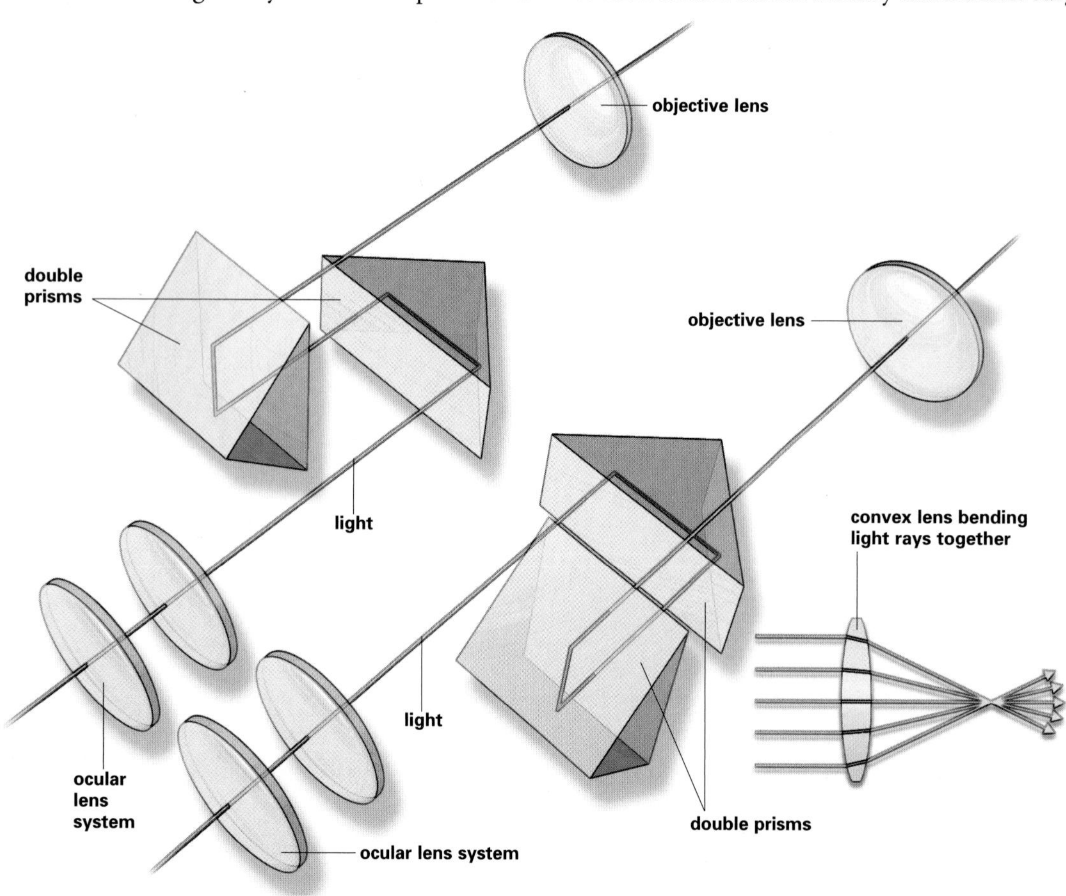

MAGNIFYING GLASS

(Right) A piece of high-tech equipment receives a visual inspection in a modem manufacturer's quality-control lab. Complex creations in convex and concave forms, lenses compensate for inadequacies in human eyes.

BINOCULAR VISION

(Left) To see, human eyes form two images of an object—one on each retina. Binoculars accommodate that by letting light enter through two large objective lenses. Reflecting prisms inside each tube fold the light's path, providing higher magnification than that afforded by field glasses, which have no prisms. The prisms also allow for a more compact binocular design; field glasses need much longer and heavier tubes for high magnification.

SEE ALSO
Film Camera · 156
Fiber Optic Cables & DSL · 210
Light & Electron Microscopes · 254

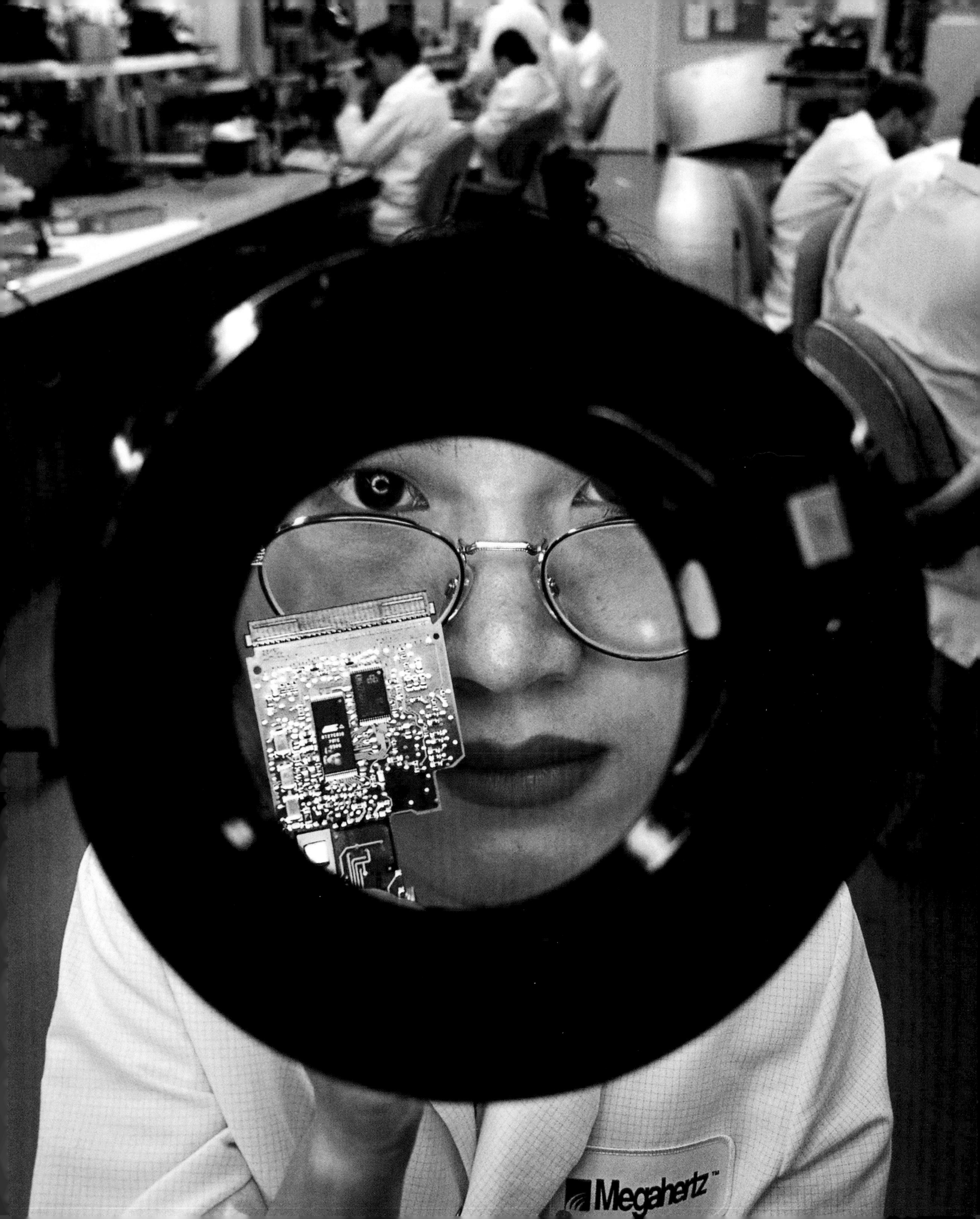
Megahertz™

NIGHT VISION

WHAT DOES IT TAKE TO SEE IN THE DARK? Humans and other mammals need the rods in the retina, millions of tiny cylindrical elements that contain rhodopsin, a purple pigment that can detect dim light. We also have millions of retinal cones, light-sensitive cells that enable us to read fine print. Cats and owls, however, have only rods, a choice of nature that makes them nocturnal creatures capable of seeing far better than humans in low light.

For a human eye to see in a dark movie theater, its rods have to rejuvenate rhodopsin, which is so chemically altered by bright outside light that the rods temporarily lose their sensitivity to light. While rhodopsin's response allows the eye to withstand the sunlight, it does not help much when you enter a dark screening room. Gradually, however, more rhodopsin is produced, increasing sensitivity to the theater's low light. Vitamin A, a lack of which can result in night blindness, is essential to the production of rhodopsin.

Medical science may someday create a pill that allows us to see better in the dark, but until that time, technology must fill the bill. Night-vision binoculars and scopes are electro-optical instruments that are incredibly sensitive to a broad range of light, from visible through infrared. Light that enters a lens in a night-vision scope reflects off an image intensifier to a photocathode and is converted to an electronic image. Amplified on a viewing screen, the image reveals much more than a night scene observed through a conventional scope.

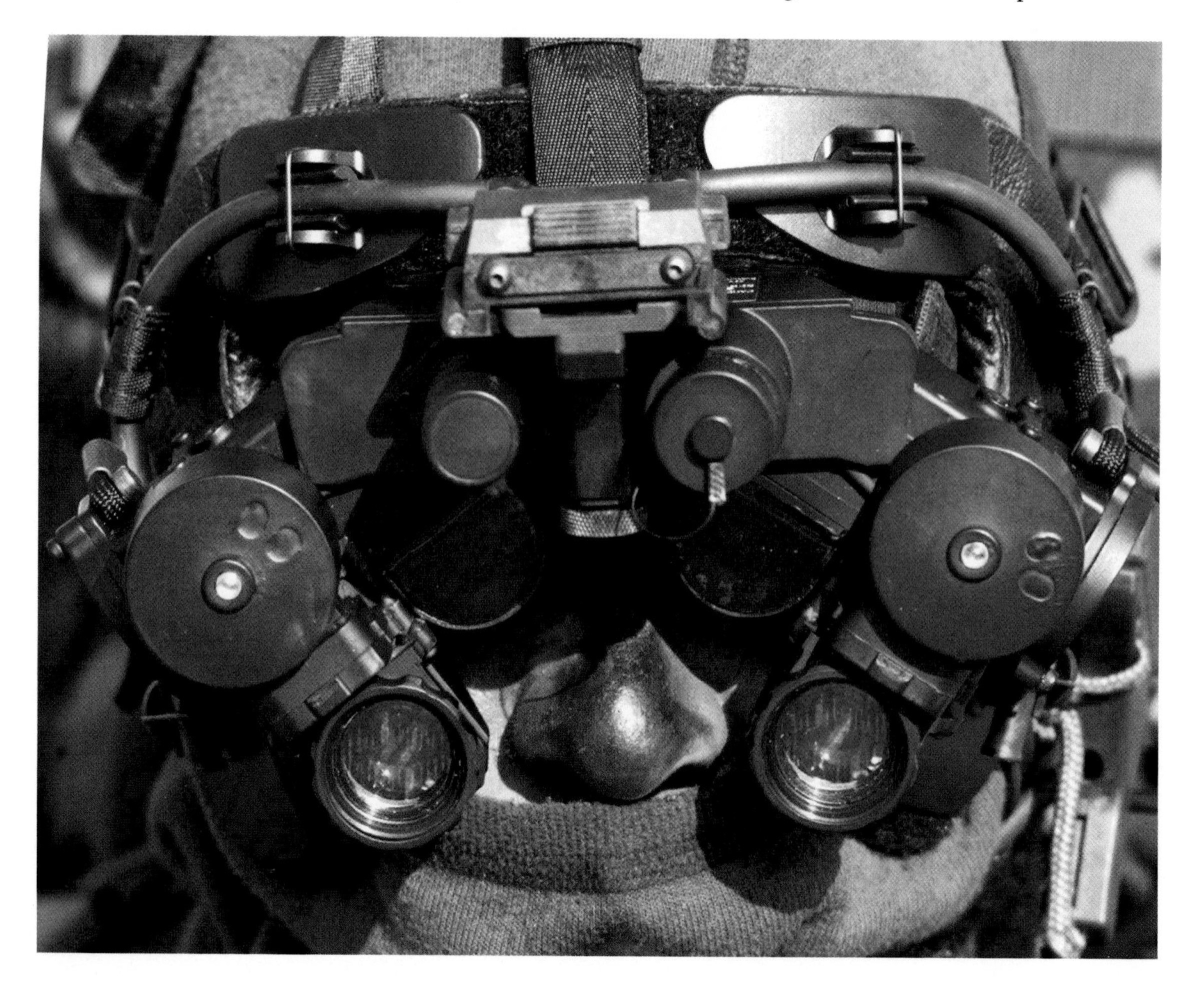

FORESIGHT

(Left) Training for anti-guerrilla engagements, a Czech commando tests his night vision binoculars. Such devices, including night goggles, require some ambient light, perhaps from the stars or from a focused-beam infrared source. Like cameras, the devices have various image magnifications, and are especially valuable during night operations involving close air support of ground troops.

ELECTRO-OPTICAL NIGHT VISION

(Right, bottom) This process works by intensifying and increasing available light. As faint light enters a photocathode, its photons accelerate through a vacuum. Next, a microchannel plate multiplies the photons thousands of times. The device converts them to an electronic image and focuses it on a phosphor screen for viewing. Amplified green light then enhances the image's visibility.

SEE ALSO
Fiber Optic Cables & DSL • 210
Optical Lenses • 250

NIGHT TRAINING

(Right) Putting their faith in night vision equipment, Israeli soldiers prepare for combat. But even with the technology that enables soldiers to peer through the darkness during nightly ambushes, such an operation remains hazardous.

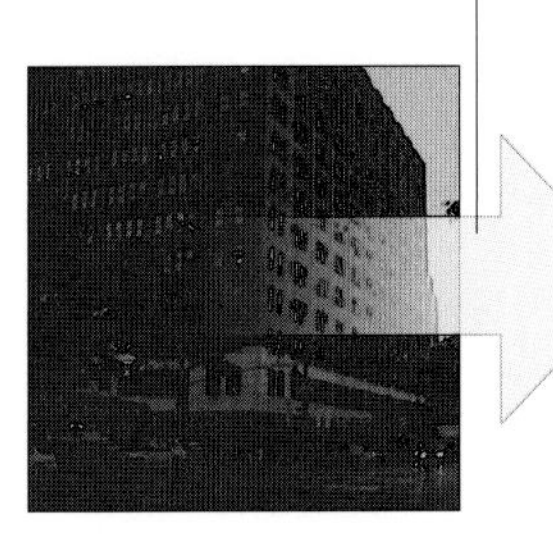
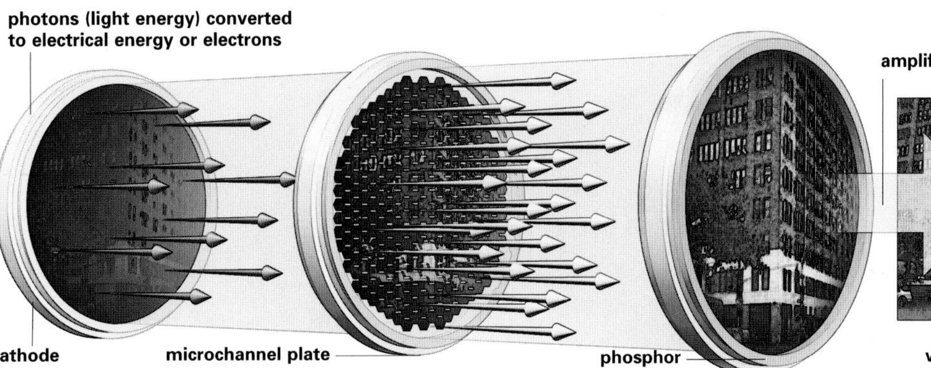

LIGHT & ELECTRON MICROSCOPES

"WHERE THE TELESCOPE ENDS," WROTE Victor Hugo in the novel *Les Misérables*, "the microscope begins. Which of the two has the grander view?"

As nearly everyone knows, microscopes are instruments that give us magnified images. But when Hugo penned those words in the 19th century, he was speaking of the simple light microscope with its then astonishing, and now very limited, view. Today, the telescope has not ended at all, and the microscope has long said good-bye to its beginning.

Perhaps the most familiar microscope is the ordinary compound microscope, which has a pair of convex lenses of short focal length on the lower (objective) end of a tube and another pair at the eyepiece. When a specimen is placed at the objective end and is illuminated with light reflected from a mirror, the magnified image is magnified again at the eyepiece end. Focusing is done by moving the objective lenses nearer to, or farther from, the specimen.

While the light microscope can identify the form and structure of extremely tiny organisms, the electron microscope reveals far more detailed information about their surface and inner workings. Under an electron microscope, a simple bacterium bares its very soul and becomes an intricate cutaway of the complexity of life.

With a magnification power of up to hundreds of thousands of times, the electron microscope can see objects that are among the most invisible of the invisible: atoms. Armed with an electron gun, this nonlight microscope focuses a beam of electrons through a vacuum and over the surface of a specimen. A signal is generated, projected onto a fluorescent screen, and then photographed. The microscope uses magnetic lenses to focus and change magnification. These produce a field that acts on the electron beam in the same way a glass lens works on light rays. The exceptionally high resolution is due to the shorter wavelength associated with electron waves.

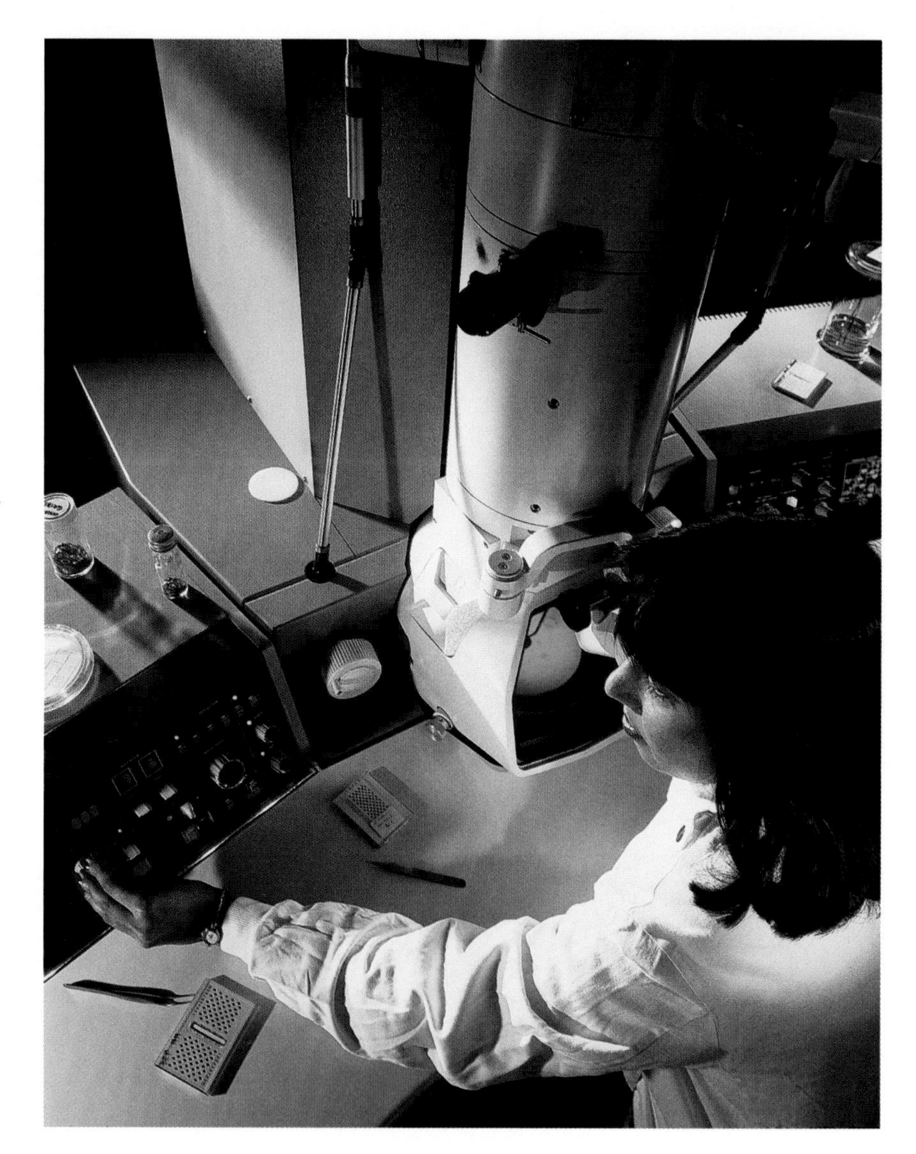

TV TECHNOLOGY

An electron microscopist gets finer details than through the more familiar eyepiece and tube of a light microscope. Equipped with a monitor and using electron beams, not light, the electron microscope draws more on television technology than on the lens science behind conventional microscopes. Indeed, electromagnets, rather than glass, function as the lens. Electrons pass through thinly sliced specimens coated with gold or water vapor to improve the image.

SEE ALSO
Television · 148
X-ray, CT, & MRI · 192
Fiber Optic Cables & DSL · 210
Optical Lenses · 250

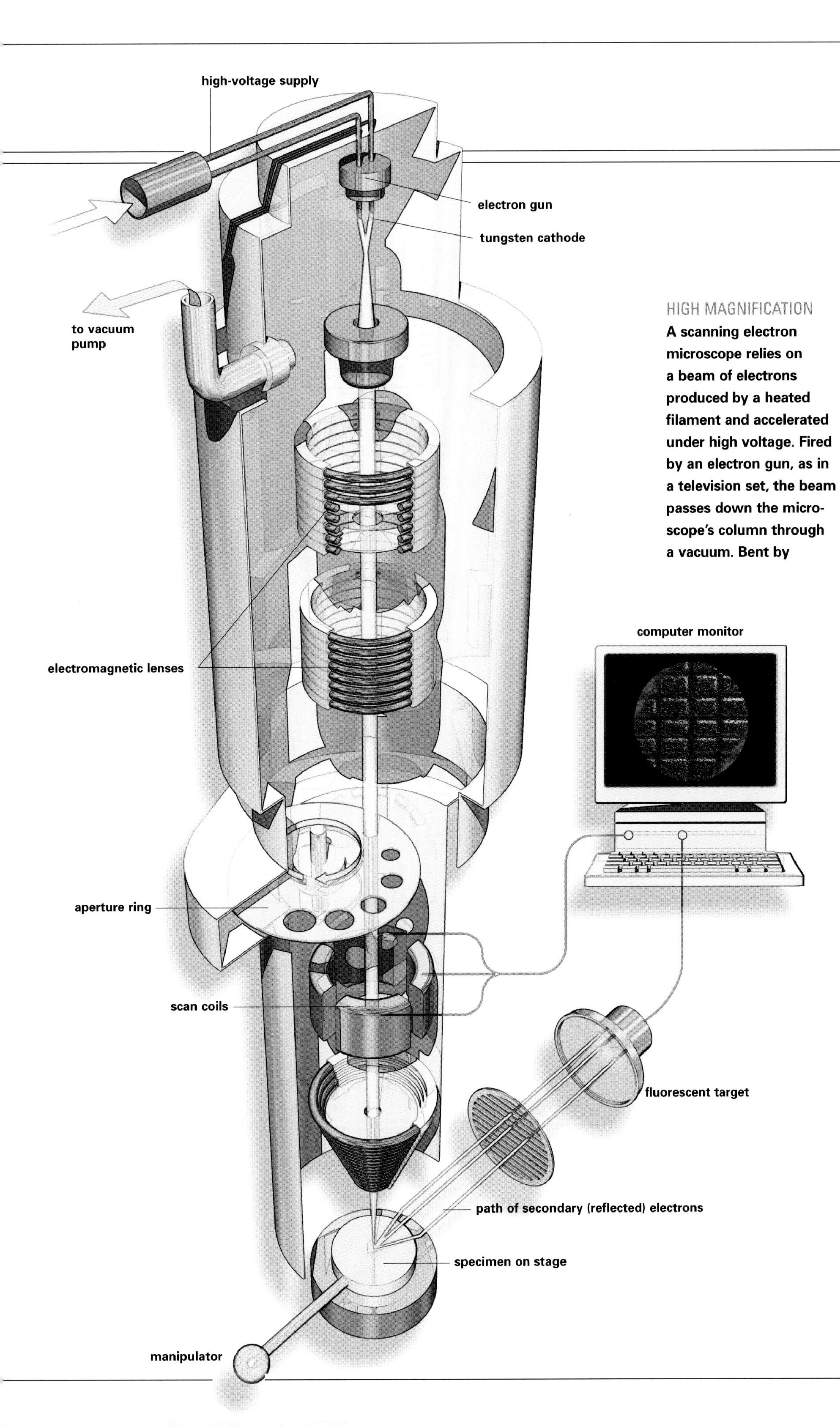

HIGH MAGNIFICATION

A scanning electron microscope relies on a beam of electrons produced by a heated filament and accelerated under high voltage. Fired by an electron gun, as in a television set, the beam passes down the microscope's column through a vacuum. Bent by electromagnets, much as glass lenses bend light, the beam focuses on a specimen cut thin to allow the electrons to penetrate. Scanning coils sweep the beam in a gridlike pattern, training it on specific points. As the beam passes through the specimen, it creates electrical signals that appear as images on a monitor. For thick specimens, microscopists can use higher accelerating voltages. High magnification—by hundreds of thousands of times—and different models may be used, depending on the specimen.

TELESCOPES

GALILEO DID NOT INVENT THE TELESCOPE, but he was probably the first person to use one for making serious astronomical observations. With his 17th-century device, he found moons around Jupiter, mountains on Earth's moon, and spots on the sun. Galileo's telescope was a simple arrangement of lenses in a tube that bent light at odd angles, thus distorting color and producing images of poor quality. Later, Isaac Newton rectified that, using mirrors to reflect light rather than bend it. Modern astronomers often view an image on a screen or as a photograph, without actually looking through an eyepiece.

Telescopes in observatories can have gigantic mirrors of polished glass having diameters of up to 32.2 feet. Spun-cast in rotating furnaces from tons of raw glass, they are linked to computers and photodetectors that, in turn, link multiple telescopes for greater viewing power. The mirrors have enormous light-gathering power and can view wide areas of the sky for large-scale surveys of the faintest objects in deep space. Indeed, instruments now under development will allow simultaneous observation of the spectra from as many as 300 galaxies. Just as important, they are distortion-free, cast as they are from vastly improved glass mixtures and finely polished by beams of electrically charged atoms. Computers also eliminate atmospheric distortion by analyzing light and correcting it if necessary. New optical detectors promise even more: Their sensitivity allows them to clock the arrival of a single light particle and to measure its energy with exceptional precision—all through the infrared, optical, and ultraviolet segments of the spectrum.

SKY DOME

(Right) The open dome of one of the twin Keck Telescopes at twilight atop Hawaii's 13,600-foot-high Mauna Kea, a dormant shield volcano. Inside the dome, the telescope's 32.2-foot-diameter mirror contains 36 hexagonal segments, each of which can be aimed by computer. The twin Kecks are the world's largest optical and infrared telescopes, both able to probe the universe's deepest regions with phenomenal power. Each instrument is eight stories tall and weighs 300 tons.

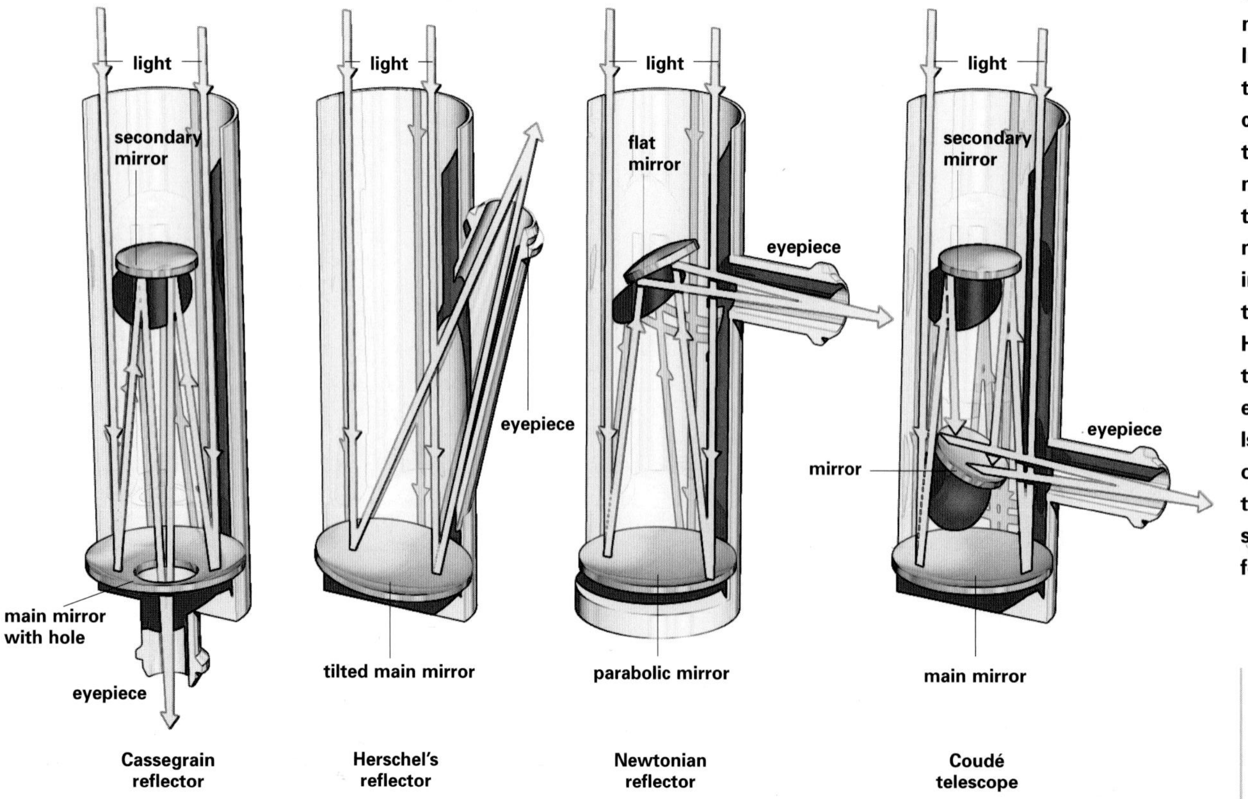

REFLECTING TELESCOPES

(Left) These telescopes use curved mirrors to magnify distant objects. In the Cassegrain reflector, a 17th-century creation, light passes through a hole in the main mirror and focuses the image on a secondary mirror. To view the image, one peers through the eyepiece. The 1773 Herschel reflector used a tilted main mirror with an eyepiece looking into it. Isaac Newton's model, only six inches long, had two mirrors; Coudé versions use three mirrors to focus images to the side.

SEE ALSO
Fiber Optic Cables & DSL • 210
Optical Lenses • 250
Space Telescopes • 258

SPACE TELESCOPES

While ground-based observatories keep expanding their ability to see into deep space, there is, as the saying goes, nothing like being there. Orbiting observatories, for example, can see into radiation wavelengths difficult to image by Earth-based instruments, or examine high-energy processes in the nuclei of galaxies and in the vicinity of black holes. One of the most powerful space telescopes bears the name of Edwin Powell Hubble, the U.S. astronomer who discovered that certain nebulae are really galaxies outside our own Milky Way. Hubble, launched aboard the space shuttle *Discovery* in 1990, has sent back images that prove, for example, that the entire universe is expanding, rushing outward from what may have been the big bang of creation. The Hubble was designed to gather light from a large, concave primary mirror and reflects it off another mirror into an array of sensors. Sensitive instruments detect x-rays, infrared light, and ultraviolet light, revealing the makeup of far-off celestial structures and systems. By focusing on a seemingly empty bit of sky, Hubble even discovered what may be a "construction site for galaxies," a vast birthplace of stars, 12 billion years in the past, that had been invisible to astronomers.

Now a replacement, the new James Webb Space Telescope, named after NASA's second administrator, is scheduled for launch aboard an Ariane 5 rocket in 2011 for a three-month journey into orbit. It will have a larger mirror to give it more light-gathering power, will operate much farther from Earth—a million miles away—and will see deeper in space.

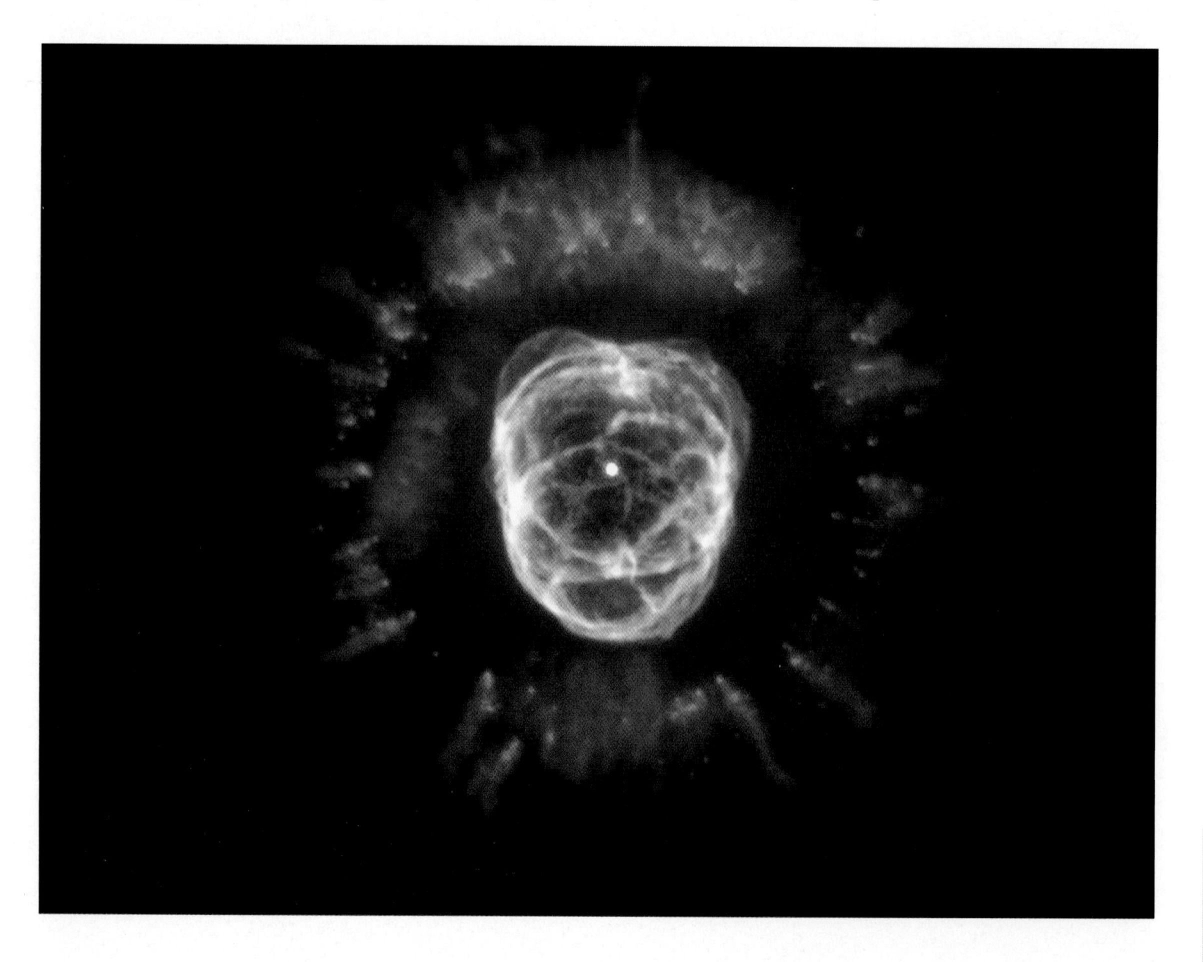

STELLAR RELIC

(Left) Captured by the Hubble Space Telescope and dubbed the Eskimo Nebula because it resembles a face framed by a fur parka, the glowing disk is what's left of a dying sunlike star 5,000 light years from Earth in the constellation Gemini. The "parka" is actually material embellished with a ring of comet-shaped objects.

THE BIGGER PICTURE

(Right) With its huge mirrors, cameras, and spectrographs seeking, and sometimes finding, distant glimmers from the big bang, the Hubble Space Telescope converts faint starlight into limelight. It orbits 375 miles above Earth, providing images free of the distortion seen in views from Earth-based telescopes not equipped with computer-controlled adaptive optics. The telescope's imaging spectrographs can read light emitted by gases, depicting it as red for nitrogen, green for hydrogen, and blue for oxygen.

SEE ALSO
Fiber Optic Cables & DSL • 210
Telescopes • 256

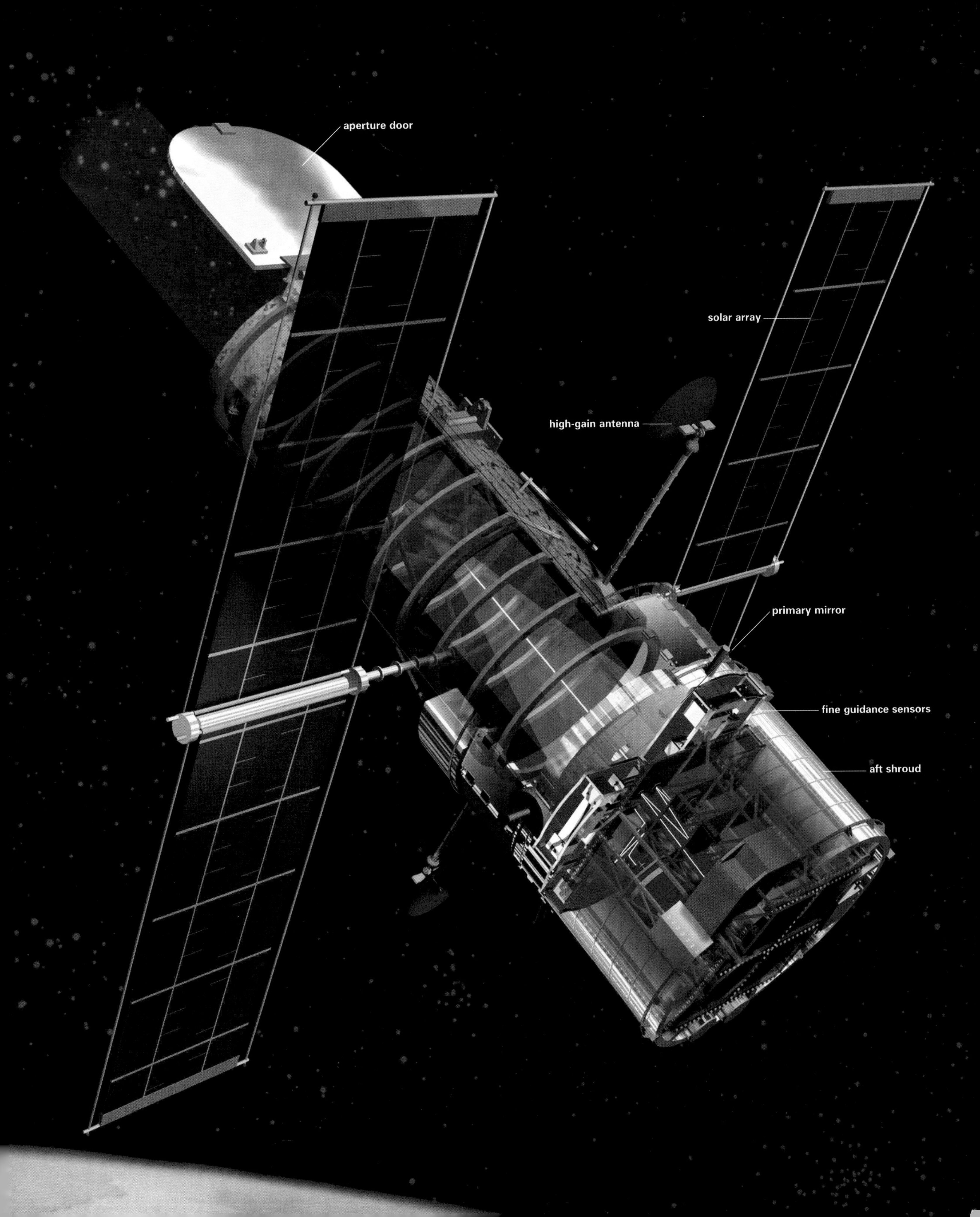
aperture door
solar array
high-gain antenna
primary mirror
fine guidance sensors
aft shroud

MILITARY TECHNOLOGY

TODAY, WARFARE HAS EVOLVED INTO SOMEthing far more advanced and precise in its power to detect an enemy target and destroy it than in the past. It is driven by a technological armamentarium of computers, satellites that spy and guide tank columns, guided missiles, lightning-speed aircraft, ships and submarines bristling with electronic gear and devastating firepower, and the deterrent capability of arsenals of nuclear weapons.

Stealth bombers with their unique shape and special surface materials, streak through the skies, able to evade radar; air-to-air missiles with electronic heat-seeking systems home in on the exhaust heat of enemy aircraft, laser-guided bunker-busting bombs penetrate deep into underground concrete fortifications before their delay-fuses initiate the explosion, electronic bombs short-circuit a city's entire power grid, or a command center, without casualties or property destruction. Much of the technology operates quietly: Battlefield robots seek out booby traps; electronic jammers confuse an enemy's remote-controlled missiles; and computers perform data encryption and decryption, the process of hiding data as ciphertext and then decoding it into plaintext.

Perhaps one of the best-known pieces of military technology is the long-range Tomahawk cruise missile, a torpedo-shaped package that carries a thousand-pound bomb and sophisticated guidance equipment programmed to reach a precise target. Able to fly at up to a thousand miles an hour and at extremely low altitude, cruise missiles are used to attack land targets after launch from booster rockets on surface ships and submarines. After the rocket fuel is burned off, the missile's wings and tail fins unfold, and a turbofan engine, like that which drives a commercial jetliner, takes over.

On board is a digital "scene-matching" system that identifies the territory over which the missile flies and compares it with a map of the terrain that is stored in the system's memory banks. A link to a network of global positioning satellites, an onboard camera and various guidance systems, help the distant controllers track the missile and find its target. Newer systems can reprogram the missile while in flight to strike many preprogrammed alternative targets or allow it to "loiter" over a target area; its on-board camera then lets commanders assess target battle damage.

On the ground, the military relies on technology as well. There is the Bradley Fighting Vehicle System, named after the famed World War II general, Omar Bradley. Heavily armored and armed, a Bradley is a high-tech, tracked weapons platform that also carries infantry to the battlefield. With thermal imaging systems that see in the dark and through fog with television-like clarity, incredibly sensitive targeting sensors, and laser range-finders, the vehicle is especially useful in supporting dismounted soldiers and locating attackers. Ground-based remote-sensor systems also detect troop movement through ground vibrations and identify the direction in which soldiers and their vehicles are moving.

DISTANT WATCHERS

(Left) Far from the scene of battle, U.S. Navy personnel aboard the destroyer *John Paul Jones* gather and employ combat information during the war on Afghanistan in 2001. Cruise missiles were launched from the vessel, their paths clearly tracked on illuminated screens. The military's ability to see with sophisticated instruments what the eye cannot observe directly is analogous, albeit with far different goals, to what space telescopes and electron microscopes can accomplish.

POWER UNLEASHED

(Right) A Tomahawk land-attack missile controlled by computers and satellite systems roars from its berth on the guided missile cruiser U.S.S. *Anzio* during the 2003 war on Iraq. Extraordinarily accurate, Tomahawks can strike relatively small targets and fly over evasive routes generally out of reach of radar systems.

SEE ALSO

Satellite Communications · 218
Light & Electron Microscopes · 254
Space Telescopes · 256

GLOSSARY

Airfoil: The curved surface of a wing or propeller that produces lift as it moves through the air.

Alloy: A substance composed of two or more elements, usually two or more metals or a metal and a nonmetal melted together.

Amplitude: The strength and size of a radio wave.

Aneurysm: An abnormal bulging of a blood vessel.

Anneal: To heat glass, metal, or an alloy and then let it cool gradually to room temperature, thus making it less brittle.

Aquaculture: The breeding and raising of fish and seafood under precise conditions to improve food production and increase the ability to stock streams, ponds, and lakes.

Aqueduct: An artificial channel or structure that conveys water.

Arch: A curved construction element that spans an opening and supports the weight above it.

Atomic clock: A precision clock that operates by measuring the natural vibration frequencies of atoms or molecules.

Bar code: Coded information that identifies a labeled object. The bars and spaces (and sometimes numerals) are designed to be scanned and read by a computer.

Bessemer process: A steelmaking process that removes impurities from pig iron by forcing a blast of air through molten metal.

Binary code: A base-two number system that uses 1s and 0s instead of the standard base-ten, or decimal, system.

Bobbin: A spool on which weft yarn is wound for weaving, or on which thread is wound for use in a sewing machine.

Buoyancy: The lifting force of a fluid exerted on a body immersed in it.

Buttress: A projecting masonry structure that supports or strengthens a wall or building.

Caisson: A watertight chamber used in the construction of foundations, bridges, and tunnels.

Cam: A sliding or rotating device used to change rotary motion to linear motion, or vice versa.

Cast iron: A hard and brittle iron alloy consisting of carbon and silicon cast in a mold.

CD-ROM (compact disc read-only memory): A device for storing computer data in digital form.

Celluloid: A tough, flammable thermoplastic composed of nitrocellulose and camphor.

Centrifugal force: The force that presses objects away from the center of rotation.

Chlorofluorocarbon: A compound containing carbon, chlorine, fluorine, and sometimes hydrogen. It is used in refrigerants, cleaning solvents, aerosol propellants, and plastic foams.

Clepsydra: An instrument that measures time by the fall or flow of water from one marked container to another; also known as a water clock.

Combine: A multifunction farming machine that cuts, threshes, cleans, and gathers cereal crops.

Compression: A force that reduces or shortens a material by pressing or squeezing it together; opposite of tension.

Compressor: The part of a cooling system in which the coolant, in its gaseous state, is compressed by a pump to heat it.

Concrete: A strong building material made by combining sand, gravel, and rock with a cementing material and water.

Condenser: In a cooling system, the part in which the coolant releases heat and changes from a gas or vapor into a liquid.

Cone: A photosensitive receptor cell in the eye's retina.

Contour plowing: Plowing across the fall of the slope rather than with it. This process helps control water runoff and soil erosion.

CPU (central processing unit): The primary chip in a computer. It not only controls the operation of other components but also interprets and executes instructions.

Curtain wall: A building's nonbearing exterior wall.

Defibrillator: An electronic device that uses electric shock to restore rhythm to a rapid and irregular heartbeat.

Density: The weight of an object or a substance in relation to its volume.

Dialysis: A process that uses a solution and a semipermeable membrane to separate waste products from the blood.

Digital: Relating to data in the form of numerical digits.

Diode: An electronic component that lets current flow through it in only one direction.

Drag: The force with which air or water resists the motion of a moving body.

Electricity: A natural entity composed of electrons and protons (or possibly electrons and positrons). It is observable in bodies electrified by friction, in lightning, and in the aurora borealis; it is generally used in the form of electric current.

Electromagnet: A magnet using electricity to increase its power.

Electromagnetic radiation: The form in which energy moves through space or matter.

Electromagnetic spectrum: The full range of light energy from the sun, consisting of electromagnetic waves distinguished by their frequency, or wavelength.

Electron: A subatomic particle with a negative charge.

Endoscope: A flexible instrument for seeing inside organs of the human body.

Energy: The capacity to do work.

Escapement: A mechanism in a clock that controls the motion of the train of wheelwork.

Evaporator: The part of a cooling system where coolant absorbs heat and changes from liquid to gas or vapor.

Filament: A metallic wire in an incandescent lamp that glows when an electric current passes through it.

Fission: The splitting of an atom's nucleus to release large amounts of energy.

Flying buttress: A projecting arched structure that carries the lateral thrusts of a roof or vault to an upright pier or buttress.

Focal length: The distance from the surface of the lens to the focal point.

Force: Any influence that exerts pressure on an object or that speeds up, slows down, or stops an object's motion.

Fossil fuel: A fuel formed from the fossilized remains of plants and animals.

Frequency: The number of wave vibrations over a given period of time.

Fuel cell: A device that continually changes the chemical energy of a fuel and an oxidant directly into electrical energy.

Fusion: A nuclear reaction in which the nuclei of light atoms, such as hydrogen, come together and form one or more heavier nuclei, thereby releasing enormous energy.

Gamma ray: A form of electromagnetism that has the shortest known wavelength. It is also the most penetrating form of radioactivity.

Gear: A mechanism using toothed wheels to convey movement from one part of a machine to another.

Generator: A device that converts mechanical to electrical energy.

Geothermal energy: Energy obtained from subsurface reservoirs of naturally occurring heat.

Gnomon: A sundial's pointer, which indicates the time of day by the position or length of its shadow.

Hard disk: In a computer, a data-storage device that uses a magnetically coated metal disk.

Harness: A frame that holds and controls a group of heddles on a loom.

Heddle: One of a set of cords or wires on a loom's harness. In the center of each is an eye through which warp yarns are threaded, allowing them to be separated for weaving.

Herbicides: Chemical preparations that are used to destroy vegetation, usually weeds.

Hertz: One hundred cycles of electromagnetic radiation per second.

Horsepower: A unit of power equal to 745.7 watts; in metric (CV), the power required to raise 75 kg one meter in one second, or 735.5 watts.

Hybrid: The product of two animals or plants having different gene compositions.

Hydrocarbons: Organic compounds containing only carbon and hydrogen.

Hydroponics: A type of soilless agriculture in which plants are grown in solution or in a moist inert medium that provides the necessary nutrients.

Hypersonic: Having a speed of at least five times that of sound.

Infrared ray: A form of electromagnetic energy whose wavelength is slightly larger than that detected by human eyes.

Internet: A worldwide communication system comprising hundreds of small computer networks linked by telephone systems.

Isotope: An atom of a chemical element having a different number of neutrons, giving it a different mass number.

Kilowatt-hour: A commercial unit—the amount of work or energy equal to one kilowatt expended in one hour.

Kinetic energy: The energy generated by an object's motion.

Laparoscope: A flexible fiber optic instrument inserted through an incision to visually examine the abdomen.

Lesion: An abnormal change in a part of the body caused by disease or injury.

Lift: The upward force that is produced by an aircraft's wing, a helicopter's rotor, or a hydrofoil's foil as it moves.

Lintel: A horizontal structural member across the top of an opening. It carries the weight of the wall above it.

Lithotripsy: A method of breaking up kidney stones. It employs high-pressure shock waves that are focused by a computer and transmitted through water.

Live load: The total weight of a structure, including its contents and occupants.

Load: The force on a structure caused by weight or wind pressure.

Maglev (magnetic levitation): A technology that uses the physical properties of magnetic fields generated by superconductors to make a vehicle float above a solid surface.

Megahertz: One million cycles of electromagnetic radiation per second.

Microprocessor: A single computer chip that has a complete central processing unit.

Microwaves: Electromagnetic radiation used in radar and microwave ovens.

Milt: Reproductive glands of male fish when filled with secretion; the secretion itself.

Monomer: Small molecule that is the basic building block of a polymer.

Optical fiber: Fine strand of glass capable of transmitting digital information in the form of light pulses.

Pantograph: A structure that carries electric current from an overhead wire to an electrically driven vehicle.

Pesticide: A chemical used for killing weeds, insects, and fungal diseases.

Petrochemical: A chemical that is derived from petroleum or natural gas.

Photocathode: A device that converts light waves to electrical charges.

Photocell/photoelectric cell: An electronic device that modifies the flow of an electric current in response to light waves.

Photodiode: A semiconducting device that converts light pulses to electronic signals.

Pier: A supporting column, pillar, or pilaster designed to take vertical loads.

Piezoelectric reaction/piezoelectricity: The electric current or electric polarity produced by applying pressure to a crystalline substance, such as quartz.

Pig iron: Iron produced in a blast furnace and used to make steel, cast iron, or wrought iron.

Pile: A long, slender column of steel, concrete, or wood driven into the ground to support the vertical load of a sructure.

Pinion: A gear having a small number of teeth that mesh with teeth on a larger gear.

Pitch: The angle at which a helicopter's rotor blades slice through the air.

Pivot: A shaft or pin on which a lever turns.

Pixel: The acronym for "picture element," the smallest unit, or dot, of color and brightness in an electronic image.

Polymer: A huge molecule made up of repeated sequences of smaller molecules.

Polymerization: A chemical process in which individual monomer molecules bind together to form a polymer chain.

Power: The energy to perform work.

Powerhouse: A section of a power plant where the turbines that generate electricity are housed.

Prism: A transparent object with nonparallel sides used to disperse a wave of light.

RAM (random-access memory): Memory space in a computer where the user can alter and temporarily store information.

Reed: A swinging comblike device set into a loom across the warp to push each new line of weft tightly against the previous lines.

Rhodopsin: A purple pigment used by the eye's rods to see in the dark.

ROM (read-only memory): Memory space on a computer for permanently storing information and programs that cannot be modified by the user.

Semipermeable: Having pores or openings that allow passage of small molecules but not large molecules.

Shuttle: A device used to carry a bobbin of weft back and forth between the warp threads of a loom.

Slag: Impurities left after the smelting of ore.

Spin-cast: A method employed to create very large mirrors, such as those in telescopes.The entire furnace spins while heating the glass for a mirror.

Superconductivity: The disappearance of electrical resistance in a conductor, especially when the substance is cooled to extremely low temperatures. Such a state allows the passage of a very large electric current and the creation of a strong magnetic field.

Supersonic: Faster than the speed of sound (about 760 mph at sea level).

Temper: To soften glass at high temperature and cool its surface quickly. Exposing the surface to such compressive stress produces very strong glass.

Tension: A force that stretches or lengthens the members of a structure; opposite of compression.

Thermoplastic: A polymer made up of individual chains that can be melted and reformed.

Thermosetting plastic: A polymer chain that undergoes a chemical reaction when heated. Called cross-linking, the reaction binds polymer chains together.

Thrust: A force that moves a vehicle or other object forward.

Tokamak: A generator with a doughnut-shaped core used in nuclear fusion.

Tomography: A technique for reassembling x-ray information by computer to obtain a three-dimensional image of an object's internal structure.

Torque: A force that produces rotation or twisting.

Transformer: A device that alters voltage.

Transistor: An electronic circuit that is composed of semiconducting material.

Transponder: An electronic transmitting-receiving device that can find, identify, and precisely locate an object or vehicle.

Turbine: A machine with a set of blades mounted on a central shaft rotated by air, water, or steam.

Voltage: The electromotive force, measured in volts, with which a source of electricity sends electrons along a circuit to form a current.

Voussoir: A wedge-shaped masonry unit used in forming an arch or vault.

Warp: A set of yarns forming the lengthwise element of a fabric.

Weft: A filling thread or yarn interlaced crosswise with the warp to create a fabric.

Wind load: The force on a structure caused by wind.

Woof: Another term for weft.

Worm gear: A gear of a worm (a screw or other spiral device) that intermeshes with a gearwheel.

Wrought iron: A tough, malleable, commercial iron that has a lower carbon content than steel or cast iron.

X-ray: Electromagnetic radiation with a very short wavelength. It can be detected by phosphor screens and photographic film.

ILLUSTRATION CREDITS

COVER, National Geographic Photographer Mark Thiessen.

FRONTMATTER: 1, Hermann/Starke/CORBIS; 2-3, Roger Ressmeyer/CORBIS; 7, D. Brewster/Bruce Coleman, Inc.; 8, Dan McCoy/Rainbow; 10, David Arky/CORBIS; 13, Onne van der Wal/CORBIS.

CHAPTER ONE: 14-15, Joel Sartore; 17, David Arky/CORBIS; 18, Rita Maas/Getty Images; 21, A. Syred/Photo Researchers, Inc.; 22, Tom Stewart/CORBIS; 24, Ezio Geneletti/Getty Images; 27, Susan Lapides; 28, Volker Steger/Siemens AG/Photo Researchers, Inc.; 29, Tom Stewart/CORBIS; 30, Nicholas Veasey/Getty Images; 31, Michael Freeman; 32, Wally McNamee/CORBIS; 34, Carlos Dominguez/CORBIS; 37, Bohemian Nomad Picturemakers/Kevin R. Morris/CORBIS; 38, Ted Mahieu/CORBIS; 39, Harald Sund/Getty Images; 41, Pete Saloutos/CORBIS.

CHAPTER TWO: 42-43, Jim Zuckerman/CORBIS; 45, Royalty-Free/CORBIS; 47, Peter Essick; 48, Bob Krist/CORBIS; 49, Lester Lefkowitz/CORBIS; 50, Chinch Gryniewicz, Ecoscene/CORBIS; 51 William James Warren/CORBIS; 52, Derek Croucher/CORBIS; 55, Roger Ressmeyer/CORBIS; 56, Flip Chalfant/Getty Images; 57, Richard T. Nowitz/CORBIS.

CHAPTER THREE: 58-59, Erik Leigh Simmons/Getty Images; 61, Joel Sartore; 62, Janet Gill/Getty Images; 64, Joel Rogers/Getty Images; 65, James Porto/Getty Images; 66, Paul Chesley; 67, Varie/Alt/CORBIS; 69, Robert Essel NYC/CORBIS; 71 QA Photos; 72, Joao Paulo/Getty Images; 73 Roger Ressmeyer/CORBIS.

CHAPTER FOUR: 74-75, Sam Abell; 76, John Madere/CORBIS; 78 (both), David Woods/CORBIS; 79, Patrick Durand/CORBIS SYGMA; 81, Frederic Pitchal/CORBIS SYGMA; 82, Chris Trotman/Duomo/CORBIS; 85, Ron Slenzak/CORBIS; 86, Dan Lamont/CORBIS; 88, Richard A. Brooks/AFP/Getty Images; 89, Karl Weatherly/CORBIS; 90, Michael S. Yamashita/CORBIS; 91, Charles O'Rear; 92, Jean-Bernard Vernier/CORBIS SYGMA; 93, Chad Weckler/CORBIS; 95, James L. Amos/CORBIS; 96, Stuart Westmorland/CORBIS; 98, Liu Liqun/CORBIS; 101, Stuart Westmorland/CORBIS; 102, Tom Tracy/CORBIS; 103, David H. Wells/CORBIS; 104, Bruce Burkhardt/CORBIS; 106, Bettmann/CORBIS; 109, David Tejada/Getty Images; 110, Erik Viktor/Science Photo Library/Photo Researchers, Inc.; 111, NASA.

CHAPTER FIVE: 112-113, Peter Dean/Getty Images; 114, Digital Vision/Getty Images; 115, Arthur C. Smith III/Grant Heilman Photography; 117, James Balog/Getty Images; 118, Richard Carson/Reuters/CORBIS; 119, William James Warren/CORBIS; 120, Michael Freeman/CORBIS; 121, Joel Sartore; 122, Peter Essick.

CHAPTER SIX: 124-125, Spencer Jones/Glasshouse Images; 126, Bernd Kappelmeyer; 127, Eberhard Grames/Bilderberg/AURORA; 128, Derek P. Redfearn/Getty Images; 129, Davies & Starr; 130, Breton Littlehales; 131, Rosenfeld Images/Science Photo Library/Photo Researchers, Inc.; 132 (left), Science Photo Library/Photo Researchers, Inc.; 132 (center), Oliver Meckes/Photo Researchers, Inc.; 132 (right), James Bell/Photo Researchers, Inc.; 133, Mark Douet/Getty Images; 134, Manfred Vollmer/CORBIS; 135, David Stoecklein/CORBIS.

CHAPTER SEVEN: 136-137, Peter Ginter/Bilderberg/AURORA; 139, William Coupon/CORBIS; 141, Randy Duchaine; 143, Penny Tweedie/CORBIS; 144, Roger Ressmeyer/CORBIS; 145, TM & © Boeing. Used under license.; 146, Dr. Jeremy Burgess/Photo Researchers, Inc.; 148, Tim Boyle/Getty Images; 150, Joel Sartore; 152, Steve McCurry; 153, Steve Prezant/CORBIS; 154, Jan Nienheysen/Reuters/CORBIS; 155, John Henley/CORBIS; 157, Tim Laman; 158, Jack Fields/CORBIS; 159, Shizuo Kambayashi, AP/Wide World Photos; 161, 'TITANIC' © 1997 Twentieth Century Fox and Paramount Pictures. All rights reserved.; 163, Charlie Munsey/CORBIS; 165, Matthew Stockman/Getty Images; 166, Mark Segal; 167, Robert Landau/CORBIS; 169, Tony Wiles/Getty Images

CHAPTER EIGHT: 170-171, Bryan F. Peterson/CORBIS; 172, Richard Hamilton Smith/CORBIS; 173, Peter Essick; 174, Keith Wood/Getty Images; 175, Roger Tully/Getty Images; 176, Phil Schermeister/NGS Image Collection; 177, James L. Amos/CORBIS; 178, Digital Vision/Getty Images; 181, Charles O'Rear/CORBIS; 182, John Lawrence/Getty Images; 184, Jeff Sherman/Getty Images; 185, Richard Cummins/CORBIS; 186, Royalty-Free/CORBIS; 187, Antonio Mo/Getty Images; 188, George Steinmetz; 189, Kelly-Mooney Photography/CORBIS.

CHAPTER NINE: 190-191, Dr. Arthur Tucker/Photo Researchers, Inc.; 192, Allan H. Shoemake/Getty Images; 193, David Job/Getty Images; 194, Royalty-Free/CORBIS; 195, Lester Lefkowitz/CORBIS; 196, Collection CNRI/Phototake; 197, S. Elleringmann/Bilderberg/AURORA; 198, BSIP/Phototake; 199, Simon Fraser/Photo Researchers, Inc.; 200, ISM/Phototake; 201, Scott Camazine/Photo Researchers, Inc.; 202, BSIP/Phototake; 203, Klaus Guldbrandsen/Science Photo Library/Photo Researchers, Inc.; 204, John Zich/Time Life Pictures/Getty Images; 205, Joe McNally; 207, Tom Stewart/CORBIS.

CHAPTER TEN: 208-209, Charles O'Rear/CORBIS; 210, Davis Meltzer; 211, Lawrence Manning/CORBIS; 213, Carin Krasner/CORBIS; 215, Edward Bock/CORBIS; 216, Joel Sartore; 217, Roy McMahon/CORBIS; 218, Roger Ressmeyer/CORBIS; 221, Royalty-Free/CORBIS; 223, B.C. Moller/Getty Images; 224, Michael Malyszko/Getty Images; 226, Pablo Corral V/CORBIS; 228, James Marshall/CORBIS; 230, Mike Blake/Reuters/CORBIS; 232, Jeffrey Sylvester/Getty Images; 233, Michael W. Davidson/Photo Researchers, Inc.; 235, Samuel Ashfield/Getty Images; 236, Brownie Harris/CORBIS; 237, J. Silver/SuperStock; 238, Royalty-Free/CORBIS; 239, Justin Sullivan/Getty Images; 240, Jimmy Chin/NGS Image Collection; 243, Jon Feingersh/CORBIS; 244, Steve Raymer/CORBIS; 246, Keren Su/CORBIS; 247, Gail Mooney/CORBIS.

CHAPTER ELEVEN: 248-249, Roger Ressmeyer/CORBIS; 251, Joel Sartore; 252, Dan Materna/AFP/Getty Images; 253, Shaul Schwarz/CORBIS; 254, Geoff Tompkinson/Science Photo Library/Photo Researchers, Inc.; 257, Roger Ressmeyer/CORBIS; 258, NASA; 259, Don Foley; 260, Ruben Sprich/Reuters/CORBIS; 261, U.S. Navy/Reuters/CORBIS.

ADDITIONAL READING

Readers may wish to consult the *National Geographic Index* for related articles and books, including the following:

Suplee, Curt, *Everyday Science Explained,* 1996; *Inventors and Discoverers,* 1988; *The Builders: Marvels of Engineering,* 1992.

The following titles may also be of interest:

Bender, Lionel. *Invention.* New York: Eyewitness Books, Knopf, 1991.

Bloomfield, Louis A. *How Things Work: The Physics of Everyday Life.* New York: John Wiley & Sons, Inc., 1997.

Breger, Dee. *Journeys Into Microspace: The Art of the Scanning Electron Microscope.* New York: Columbia University Press, 1995.

Bruun, Erik, and Buzzy, Keith. *Heavy Equipment.* New York: Black Dog & Leventhal Publishers, Inc., 1997.

Clarke, Donald. *The Encyclopedia of How It Works From Abacus to Zoom Lens.* New York: A & W Publishers, 1977.

Delf, Brian, and Platt, Richard. *In the Beginning: A Nearly Complete History of Nearly Everything.* London: DK, 1995.

Facts on File, Inc. *Illustrated Dictionary of Science.* New York, 1988.

Kerrod, Robin. *The Way It Works: Man and His Machines.* London: Octopus Books, 1980.

Langone, John. *Superconductivity, the New Alchemy.* Chicago: Contemporary Books, 1989.

Lubar, Steven. *InfoCulture: The Smithsonian Book of the Inventions of the Information Age.* Boston: Houghton Mifflin, 1993.

Macaulay, David. *The New Way Things Work.* Boston: Houghton Mifflin, 1998.

Macmillan. *The Way Science Works.* New York, 1995.

Reader's Digest. *How in the World?* Pleasantville, 1990.

Ronan, Colin A. *Science Explained: The World of Science in Everyday Life.* New York: Henry Holt, 1993.

Scientific American. *How Things Work Today.* New York: Crown, 2000.

Simon and Schuster. *The Way Things Work, An Illustrated Encyclopedia of Technology.* New York, 1967.

Smolan, Rick. *One Digital Day: How the Microchip Is Changing Our World.* New York: Random House, 1998.

Time-Life. *How Things Work in Your Home.* Alexandria, 1985; *Understanding Science & Nature Series: Computer Age,* 1992, and *Machines and Inventions,* 1993.

Trefil, James. *1001 Things Everyone Should Know About Science.* New York: Doubleday, 1992.

ACKNOWLEDGMENTS

The Book Division wishes to thank the many individuals, groups, and organizations mentioned or quoted in this publication for their help and guidance.

For their help on the first *How Things Work,* we would like to express our immense gratitude to, and admiration for, our chief scientific consultant, Andrew J. Pogan, whose knowledge and assistance proved invaluable from the project's conception through its completion.

We are especially grateful to Neil W. Averitt; Edmond P. Bottegal, Joy Mining Machinery; Dennis R. Dimick; Clarence E. Hill; Brenda Hooper of Arc Second; Sandy Miller Hays, USDA ARS Information; Mark Princevalle, Nuclear Communications Services, Northeast Utilities System, Niantic, Conn.; George Sanborn, Massachusetts Transportation Museum in Boston; Christopher Stewart, Joy Mining Machinery; and Richard F. Zamboni, Frank J. Zamboni & Co., Inc.

For his help on *The New How Things Work,* we would like to thank Keith Bryan for his insightful comments on computer and internet technology.

INDEX

Boldface indicates illustrations.

THE NEW HOW THINGS WORK:
EVERYDAY TECHNOLOGY EXPLAINED
By John Langone

Published by the National Geographic Society

John M. Fahey, Jr. *President and Chief Executive Officer*
Gilbert M. Grosvenor *Chairman of the Board*
Nina D. Hoffman *Executive Vice President*

Prepared by the Book Division

Kevin Mulroy *Vice President and Editor-in-Chief*
Charles Kogod *Illustrations Director*
Marianne R. Koszorus *Design Director*
Barbara Brownell Grogan *Executive Editor*

Staff for this Book

Jane Sunderland *Project Manager*
Dale-Marie Herring *Text Editor*
Jane Menyawi *Illustrations Editor*
Peggy Archambault *Art Director*
David Seeger *Designer*
Carol Farrar Norton *Design Consultant*
Emily McCarthy *Researcher*
R. Gary Colbert *Production Director*
Lewis R. Bassford *Production Project Manager*
Meredith Wilcox *Illustrations Assistant*
Theodore B. Tucker, IV,
Cameron Zotter, *Design Assistants*
Connie Binder *Indexer*

Manufacturing and Quality Control

Christopher A. Liedel *Chief Financial Officer*
Phillip L. Schlosser *Managing Director*
John T. Dunn *Technical Director*
Alan Kerr *Manager*

Staff for Previous Edition

Martha C. Christian *Editor*
Carolinda E. Averitt *Text Editor*
Elisabeth B. Booz,
Kimberly A. Kostyal *Senior Researchers*
Joyce Marshall Caldwell,
Alexander L. Cohn,
Sew Fun Mangano *Researchers*
Mimi Harrison *Illustrations Researcher*

One of the world's largest nonprofit scientific and educational organizations, the National Geographic Society was founded in 1888 "for the increase and diffusion of geographic knowledge." Fulfilling this mission, the Society educates and inspires millions every day through its magazines, books, television programs, videos, maps and atlases, research grants, the National Geographic Bee, teacher workshops, and innovative classroom materials. The Society is supported through membership dues, charitable gifts, and income from the sale of its educational products. This support is vital to National Geographic's mission to increase global understanding and promote conservation of our planet through exploration, research, and education.

For more information, please call
1-800-NGS LINE (647-5463)
or write to the following address:

National Geographic Society

1145 17th Street N.W.

Washington, D.C. 20036-4688 U.S.A.

Visit the Society's Web site at
www.nationalgeographic.com.

Library of Congress Cataloging-in-Publication Data

Langone, John, 1929-
The new how things work: everyday technology explained / John Langone ; art by Pete Samek, Andy Christie, and Bryan Christie.
p. cm.
Rev. ed. of: National Geographic's how things work. c1999.
Includes index.
ISBN 0-7922-6956-X
1. Technology--Popular works. 2. Inventions--Popular works. I. Title: How things work. II. Langone, John, 1929- National Geographic's how things work. III. Title.

T47.L2923 2004
600--dc22

2004050438

Author John Langone, a veteran science journalist, was a staff writer for *Discover* and *Time* magazines, a reporter and writer for United Press International, and science editor at the *Boston Herald.* He was a Kennedy Fellow in Medical Ethics at Harvard, a Fellow at the Center for Advanced Study in the Behavioral Sciences at Stanford, and a Fulbright Fellow at the University of Tokyo, where he researched the impact of science and technology on the Japanese. He writes the weekly "Books on Health" column in the Science Times section of *The New York Times.* This is Langone's 25th book.